Filamentary A15 Superconductors

CRYOGENIC MATERIALS SERIES

Nonmetallic Materials and Composites at Low Temperatures
Edited by A. F. Clark, R. P. Reed, and G. Hartwig

Filamentary A15 Superconductors
Edited by Masaki Suenaga and A. F. Clark

A Continuation Order Plan is available for this series. A continuation order will bring delivery of each new volume immediately upon publication. Volumes are billed only upon actual shipment. For further information please contact the publisher.

Filamentary A15 Superconductors

Edited by
Masaki Suenaga
Brookhaven National Laboratory
Upton, New York

and
Alan F. Clark
National Bureau of Standards
Boulder, Colorado

PLENUM PRESS · NEW YORK AND LONDON

Library of Congress Cataloging in Publication Data

Topical Conference on A15 Superconductors, Brookhaven National Laboratory, 1980
  Filamentary A15 superconductors.

  (Cryogenic materials series)
  Includes bibliographies and index.
  1. Superconductors–Congresses. I. Suenaga, Masuka. II. Clark, A. F. III. Title.
IV. Series.
TK454.4.S93S95 1980   621.3815'2   80-24312
ISBN 0-306-40622-5

Proceedings of the Topical Conference on A15 Superconductors, sponsored by the ICMC.
held at the Brookhaven National Laboratory, Upton, New York, May, 1980.

© 1980 Plenum Press, New York
A Division of Plenum Publishing Corporation
227 West 17th Street, New York, N.Y. 10011

1980
INTERNATIONAL CRYOGENIC MATERIALS
CONFERENCE BOARD

Office: National Bureau of Standards
Boulder, Colorado, U.S.A.

A. F. Clark, Chairman
National Bureau of Standards
Boulder, Colorado, U.S.A.

R. W. Boom
University of Wisconsin
Madison, Wisconsin, U.S.A.

E. W. Collings
Battelle Memorial Institute
Columbus, Ohio, U.S.A.

D. Evans
Rutherford Laboratory
Chilton, Didcot, England

G. Hartwig
Nuclear Research Center Karlsruhe
Institute for Technical Physics
Karlsruhe, Federal Republic of Germany

T. Horiuchi
Kobe Steel Ltd.
Kobe, Japan

J. W. Morris, Jr.
University of California
Berkeley, California, U.S.A.

R. P. Reed
National Bureau of Standards
Boulder, Colorado, U.S.A.

M. Suenaga
Brookhaven National Laboratory
Upton, New York, U.S.A.

K. Tachikawa
National Research Institute for Metals
Tokyo, Japan

K. A. Yushchenko
E. O. Paton Institute of Electrowelding
Kiev, USSR

PREFACE

Compound superconductors with the A15 structure carry the promise of the second generation in practical superconductivity. They will provide higher operating magnetic fields at higher temperatures than the preceding alloy superconductors. To fulfill this promise, their brittle nature must be accommodated in a filamentary structure. Achieving this has been no simple task and imaginative research and clever production techniques have led to many usable conductor configurations. In addition, several new and exciting possibilities are being proposed; for example, in situ processing promises easier production with improved strain tolerance. It is timely, therefore, to take measure of what we have achieved and to assess our understanding so that we may choose, with some confidence, paths for future research and potential applications.

To meet these needs, the International Cryogenic Materials Conference Board has sponsored this special topic conference on Filamentary A15 Superconductors bringing together superconductivity researchers, superconducting wire producers, and high field magnet users to discuss current research problems. That the information exchange was intensive and successful is evidenced by the excellent papers in this volume. In order to capture some of the synergistic wisdom generated in discussions at the conference several people were asked to assemble and interpret the comments and concerns of the fusion, high energy physics, solid state physics, and metallurgy groups. These reports and a conference summary are also included in the proceedings in an attempt to preserve at least a small part of the invaluable "back hall" component of any successful conference. These reports are particularly helpful in identifying research needs. We feel that all present contributed to the commitment of successful applications of filamentary A15 superconductors based on solid understanding and added insights to the potential, not only of these conductors, but to the future of all superconductivity as well.

vii

The editors wish to express their sincere appreciation to L. C. Arns who worked hard to produce a uniform and exceptional proceedings to this conference.

M. Suenaga
A. F. Clark

CONTENTS

MECHANICAL PROPERTIES

MULTIPLY CONNECTED SUPERCONDUCTORS

GROUP COMMENTS

INDEXES

CONFERENCE SUMMARY

FILAMENTARY A15 SUPERCONDUCTORS

J. E. Evetts

Cambridge University
Cambridge, UK

This conference is exceedingly timely: the first generation
multifilamentary A15 conductors have successfully proved themselves
in the market place and attention is now turning with confidence to
the future. Renewed consideration is being given to every aspect
of conductor design. The immediate objectives are on the one hand
to develop large volume production of conductor for medium field
applications (8-12 T), and on the other hand, to explore systemat-
ically the exciting and unique prospects A15 materials offer for
the technological exploitation of superconductivity at high fields
(14-18 T and higher). The design of a five or six component composite turns on an extravagant blend of electrodynamic, mechanical
and metallurgical considerations, it is hardly surprising therefore
that conductor development has tended to proceed on semi-empirical
lines. The explicit aim of this meeting was to bring together producers, users and academics concerned with multifilamentary A15
conductors in the expectation that the exchange of information and
ideas would highlight important issues and stimulate a more con-
certed cross-disciplinary approach to understanding complex materials. This objective was reinforced by devoting the fourth session of the meeting to reports by chairmen of specialist groups
appointed by the Organizing Committee to monitor the previous three
sessions.

The first of the sessions comprised national and company over-
views and briefly surveyed all current research and development
work. The range of experimental conductors receiving serious attention gave emphasis to the conflicting requirements of different
applications. As the demand for conductor increases there is likely
to be a proliferation of conductor types to suit specific requirements. The trend towards more complex alloys with ternary and
quaternary additions seems irresistible; as a consequence the

metallurgical effects observed become more intricate and their quantitative interpretation more difficult.

The theme of the second session, mechanical properties, reflects the great effort that has gone into understanding the variation of superconducting properties with strain. Since in certain field regimes intrinsic conductor strain can lead to a fourfold degradation in critical current, a quantitative description of the effect of strain is a prerequisite to the development of accurate models for flux pinning in these materials. The observation that the shape of the critical current versus reduced field curves scales directly with strain leads to the conclusion that the various different shaped critical current curves commonly observed do not depend directly on the strain sensitivity of the particular composite components and conductor geometry.

The session on multiply connected conductors was valuable for two reasons. Firstly, it served to emphasize the wide range of conductor routes that merit serious investigation. Secondly, the remarkable microstructures exhibited by in-situ materials raise fundamental questions for the interpretation of their properties. Furthermore on reflection it is clear that many of the same questions should also be raised in the case of more conventional conductors. The practice of carrying over concepts and ideas developed for bulk superconductors must be questioned. For instance the concept of a critical current density becomes difficult to justify on a scale less than the pinning penetration depth. Also in the case of sub-micron filaments the surface pinning contribution can be appreciable, and the bulk pinning summation, involving only a few pinning sites, cannot be treated using a conventional statistical theory.

Before the final session, a small poster session (9 papers) was provided to cover pertinent subjects to the conference but could not be included in the oral sessions. These papers discussed, for example, composition profiles in Nb_3Sn layers, ac losses in in-situ processed Nb_3Sn wires, flux pinning in bronze processed Nb_3Sn, etc.

In the final session the group chairmen reported to the Conference, describing those aspects of the subject that had appeared of particular interest or importance to their specialist groups. Both specific and general problems were discussed with emphasis on gaps in our present understanding and on the likely future requirements for conductors for different applications. (These reports are included in this volume.)

These proceedings represent a critical overview of the subject at a crucial stage in the development of a new generation of materials. The interaction between disciplines was particularly

valuable. When research on a project involves progress in a very
broad front there is a tendency for each specialist area to make
local progress on premises that derive from the stage before last
in related areas. Presented with this situation in the development
of multifilamentary A15 conductors this meeting has enabled the
different disciplines to cross relate and set a new common level
of understanding.

Full recognition must be extended to the Organizing Committee,
firstly, for its appreciation of the need for a topical conference
in this area and, secondly, for the clarity and balance of the
program which, without doubt, contributed greatly to the success of
the conference.

DEVELOPMENTS OF A15 FILAMENTARY COMPOSITE SUPERCONDUCTORS

IN JAPAN

K. Tachikawa

National Research Institute for Metals
1-2-1, Sengen, Sakura, Niiharigun
Ibaraki 305, Japan

INTRODUCTION

At the National Research Institute for Metals, the so-called surface diffusion process (SDP) to produce V_3Ga tape was invented in 1967.[1] In this process, a vanadium substrate is passed continuously through a molten gallium bath in which some copper is added. The copper acts as a catalyst for enhancing the diffusion reaction between the vanadium substrate and the gallium to form V_3Ga phase.[2] The critical current density versus magnetic field, J_c-H, curve of V_3Ga is convex upward in fields above 10 T, and the V_3Ga exhibits a higher J_c at fields above 13 T than any other practical superconductors. The V_3Ga tape was used to construct the 17.5 T superconducting magnet, which established the highest field record in the world in the superconducting state.[3] The invention of the surface diffusion process led to the following success of the so-called composite diffusion process, which enables the fabrication of multifilamentary type-V_3Ga superconductors.[4] The multifilamentary V_3Ga wires were expected to be much more stable than the tapes and to be useful for applications in time varying fields.

FABRICATION OF MULTIFILAMENTARY V_3Ga WIRES AND THEIR SUPERCONDUCTING PROPERTIES

In the composite diffusion process (CDP) to fabricate multi-filamentary V_3Ga wires, a composite of a Cu-Ga solid solution alloy matrix containing 15-20 at.% Ga and vanadium cores is fabricated into a thin wire, and then heat treated at a temperature between 600-700°C. In the heat treatment, gallium in the Cu-Ga

alloy matrix selectively diffuses with vanadium, and only V$_3$Ga
layers are formed around the vanadium cores. When the gallium
concentration in the Cu-Ga alloy exceeds about 40%, the compounds
richer in gallium appear. Figure 1 shows an X-ray microanalysis
line scanning chart taken on the cross-section of the CDP V$_3$Ga
tape.[5] Very little copper (< 0.2 at.%) is incorporated within the
V$_3$Ga layer. The copper acts as a mother metal of gallium and enhanced the formation of V$_3$Ga. The effects of metallurgical variables on superconducting properties and workability of the CDP
V$_3$Ga have been reported in detail in references 6-8.

The higher transition temperature, T$_c$ is obtained after the
heat treatment for the specimen with higher gallium concentration
in the matrix as shown in Fig. 2. However, the maximum gallium
concentration in the Cu-Ga matrix workable at room temperature is
limited up to about 21 at.%. The highest T$_c$ is obtained in the
V/Cu-20 at.% Ga composite after the heat treatment at 625-650°C
for 100-200 h.[7] The dependences of the upper critical field, H$_{c2}$
on the matrix composition and on the heat treatment condition are
similar to those of T$_c$. The CDP V$_3$Ga shows about the same T$_c$, but

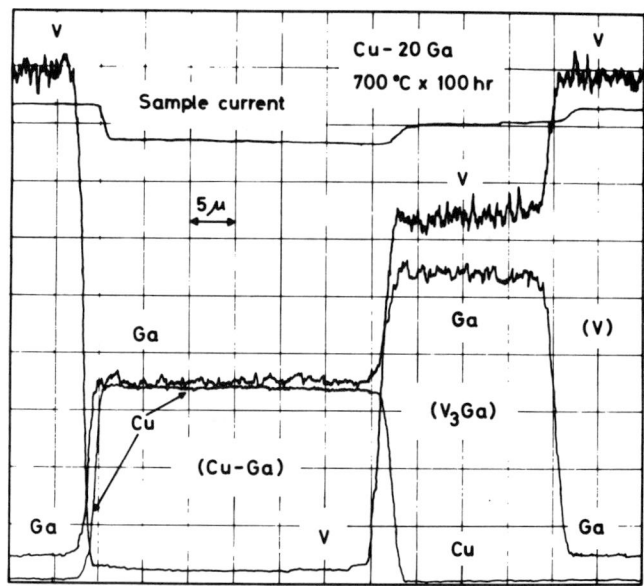

Fig. 1. An XMA line scanning chart taken on the cross-section of
the CDP V$_3$Ga, which shows the concentration profiles of
vanadium, gallium and copper in the vanadium core, in the
Cu-Ga matrix and in the V$_3$Ga.[5]

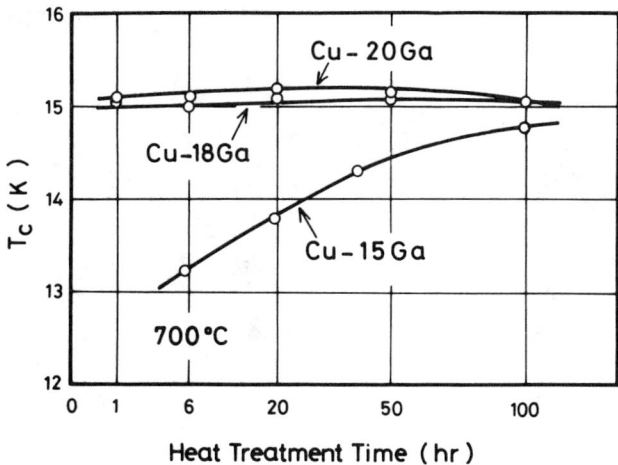

Fig. 2. Transition temperature versus heat treatment time at
         700°C for the V/Cu-Ga composite with different gallium
         concentration in the matrix.[7]

0.5-1.0 T lower $H_{c2}(4.2 K)$, as compared to those of the SDP V_3Ga.

    J_c of the CDP V_3Ga tape is greatly affected by both the Cu-Ga
matrix composition and the heat treatment condition. A J_c of over
$1x10^5$ A/cm^2 at 4.2 K and 17 T, which is equivalent to that of the
SDP V_3Ga tape, is obtained for the specimen with Cu-20 at.% Ga
matrix heat treated at 625°C. Specimens heat treated at higher
temperatures or for longer time show peak effect at a magnetic
field of about 17 T.

    Figure 3 shows scanning electron micrographs taken on the
cross-section of the CDP V_3Ga layer.[8] At 625°C, a little elongated V_3Ga grains with small diameters are formed. At heat treatment temperatures above 650°C, columnar V_3Ga grains with larger
diameters are grown perpendicular to the vanadium core. Crystallographic textures in the V_3Ga and Nb_3Sn layers formed by the SDP
and the CDP have been also studied by pole figure techniques.[9]

    The J_c at magnetic fields below 10 T is strongly related to
the V_3Ga grain size. Figure 4 shows the relation between the J_c
at 6.5 T and the V_3Ga grain size.[8] The result indicates that J_c
is nearly inversely proportional to the diameter of the V_3Ga
grains. Above the peak field, however, J_c is almost independent
on the heat treatment condition, i.e., the V_3Ga grain size. J_c

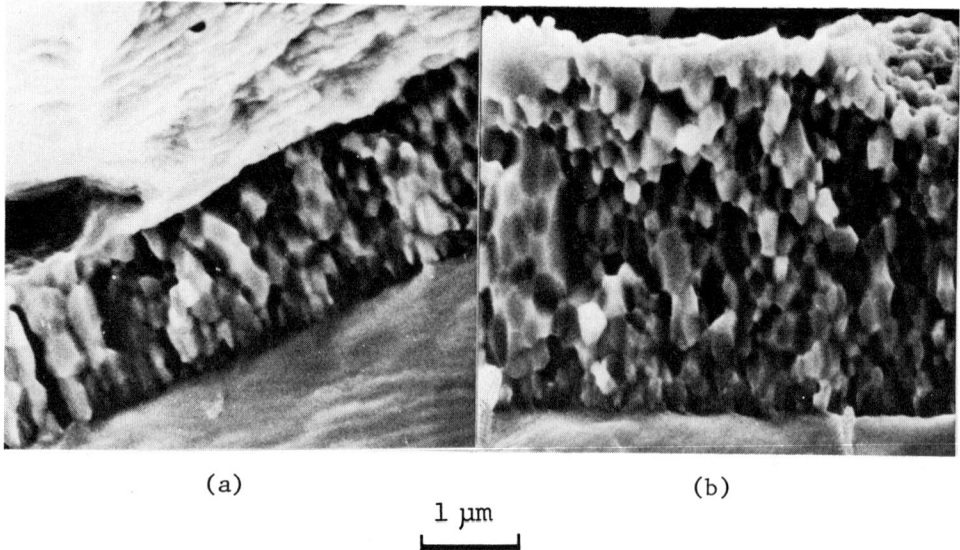

(a)   (b)

1 μm

Fig. 3.   Scanning electron micrographs taken on the cross-section
          of the V_3Ga layer heat treated at 625°C.[8]   (a) for 20 h;
          (b) for 100 h.

of the CDP V_3Ga tape measured in field perpendicular to the tape
surface is considerably larger than that measured in parallel
field.   This anisotropy in J_c seems to be related to the shape
anisotropy of the V_3Ga grains.   However, above the peak field,
$J_c(H_\perp)$ is nearly identical to $J_c(H_\parallel)$.[8]

     The multifilamentary V_3Ga wire with excellent stability and
high field performances has been commercially produced by the CDP
in Japan.[10]   A cross section of the stranded-type V_3Ga wire is
shown in Fig. 5(a).   This wire shows an overall J_c of $3.3 \times 10^4 A/cm^2$
at 15 T.[10]   The multifilamentary V_3Ga wire is quite stable under a
rapid field change even over 20 T/sec.[11]   A compact 13 T magnet
shown in Fig. 6 has been recently wound using the multifilamentary
V_3Ga tape with larger current carrying capacity of which cross-
section is shown in Fig. 5(b).   The magnet is much stable and
does not quench even if the power source is turned off at 13 T.

IMPROVEMENTS IN HIGH-FIELD PERFORMANCES OF COMPOSITE-PROCESSED
A15 SUPERCONDUCTORS

     The effects of additional element on the superconducting
properties of the CDP V_3Ga have been investigated in NRIM.   The
addition of aluminum to the Cu-Ga matrix increases I_c, J_c and the

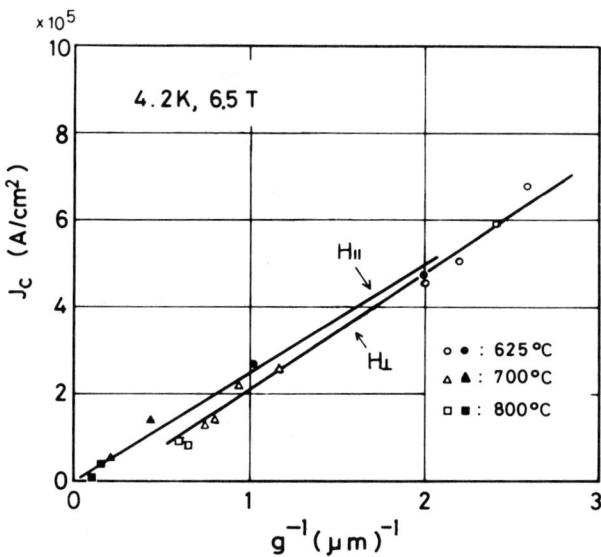

Fig. 4. Critical current density versus reciprocal of the V_3Ga
         grain size, g which is average grain diameter and average
         grain length for $J_c(H_\perp)$ and $J_c(H_\parallel)$, respectively.[8]

formation rate of V_3Ga.[12] The addition of 0.3-0.5 at.% Mg to the
Cu-Ga matrix significantly increases J_c in entire field range and
slightly increases H_{c2}.[13] The magnesium addition also increases
the formation rate of V_3Ga. The magnesium or aluminum addition
to the matrix produces finer and more isotropic V_3Ga grains. The
addition of 3-6 at.% Ga to the vanadium core increases H_{c2} by about
2 T, while it does not appreciably change T_c.[13] The simultaneous
addition of gallium to the core and magnesium to the matrix, which
is considered to be the most effective combined addition, achieves
enhancements both in H_{c2} and J_c. Typical I_c and J_c versus magnetic field curves obtained for these samples are shown in Fig. 7.
The core and the matrix compositions of the specimens are given
in the figure caption. A J_c value of over 1×10^5 A/cm^2 is obtained
at 20 T even for relatively thick V_3Ga layers of 5-10 μm, which in
turn might achieve a significant improvement in the overall J_c of
the multifilamentary V_3Ga wire in high magnetic fields.

     The addition of small amounts of magnesium to the Cu-Sn
matrix also produces much finer Nb_3Sn grains in the CDP Nb_3Sn.[14]
This grain refinement enhances the formation of Nb_3Sn, where the
grain boundary diffusion of tin is predominant, and produces
about twice as thick Nb_3Sn layer. The grain refinement also

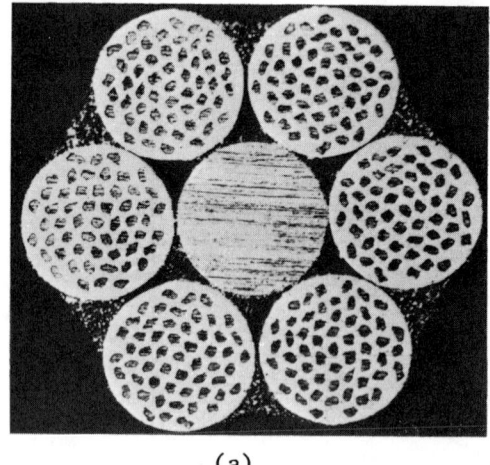

(a)

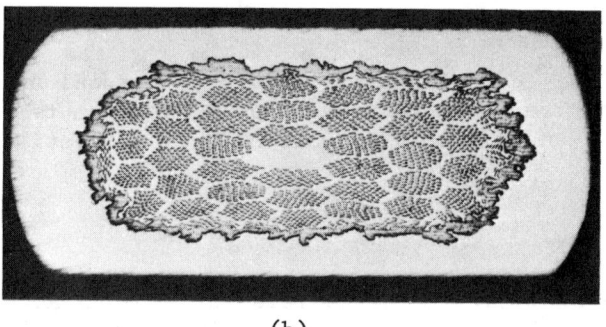

(b)

Fig. 5.   Photomicrographs of the cross-section of the multifila-
          mentary V_3Ga conductors. (a) 6 stranded wire. Overall
          diameter: 0.36 mm, core diameter: 10 μm, core number:
          55x6.[10] (b) Multifilamentary tape (V/Cu-Ga/Nb/Cu, be-
          fore final rolling). Final dimension: 0.16x5.0 mm, core
          diameter: 6 μm, core number: 2574.

increases J_c, and significantly increases current-carrying
capacity of the CDP Nb_3Sn. However, no appreciable changes in
T_c and H_{c2} are obtained by the magnesium addition to the CDP
Nb_3Sn.

     The effects of the hafnium addition to the niobium core, and
that of the simultaneous addition of hafnium to the core and gallium to the matrix have been studied in NRIM.[15] The hafnium addition to the niobium core and the substitution of gallium for
tin in the matrix causes only minimum deterioration in the
workability of the composite, and the preliminary 19-core wires

Fig. 6. The 13 T superconducting magnet with high stability.
Outer diameter 250 mm, clear bore 30 mm, height 250 mm.

have been fabricated without any problem. Figure 8 shows Nb_3Sn layer thickness versus heat treatment time at 800°C for the single core composite tape specimens with different core and matrix compositions. The growth rate of the Nb_3Sn layer in the composite specimens with a Nb-2 or 5 at.% Hf alloy core is about 2-3 times greater than that of the composite specimen with a pure niobium core and a matrix of the same composition. The Nb_3Sn growth rate in the Cu-5 at.% Sn-4 at.% Ga matrix specimen is less than half of that in the Cu-7 at.% Sn matrix specimen. However, the decrease of the Nb_3Sn growth rate is compensated by the hafnium addition to the core, as can be seen in Fig. 8. The scanning electron microscopy on the specimens indicates that the hafnium addition to the core produces a thick and fine-grained Nb_3Sn layer. On the other hand, the gallium addition to the matrix causes appreciable Nb_3Sn grain coarsening. The X-ray microprobe analysis

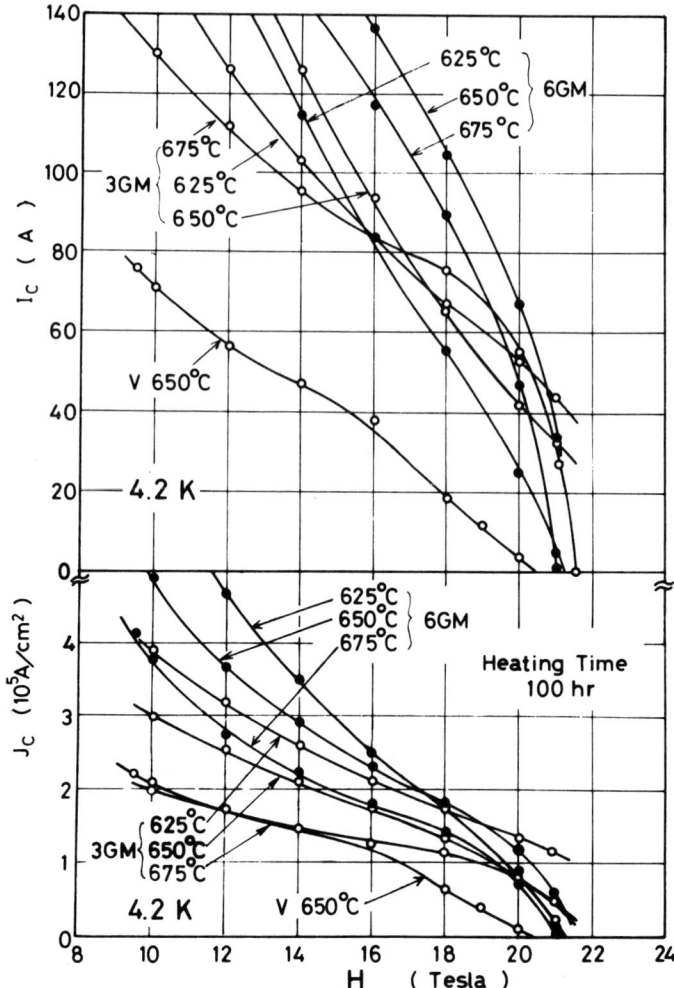

Fig. 7. I_C and J_C versus magnetic field curves at 4.2 K for the improved CDP V_3Ga specimens. The thicknesses of the V_3Ga layers in the 3GM and 6GM specimens heat treated at 625°C and 650°C are about 6 μm and 9 μm, respectively.[13]

| Specimen | Composition (at.%) | |
Abbreviation	Core	Matrix
V	pure V	Cu–19Ga
3GM	V–3Ga	Cu–19Ga–0.5Mg
6GM	V–6Ga	Cu–19Ga–0.5Mg

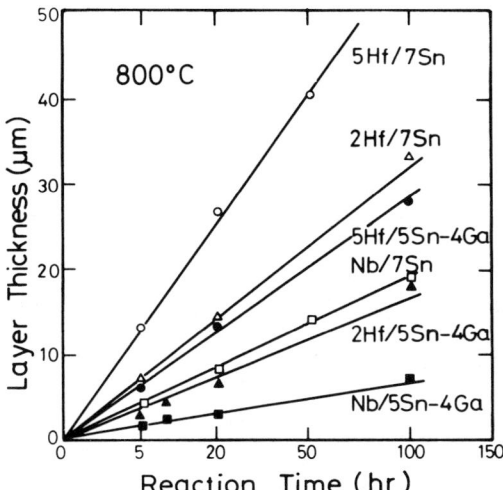

Fig. 8. Nb_3Sn layer thickness versus square root of the reaction
time at 800°C for the CDP Nb_3Sn tape specimens with
hafnium addition to the core and/or gallium addition to
the matrix.[15] The abbreviation, for example, 5 Hf/7 Sn
means Nb-5 at.% Hf/Cu-7 at.% Sn composite.

indicates the presence of about 1.5 at.% Ga and 0.6 at.% Hf in
the Nb_3Sn layer formed in the Nb-2 at.% Hf/Cu-5 at.% Sn-4 at.% Ga
composite specimen reacted at 800°C for 100 h.

    The hafnium addition to the core as well as the gallium ad-
dition to the matrix increases T_c of the CDP Nb_3Sn by 0.4-0.5 K
and $H_{c2}(4.2$ K) by 2.0-4.0 T. The simultaneous addition of hafnium
to the core and gallium to the matrix increases T_c by 0.6-0.9 K
and $H_{c2}(4.2$ K) by 6.0-7.0 T. In the Nb-5 Hf/Cu-5 Sn-4 Ga and
Nb-5 Hf/Cu-3 Sn-9 Ga composite specimens heat treated at 800°C
for 20 h, $H_{c2}(4.2$ K)'s of 25-26 T are obtained. Figure 9 shows
critical current, I_c versus magnetic field curves at 4.2 K for
the specimens with different core and matrix compositions. The
data presented in Fig. 9 indicate that the hafnium addition to
the niobium core shifts the I_c versus H curve to higher field
without changing the shape. The gallium addition to the matrix,
on the contrary, changes the I_c versus magnetic field curve of
Nb_3Sn from convex downwards to convex upwards in fields about 12T,
which is similar to that of V_3Ga. A detailed study on the mechanism of this effect shall be interesting. The simultaneous addition of hafnium to the core and gallium to the matrix is most

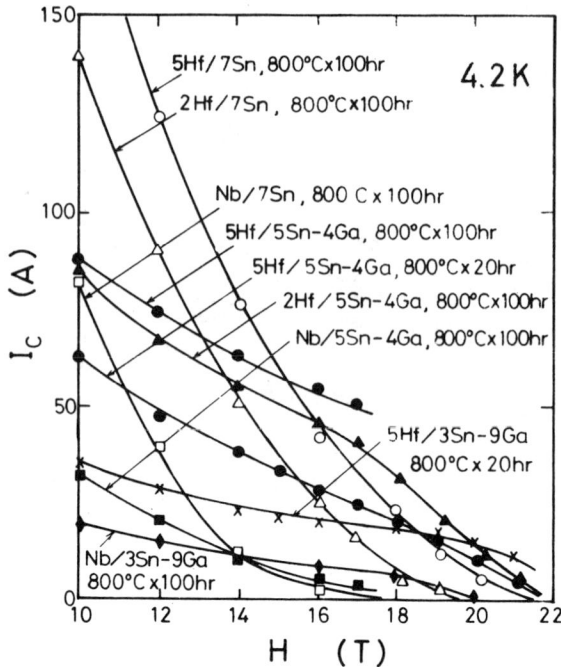

Fig. 9. Critical current versus magnetic field curves at 4.2 K
         for the improved CDP Nb₃Sn specimens with different core
         and matrix compositions.[15]

effective in increasing the current-carrying capacity of CDP Nb₃Sn
in high fields (> 15 T).

     Figure 10 shows J_c versus magnetic field curves at 4.2 K for
the 19-core wire specimens. J_c's of over 1×10^5 A/cm² are obtained
at 17 T for the Nb-5 Hf/Cu-3 Sn-9 Ga wire specimens reacted at
700°C. The simultaneous addition of hafnium to the core and gallium to the matrix may increase the maximum field generated by the
multifilamentary Nb₃Sn by nearly 5 T. This improvement provides
a larger safety margin for generating high magnetic fields with
the CDP Nb₃Sn conductors. The measurements of stress effects on
I_c of the Nb-Hf/Cu-Sn-Ga single core tapes show that no degradation occurs in I_c until the strain reaches nearly 1.0% in the
Nb-5 Hf/Cu-5 Sn-4 Ga composite, while the irreversible degradation occurs in I_c at 0.7% strain in the Nb/Cu-7 Sn composite.[16]

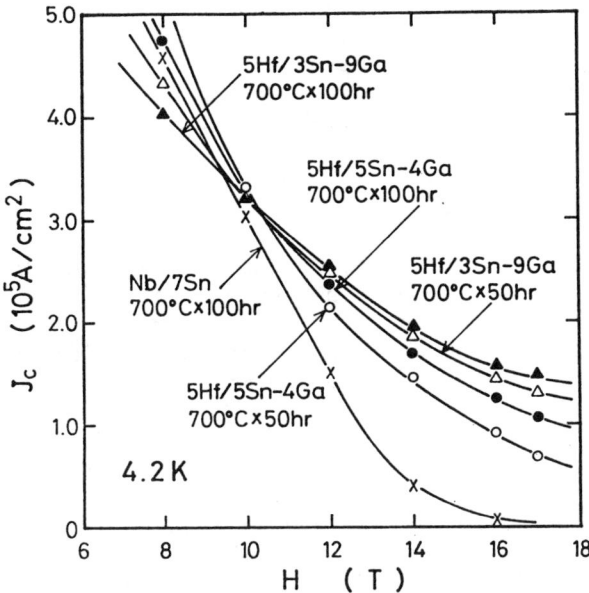

Fig. 10: J_c versus magnetic field curves at 4.2 K for the 19-core
         CDP Nb_3Sn wire specimens with different core and matrix
         compositions.

The external diffusion of gallium into the Nb-Hf/Cu-Sn composite
pushes the tin in the matrix towards the niobium core, and ac-
celerates the formation of Nb_3Sn layer.[16] The external diffusion
of gallium also improves the high-field performances of the CDP
Nb_3Sn.

     In NRIM, the formation of V_3Si[5,17] and that of V_3Ge[18] by the
CDP have been also studied. The aluminum addition to the vanadium core in the V/Cu-Ge composite enables the formation of ternary $V_3(Ge,Al)$ phase. T_c and $H_{c2}(4.2$ K), especially the latter,
of the V-Al/Cu-Ge composite increase significantly with aluminum
concentration in the vanadium core from 7 K and 2 T at 0 at.% Al
to 12.5 K and 18 T at 23 at.% Al, respectively.[20]

DEVELOPMENTS OF COMPOSITE-PROCESSED A15 CONDUCTORS IN JAPANESE
RESEARCH GROUPS OTHER THAN NRIM

In the governmental institutes, the development of multifilamentary Nb_3Sn conductors, in which the Nb_3Sn layer is formed
by the solid-liquid reaction between the niobium tubes and the
Sn-Cu alloy core containing 5 at.% Cu, is being carried out at
the Electrotechnical Laboratory.[20] The development is proceeded
in collaboration with Sumitomo Electric Industries,[21] and a small
test magnet wound by this Nb_3Sn conductor has generated 12 T with
a back-up field of 10 T.

At the Japan Atomic Energy Research Institute (JAERI), a
10T-150 mm bore Nb_3Sn magnet has been successfully operated in a
back-up field of 8 T generated by the 240 mm-bore Nb-Ti magnet.[22]
The conductor is the composite-processed multifilamentary Nb_3Sn
tape with a cross-section of 1.3x8.5 mm^2 containing 42037 filaments, and carries 816 A at 10.2 T.

In the industries, Furukawa Electric Co. first commercialized
the multifilamentary V_3Ga wires in the world,[10] and succeeded to
construct high stability 10 $T^{[23]}$ and 13 T magnets. Furukawa also
developed Nb_3Sn 12-strand compacted cable with medium current
capacity for the field windings of the superconducting synchronous machine. Furukawa is now developing Nb_3Sn conductors with
large current capacities.[24]

At Hitachi Ltd. and Hitachi Cable Co., large scale multifilamentary Nb_3Sn conductors which are designed to carry 1 kA at
11-13 T and 10 kA at 13 T have been developed.[25] Hitachi prepared
the Nb_3Sn conductor for the 10 T-150 mm bore magnet of JAERI. The
Hitachi group is also developing hydrostatic extrusion techniques
for the production of large scale multifilamentary Nb_3Sn conductors.[26]

Mitsubishi Electric Co. developed the internal tin diffusion
process for the fabrication of multifilamentary Nb_3Sn conductors.[27] Mitsubishi has recently developed an improved process in
which a multifilamentary Nb/Cu composite enveloping tin at the
center is drawn into a wire and then heat treated.[28] The copper-
stabilized conductor which has a cross-section of 1.4x10 mm^2 with
35280 Nb_3Sn filaments and a current capacity of about 1 kA at 8 T
has been developed.

Toshiba R&D Center, and Showa Electric Wire and Cable Co.
have developed the niobium tube method to fabricate multifilamentary Nb_3Sn conductors.[29] This method starts with a niobium
tube component containing copper-sheathed tin inside and high
conductivity copper outside. A bundle of these component tubes,
which is encased in a copper jacket, is drawn and heat treated.

Toshiba has constructed several test magnets by the wind and react technique. The 1×2 mm^2 conductor with 258 filaments carries about 200 A at 10.7 T.[30]

At Nihon University, high H_{c2} multifilamentary $Nb_3(Sn,In)$ wires have been fabricated in collaboration with materials research group of FBNML (MIT).[31] In this process, niobium rods with 19 holes are impregnated with a ductile Sn-Cu-In alloy and reduced to final size wires without intermediate annealing. Addition of indium and copper substantially improves H_{c2} and high-field J_c. The 19 core wire reacted at 900°C has an overall J_c of 10^4 A/cm^2 at 18 T and an extrapolated H_{c2} of 25.5 T at 4.2 K.

CONCLUSION

Multifilamentary V_3Ga and Nb_3Sn wires have been developed by the composite process. Small scale high-field superconducting magnets with excellent stability have been successfully constructed. These results indicate that the composite-processed multifilamentary A15 wires are promising for variety of high-field applications. Recent improvements in high-field performances of the composite-processed V_3Ga and Nb_3Sn conductors make the construction of 20 T superconducting magnet by the stable multifilamentary wires feasible in the near future.

Developments and testings of multifilamentary Nb_3Sn conductors with large cross-sectional area and current capacity are being proceeded in many Japanese governmental and industrial research groups. For large scale applications of composite-processed A15 conductors, improvements in stress properties are required. Future progresses might be expected on this problem through the effect of additional elements reinforcing the core and the matrix of the composite. The in situ processed filamentary A15 conductors show relatively superior mechanical properties over the composite-processed A15 conductors. On the other hand, the composite-processed A15 conductors show appreciably better high-field performances than the in situ processed A15 conductors. From the viewpoint of fundamental research, mechanisms of the effects of third element addition on the growth kinetics, the grain size and the H_{c2} enhancement, as well as mechanism of the flux pinning, in A15 phase are still remaining subjects.

REFERENCES

1. K. Tachikawa and Y. Tanaka, Japan. J. Appl. Phys. 6:782 (1967).
2. Y. Tanaka, K. Tachikawa, and K. Sumiyama, J. Japan. Inst. Metals 34:835 (1970).

3. W. D. Markiewicz, E. F. Mains, R. M. Vankeuren, R. E. Wilcox,
    C. H. Rosner, H. Inoue, C. Hayashi, and K. Tachikawa,
    IEEE Trans. on Magnetics MAG-13:35 (1977).
4. K. Tachikawa, Proc. 3rd ICEC, Iliffe Science and Technology
    Pub. Ltd., Surrey (1970).
5. K. Tachikawa, Y. Yoshida, and L. Rinderer, J. Materials Sci.
    7:1154 (1972).
6. Y. Tanaka and K. Tachikawa, J. Japan. Inst. Metals 40:502
    (1976).
7. Idem, ibid. 40:509 (1976).
8. Y. Tanaka, K. Itoh, and K. Tachikawa, ibid. 40:515 (1976).
9. K. Togano and K. Tachikawa, J. Appl. Phys. 50:3495 (1979).
10. Y. Furuto, T. Suzuki, K. Tachikawa, and Y. Iwasa, Appl. Phys.
    Letters 24:34 (1974).
11. K. Itoh and K. Tachikawa, Appl. Phys. Letters 26:67 (1975).
12. Y. Yoshida, K. Tachikawa, and Y. Iwasa, Appl. Phys. Letters
    27:632 (1975).
13. K. Tachikawa, Y. Tanaka, Y. Yoshida, T. Asano, and Y. Iwasa,
    IEEE Trans. on Magnetics MAG-15:391 (1979).
14. K. Togano, T. Asano, and K. Tachikawa, J. Less-Common Metals
    68:15 (1979).
15. H. Sekine and K. Tachikawa, Appl. Phys. Letters 35:472 (1979).
16. idem, to be presented at 1980 Appl. Super. Conf., Santa Fe.
17. Y. Yoshida and K. Tachikawa, J. Japan. Inst. Metals 37:558
    (1973).
18. K. Tachikawa, R. J. Burt, and K. T. Hartwig, J. Appl. Phys.
    48:3623 (1977).
19. K. Tachikawa, H. Sekine, and K. Togano, IEEE Trans. on
    Magnetics MAG-15:762 (1979).
20. Y. Aiyama, presented at Japan-USA Workshop on High-Field
    Superconducting Materials for Fusion, Tokyo (1980).
21. M. Nagata, M. Yokota, and M. Kawashima, ibid.
22. E. Tada, M. Nishi, T. Ando, and S. Shimamoto, ibid.
23. M. Ikeda, Y. Furuto, Y. Tanaka, I. Inoue, and J. Tanii,
    Proc. 5th Int. Conf. on Magnet Technology, Rome (1975).
24. Y. Furuto, Y. Tanaka, S. Meguro, and T. Suzuki, presented at
    Japan-USA Workshop on High-Field Superconducting Materials
    for Fusion, Tokyo (1980).
25. N. Tada, K. Aihara, Y. Hotta, Y. Ishigami, and H. Moriai,
    IEEE Trans. on Magnetics MAG-15:810 (1979).
26. N. Tada, K. Aihara, Y. Ishigami, and H. Moriai, presented at
    Japan-USA Workshop on High-Field Superconducting Materials
    for Fusion, Tokyo (1980).
27. Y. Hashimoto, K. Yoshizaki, O. Taguchi, and M. Tanaka, Proc.
    7th ICEC, IPC Sci. and Technology Press, London (1979).
28. Y. Hashimoto, K. Yoshizaki, O. Taguchi, and M. Iwamoto,
    presented at Japan-USA Workshop on High-Field Supercon-
    ducting Materials for Fusion, Tokyo (1980).
29. Y. Koike, H. Shiraki, S. Murase, E. Suzuki, and M. Ichikawa,
    Appl. Phys. Letters 29:384 (1976).

30.   O. Horigami and E. Suzuki, presented at Japan–USA Workshop
      on High–Field Superconducting Materials for Fusion,
      Tokyo (1980).
31.   R. Akihama, K. Yasukochi, and R. Roberge, IEEE Trans. on
      Magnetics MAG–15:629 (1979).

DEVELOPMENT OF A15 MULTIFILAMENTARY SUPERCONDUCTORS[+]

H. Hillmann,* H. Pfister,** E. Springer,* M. Wilhelm,**
and K. Wohlleben**

*Vacuumschmelze GmbH, Development Department
D-6450 Hanau, FRG
**Siemens AG, Research Laboratory
D-8520 Erlangen, FRG

INTRODUCTION

The superconducting properties of bronze processed Nb_3Sn multifilamentary conductors are dependent on the structure and the diffusion treatment of the conductor. The tin concentration of the bronze, the volume ratio bronze to niobium, the filament diameter and the filament distribution are significant. The current carrying capacity can be increased considerably by the suitable addition of third elements. On the other hand conductor properties are sensitively influenced by diffusion temperature and diffusion time. High diffusion temperatures favor a high critical temperature T_c and a high critical flux density B_{c2}, low diffusion temperatures promote a better flux pinning by the increasing number of grain boundaries. As a consequence optimal diffusion conditions for high critical currents at high fields, and high critical currents at low fields, are different. Finally the influence of mechanical stress on the superconducting properties has to be regarded. Depending on the composition of the conductors different compressive stresses are acting on the Nb_3Sn layers.[1-10]

[+]Part of this work has been supported by the technological program of the Federal Department of Research and Technology of the FRG.

FABRICATION OF CONDUCTORS

Bronze Technology

Commercial copper-tin alloys with a tin content up 10 wt% are unsuitable for the bronze process. The tin content of the bronze has to be as large as possible. Therefore in accordance with the phase diagram a saturated α solution of 13.5 wt% Sn was chosen.[1,2] In numerous commercial tin bronzes phosphorus is used as a deoxidizer. In bronze-niobium compound material phosphorus prevents diffusion of the tin atoms. Therefore phosphorus free ingots up to 400 kg by weight have been molten under vacuum and have been rolled to rods of 30 mm φ. Because of serious work hardening, recovery by frequent annealing steps was necessary. Great care had to be taken to get homogeneous material because of low melting eutectics in the phase diagram.

Composite Technology

Two techniques have been applied for fabricating niobium-bronze composites: a conventional tube technique and an extrusion technique. Figure 1 shows the principle of the two fabrication techniques.[2,5]

Conventional Tube Technique

For the conventional tube technique bronze rods have been worked down to a diameter of about 30 mm. Axial holes were drilled and filled with niobium rods. These bronze-niobium composites were deformed and after being reduced to a suitable diameter bundled and passed into a bronze tube which again was worked down. This process can be repeated. Composites containing 10,000 to 70,000 filaments have been manufactured by this way. Figure 2a and b shows the cross section of the 70,000 filament conductor.[2]

Extrusion Technique

To produce extrusion billets both the drilling technology and the hexagonal tube technology have been applicated. In order to reduce the overall area ratio of substrate to niobium, the hexagonal tube technique was preferred especially for the fabrication of high filament numbered conductors. Hexagonal tubes have been stacked in a bronze can with a packing density in excess of 90%. The cans were evacuated and sealed with bronze lids by electron beam welding. Billets with diameters up to 175 mm have been extruded. We mainly applicated indirect extrusion to achieve a

laminar flow during the extrusion process and to prevent friction.
A typical extrusion speed of the billet was 1 mm/s, the extrusion
ratio was 12.25. Billets of 175 mm thickness have thus been extruded to a diameter of 50 mm. The length of the billets amounted
to 800 mm. Figure 3a shows the influence of the billet temperature
on the extrusion force. The decrease of the extrusion force between 700°C and 750°C to nearly one half is striking. The influence of the temperature of the extrusion container is given in
Fig. 3b.

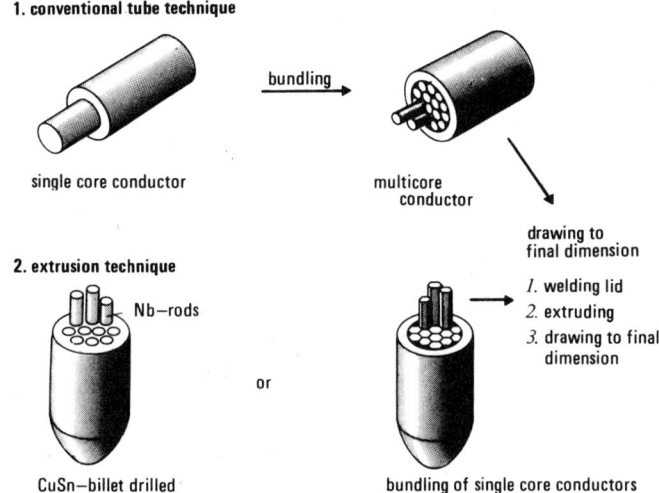

Fig. 1.   Conventional tube and extrusion technique.

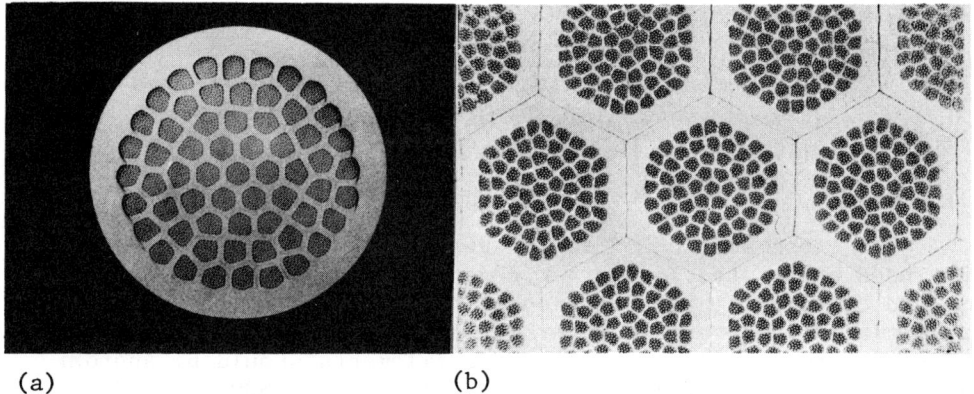

(a)   (b)

Fig. 2.   Cross section of a 70,000 filament conductor.

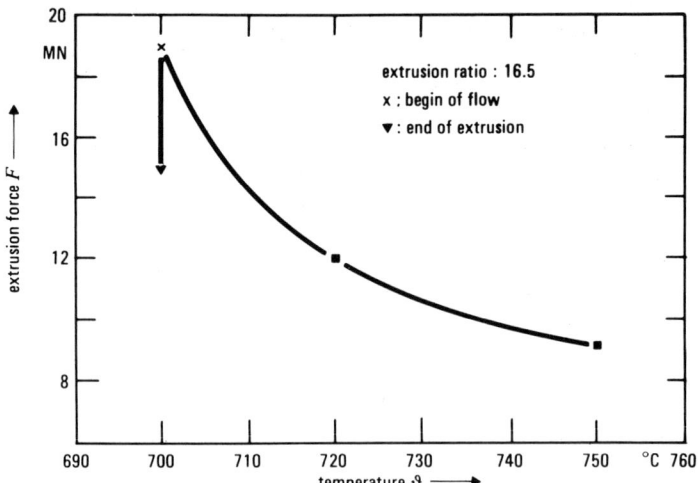

Fig. 3a. Influence of the billet temperature on the extrusion force.

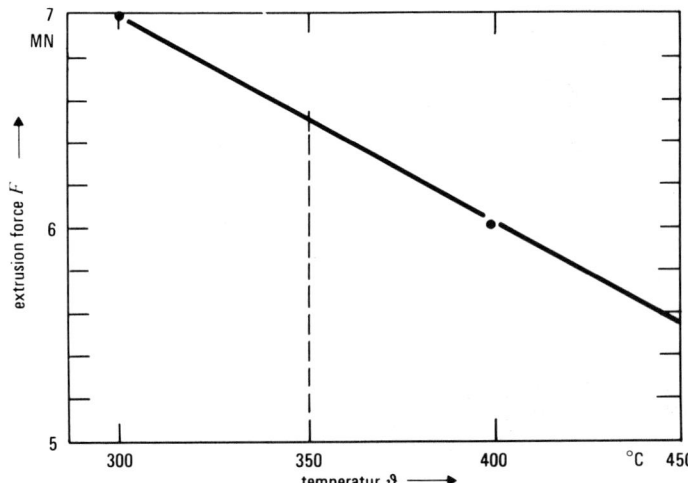

Fig. 3b. Influence of the container temperature on the extrusion force.

The cross section of the conventionally fabricated and of the extruded composites is further reduced by rolling and drawing with regard of suitable recovery annealings. According to the results of optimization studies concerning diffusion treatment and layer thickness the final diameter of the filaments should be approximately 3.5 µm, the filament number being thus 3,600 for a 0.4 mm diameter conductor, 6,000 for a 0.5 mm diameter conductor and 10,000 for a 0.7 mm diameter conductor.

Stabilization and High Current Conductors

Different stabilization techniques have been developed with respect to a reproducible and economic wire fabrication in long lengths necessary for magnet winding.

The integrated stabilization in monolithic conductors is normally provided by inserting high purity copper using Ta barriers to prevent diffusion of tin atoms into the high purity copper. Different aspects of geometrical distribution of stabilizing material have been investigated.

Figure 4 shows a conductor with a central copper stabilization; Fig. 5 with a peripheral one. The fabrication process of the two conductors is different. In the case of the central copper core copper-tantalum has to be inserted into the extrusion billet whereas stabilization at the periphery can be provided by adding copper-tantalum after extrusion. If for peripheral stabilization bronze-niobium is inserted into the extrusion billet, the billet diameter is limited by the diameter of the Ta tube available and the mechanical stability of the Ta tube during extrusion. Co-extrusion of Cu-Ta with bronze-Nb results in better bonding of tantalum to the other components. Ta reacts very eagerly with oxygen, thus access of oxygen has to be prevented during any kind of heat treatment. The possibility of oxygen penetration increases considerably for wires which are stabilized peripherally and carry very thin copper layers, or layers which have been damaged. With wires covered by only 10 μm of copper bursting of the copper layer has been observed. The reason is penetration of oxygen down to the tantalum barrier which had been converted to tantalum oxide.

Another possibility for stabilizing bronze conductors provides the insertion of separate CuTa composite strands or a CuTa core in flat cables.[5] This procedure has been successfully tested in several magnet experiments. The advantage of this technique is a more economic fabrication. Figure 6 shows different externally stabilized flat cables. Experiments with stabilization of conductors after reacting will be reported later.

SUPERCONDUCTING PROPERTIES

Critical Current

For critical current measurement nearly one meter of the conductor was wound on a Al_2O_3 tube with 48 mm diameter. The distance between the voltage probes was approximately 15 cm, critical current is defined for a voltage drop of 1 μV that is 0.07 μV/cm. It should be noted that the magnitude of the critical current

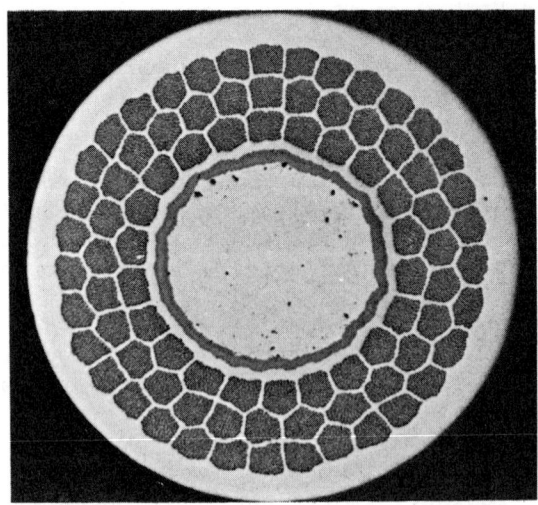

Fig. 4.   Cross section of an internally stabilized 10,000 filament
          conductor. Wire diameter 0.9 mm, critical current 295 A
          at 10 T.

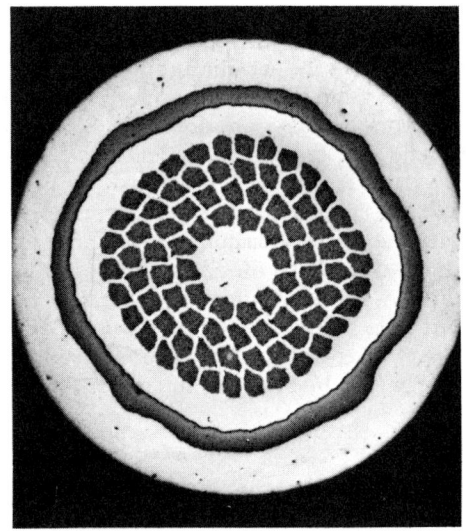

Fig. 5.   Cross section of a peripherally stabilized 10,000 fila-
          ment conductor. Wire diameter 0.9 mm, critical current
          230 A at 10 T.

depends very strongly on the material of the sample holder.[3]

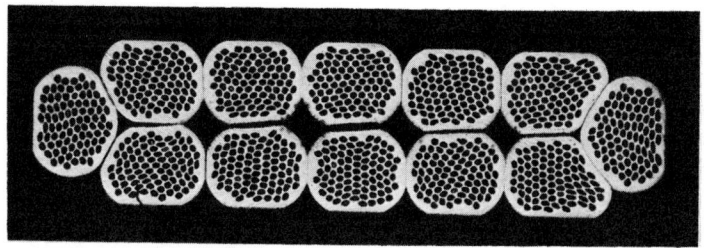

Fig. 6a. 12 strands unstabilized, wire diameter 0.66 mm, 10,000
filament per strand, total cross section 1 mm x 3.85 mm.
Critical current 2300 A at 10 T.

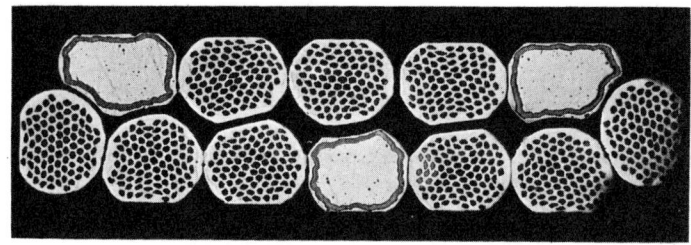

Fig. 6b. Cable as shown in Fig. 6a containing 3 CuTa strands.
Critical current 1700 A at 10 T.

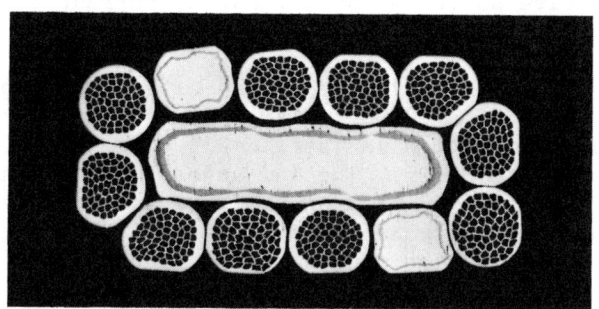

Fig. 6c. 12 strands wrapped around a CuTa core, wire diameter
0.4 mm, 3700 filament per strand, total cross section
1.1 mm x 2.25 mm. Critical current 750 A at 10 T.

Figure 7 shows the dependence of the critical currents on the
flux density for Al_2O_3 and laminated paper as sample holder. On
cooling to 4.2 K the thermally induced contraction of the wire and
the laminated paper holder is comparable whereas contraction of
Al_2O_3 is smaller. Therefore the conductor suffers tensile stress
when being wound on a Al_2O_3 holder. As a consequence the critical

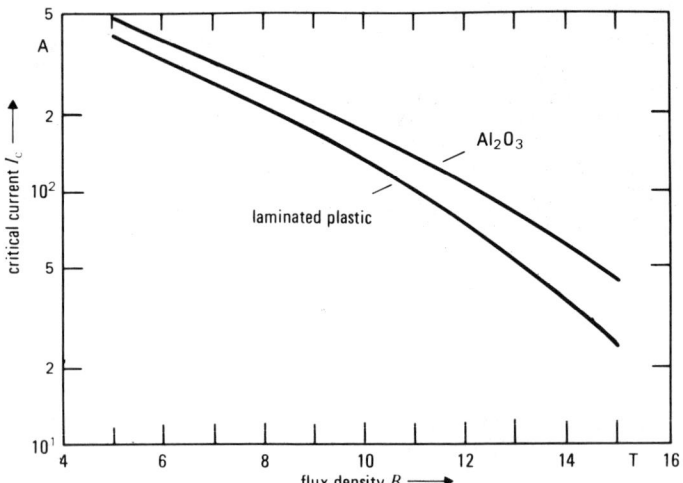

Fig. 7.   Critical current I_c versus flux density B for a 10,000
          filament Nb$_3$Sn conductor with a diameter of 0.5 mm on
          different sample holders.

currents, measured on Al$_2$O$_3$ exceed those measured on hard paper by
approximately 30% at 10 T and 50% at 15 T.   Under real conditions
in a magnet winding the conductors suffer tensile stress, therefore short sample measurements using a Al$_2$O$_3$ sample holder are
more representative as to the actual magnet current than are those
using laminated paper.

Figure 8 shows the influence of the diffusion temperature on
the current density of a 10,000 filament conductor, wire diameter
0.4 mm.   The diffusion time was sufficient to convert the Nb filaments completely into Nb$_3$Sn.   The low annealing temperature of
650°C is very favorable in the range of low flux densities.   A
high diffusion temperature is to be preferred, when high currents
are to be reached at a high flux density.   Thus the diffusion temperature for attaining optimum current density has to be chosen
deliberately.[5,7,10]

As to the diffusion time it should be rated so that thin filaments are completely converted to Nb$_3$Sn.   According to measurements
of Fig. 9, excessive diffusion time leads to a decrease of the
critical current because of grain growth.

Figure 10 shows the influence of diffusion time and temperature on the diffusion area and the maximum volume pinning force.
Additionally the product of both is inserted showing the expected
current carrying capacity of the filaments.   As is to be seen the
highest current density in Nb$_3$Sn is reached at a diffusion tempera-

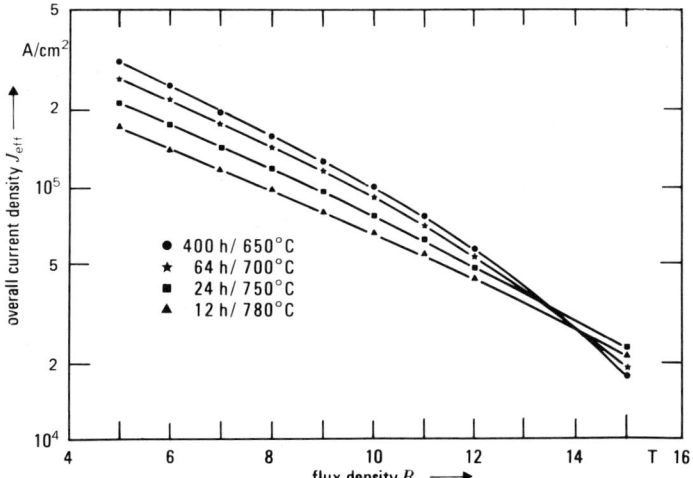

Fig. 8.   Critical current density J_{eff} versus flux density B of a
10,000 filament Nb_3Sn conductor for different diffusion
temperatures.

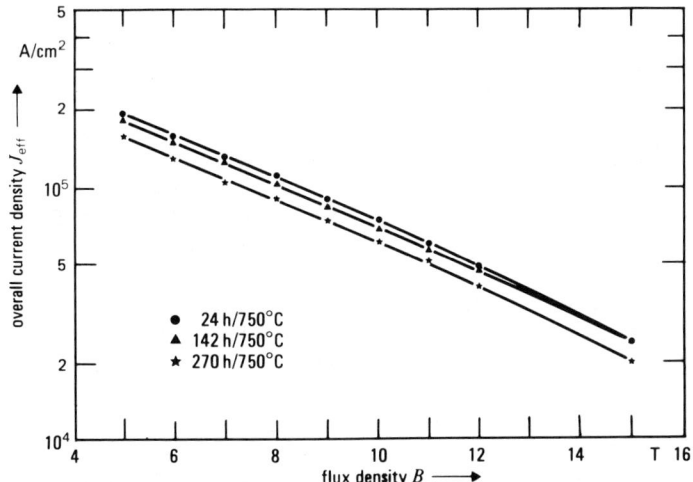

Fig. 9.   Critical current density J_{eff} versus flux density B of a
10,000 filament Nb_3Sn conductor for different times of
heat treatment.

ture of 600°C but the layer area on the other hand is too small to
attain a satisfactory overall current density.

Favorable temperatures lie between 680 and 720°C where the
increase of the diffusion area results in an increase of the
product A x F_{max} even when the layer current density decreases.[10]

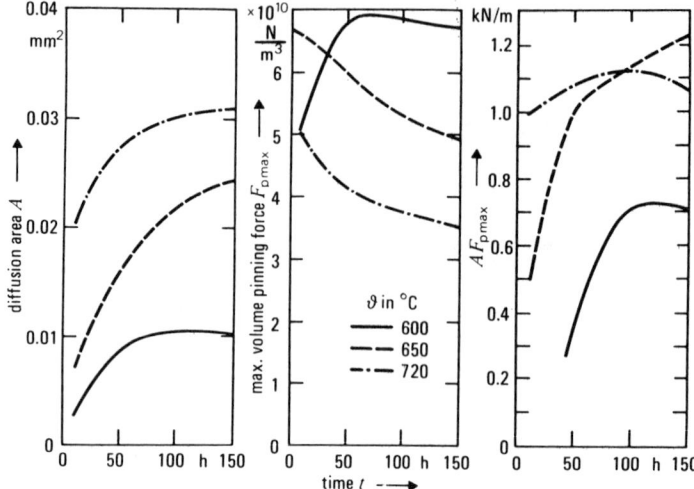

Fig. 10. Influence of diffusion time and temperature on the diffusion area and the maximum volume pinning force.

Beside the conditions during diffusion treatment the current
density is influenced by the diameter of the filaments.

Figure 11[*] shows measurement of J_{eff} vs B for wires of 0.6 mm
and 0.3 mm thickness containing 10,000 filaments, the filament
diameter being 3 µm and 1.5 µm, respectively. For low flux densities small filament diameters are favorable for high flux densities large diameters result in higher critical currents.

The highest overall current density running up to 3×10^5 A/cm^2
at 5 T, 10^5 A/cm^2 at 10 T and 0.35×10^5 A/cm^2 at 15 T has been
measured with a 0.4 mm thick conductor containing 10,000 filaments
with a diameter of 2 µm.

In addition to time and temperature of diffusion a suitable
filament distribution is important, providing permeable diffusion
paths. Because of the formation of Kirkendall voids during the
diffusion process the effective width of these channels is increasingly diminished resulting in smaller diffusion layers on the
central filaments. Therefore experiments with additional tin
reservoirs providing equal diffusion layers have been performed.

[*]The authors wish to thank G. Rupp for the high field measurements
of Figs. 11 and 15 performed at the Francis Bitter National
Magnet Laboratory.

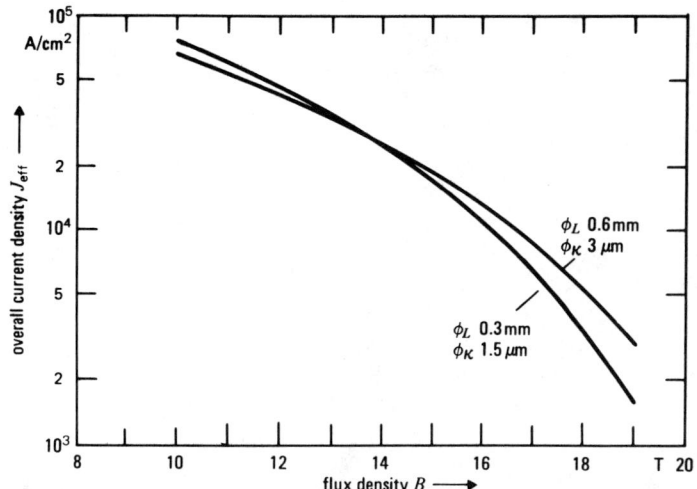

Fig. 11. Critical current density J_{eff} versus flux density B of 10,000 filament Nb_3Sn conductors with fully reacted filaments of 1.5 μm and 3 μm diameter.

Figure 12 shows conductor cross sections with and without additional tin reservoirs in the conductor center, in the subbundle layers. The latter one being stabilized with copper-tantalum at the circumference.

High current density conductors with a large aspect ratio of width to thickness have been developed in the form of cables. As already mentioned cables can be fabricated with and without external stabilization. Critical currents as high as 1700 A at 10 T, corresponding to an overall current density of 4.4×10^4 A/cm^2 have been measured in a 12-strand cable including 3 Cu stabilizers the cross section area being 1 mm x 3.85 mm (Fig. 6b).

Figure 13 shows an 8 strand internally stabilized cable. The critical current of this cable was 1710 A at 10 T, according to 4.8×10^4 A/cm^2 overall current density. The cross section was 1.2 mm x 2.95 mm, the filament number 10,000 per strand.

The Critical Temperature

The critical temperature T_c of the Nb_3Sn-bronze-composites is usually smaller than that of stoichiometrical Nb_3Sn with 18.2 K. This degradation originates from the compressive stress exerted by the bronze onto the filaments. In addition there has been found a dependence of T_c on the filament diameter and on the diffusion layer thickness.[6]

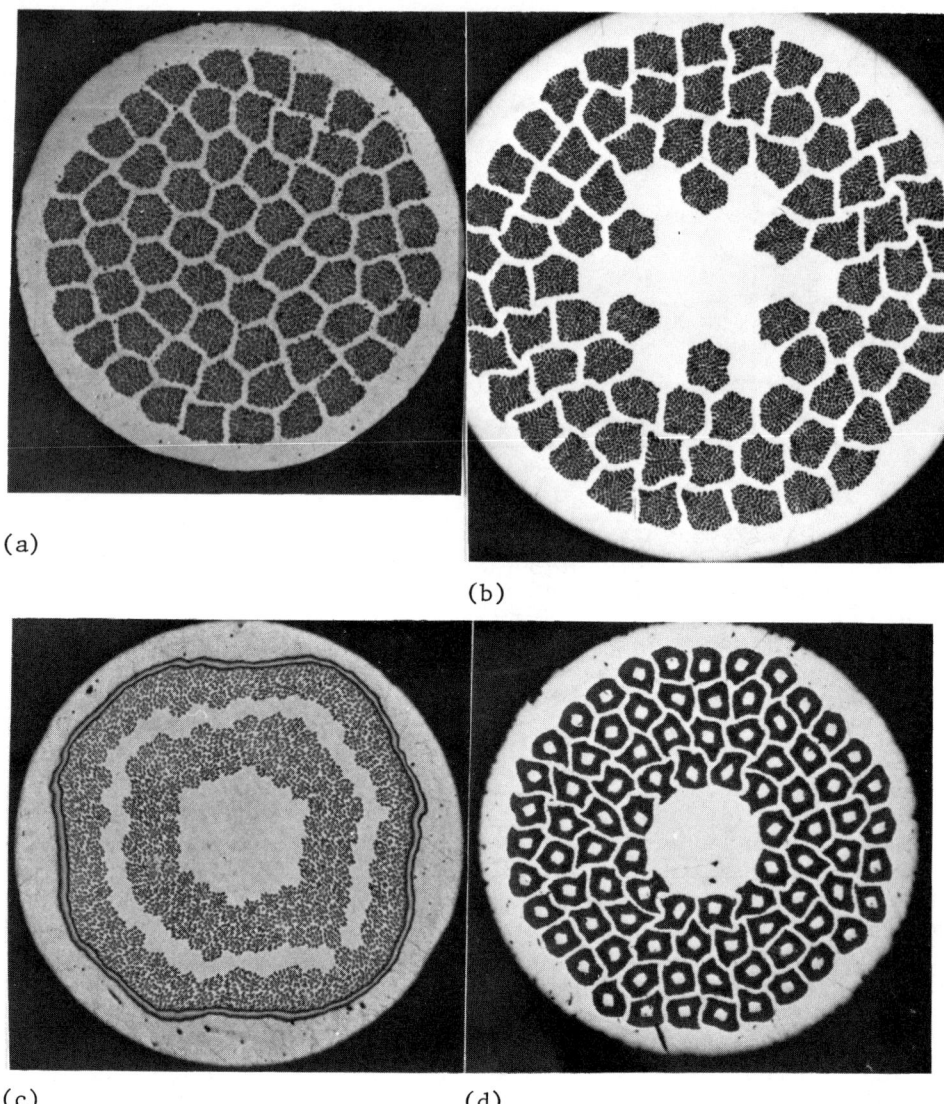

(a)

(b)

(c)   (d)

Fig. 12. Conductor cross section (a) without; (b), (c), (d) with
         internal tin reservoirs.

        Figure 14 shows T_c as a function of the Nb3Sn layer thickness
in a 1615 filament material. The niobium filaments had not been
completely reacted through. T_c lies between 16 and 17 K and
increases by one K for samples in which the entire bronze matrix
has been removed by etching thus the prestress becomes nearly zero.

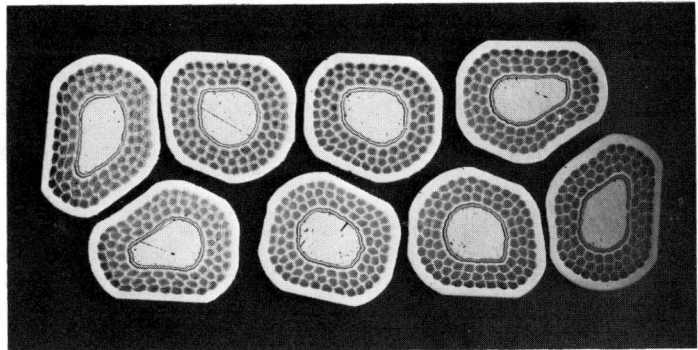

Fig. 13. Cross section of a 8 strand internally stabilized cable.

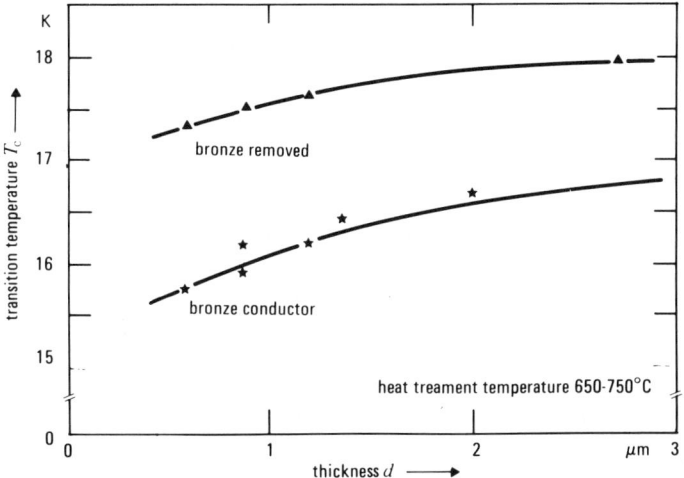

Fig. 14. Transition temperature T_c versus layer thickness d for a 1615-filament conductor before and after removal of the tin bronze.

The Critical Flux Density

The critical flux density B_{c2} is influenced by prestress.[9] In conductors containing a large amount of bronze B_{c2} is significantly decreased by prestress. Kramer extrapolation of I_c measurements on conductors without external stress leads to B_{c2} of 20 T for a volume ratio bronze to niobium of 3.5; this is 2 T lower than for a corresponding conductor with a volume ratio of only 2. If I_c measurements are performed on Al_2O_3 sample holders Kramer extrapolation leads to a critical flux density which exceeds that of samples on laminated paper holders by about 2 T as a consequence of tensile stress.

Figure 15 shows B_{c2} values extrapolated from I_c measurements between 11 T and 19 T for a 10,000 filament conductor containing filaments between 1.5 and 3 μm in diameter. These results indicate that the critical field of conductors with thick filaments exceeds B_{c2} of those with thinner filaments.

Coil Performance

For testing the electrical and mechanical performance of the multifilamentary conductors magnets have been wound with pre-reacted and nonreacted wires.[2,4,5]

Figure 16 shows the influence of bending on the current density. While decreasing the bending diameter initially an increase of the critical current is observed as consequence of tensile stress, which compensates prestress. Further reduction of the bending diameter causes remarkable I_c degradation leading finally to cracks in the Nb_3Sn layers. The bending diameter which coincides with I_c degradation is called the critical one D_c. It is proportional to the wire thickness d. The ratio D_c/d equals approximately to 100.

Figure 17 shows scanning micrographs of a strained Nb_3Sn layer. After 0.7% strain no cracks have been observed but slip steps with an angle of 45° to the filament axis. After 0.9% strain first cracks in the Nb_3Sn layer are observed, their number increased substantially for 3% deformation.

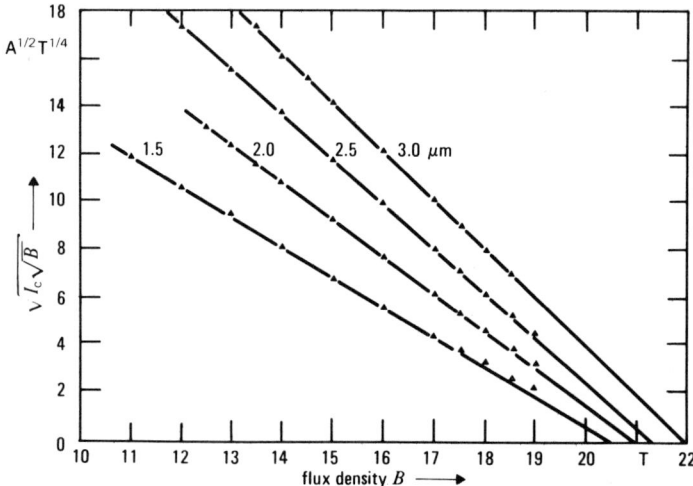

Fig. 15. Kramer function $I_c^{1/2} B^{1/4}$ for 10,000 filament Nb_3Sn conductors with fully reacted filaments of different diameter.

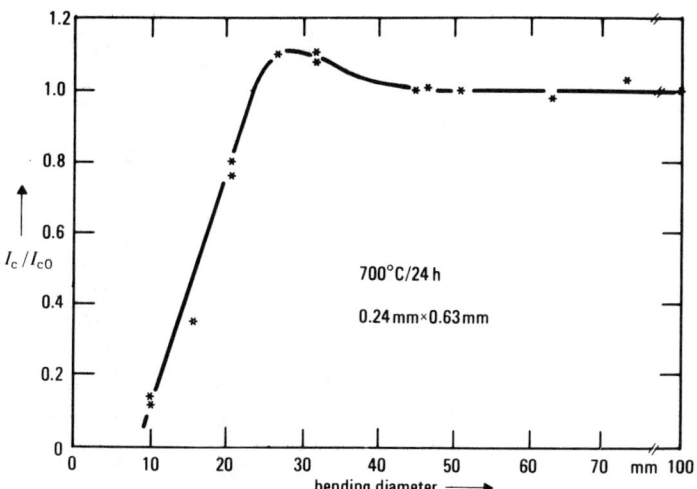

Fig. 16. Effect of bending on critical current I_c for a 1159 filament conductor with rectangular cross section.

        Table 1 shows the key dates of different solenoids. In the
case of wind and react glass and quartz fiber braid was employed
as layer insulation. Glass filaments served as turn insulation.
To fix the windings the coil was impregnated. All coils were
operated coaxially in a background NbTi coil. Solenoid 1 has been
wound with monolithic conductors of rectangular cross section without additional stabilization. Within an external field of 7.0 T
a central flux density of 13.8 T has been reached. Solenoid 2
with a free bore of 51 mm has been wound with a flat cable of 12
Nb$_3$Sn strands in the high field region of the coil and of 8 Nb$_3$Sn
strands in the low field region. The central flux density was
12.3 T within 8 T backing field. Solenoid 3 is a threefold
graduated coil with stabilized flat cables, containing three different numbers of Nb$_3$Sn- and CuTaCu-strands. The intermediate
connections are mounted on the coil flange. The solenoid reached
at the center of the 51 mm free bore diameter 13.9 T at 528 A
within a backing field of 8.0 T. Without backing field the
central field was 11.3 T at 1022 A. At 2.5 K a flux density of
15.1 T within 8.9 T backing field has been achieved.

        Figure 18 shows this Nb$_3$Sn solenoid inserted in a NbTi back-
ground coil. Solenoids 4 and 5 have been wound with prereacted
conductors. Glass filaments were used as winding insulation and
hostaphan foils as layer insulation. Both solenoids are threefold
graduated employing conductor combinations with 2.3 and 4 prereacted rectangular conductors soldered side by side between two

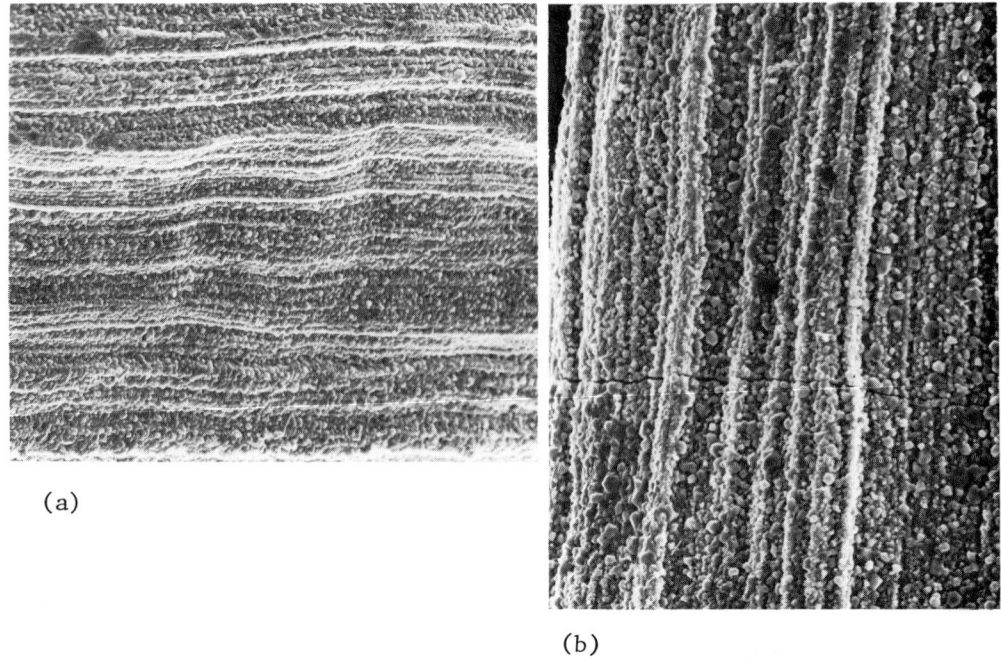

(a)

(b)

(c)

Fig. 17. SEM-micrograph showing Nb₃Sn filaments after strain of
0.7% (a), 0.9% (b) and 3% (c).

copper foils for electrical stabilization. The partial lengths
were connected by soldering. 13.0 T and 14.2 T, respectively, have
been reached with the two coils.

Table 1. Solenoids with Nb₃Sn multifilament conductors.

No.	coil dimension					type	conductor	max. central flux density at 4.2 K	
	ϕ_b	ϕ_{in}	ϕ_{out}	Z	L			B_{NbTi}	B_{tot}
	mm	mm	mm	mm	H			T	T
1	11	15	95	120	0.59	wind and react	monolith. unstab.	7.0	13.8
2	51	58	132	140	0.078	wind and react	cable stabilized	8.0	12.3
3	51	58	166	134	0.117	wind and react	cable stabilized	8.0	13.9
								0	11.3
								8.9	15.1 (2.5 K)
4	55	60	130	120	0.8	prereacted conduc.	monolit. stab. soldered	7.8	13.0
5	51	60	133	140	0.33	prereacted conduc.	monolith. stab. soldered	8.5	14.2

ϕ_b free bore diam.
ϕ_{in} internal coil diameter
ϕ_{out} external coil diameter
Z coil length
L inductivity of the coil

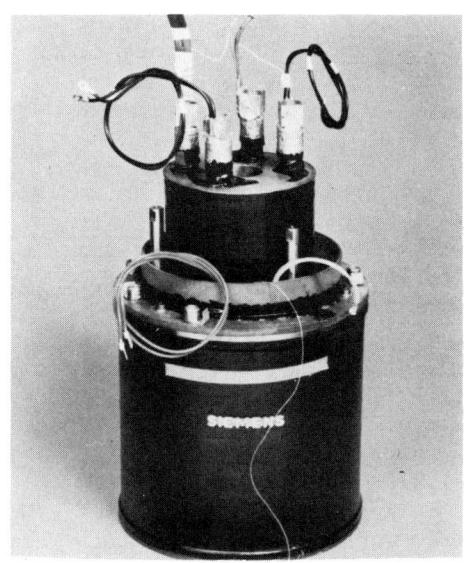

Fig. 18. Nb₃Sn coil with a free bore of 51 mm diameter inserted in a NbTi coil. Central flux density 13.9 T at 4.2 K and 15.1 T at 2.5 K.

The excellent electrical stability of these Nb₃Sn multicore conductors which can be excited with rates of more than 1 T/s is remarkable. The training of the solenoids is short. Commonly

after the second step the short sample current densities are
reached in some cases even higher values.

SUMMARY

Nb_3Sn multifilamentary superconductors have been developed by
the bronze process. Composites were fabricated by bundling and
extruding hexagonal niobium rods sheathed in jackets of phosphorus
free high tin content bronzes and by successively drawing. The
current carrying capacity has been optimized with respect to the
heat treatment to size and distribution of the filaments and to
the ratio of bronze matrix to niobium. Optimal heat treatment
and filament size are different for high and for low field application. Copper stabilized conductors have been fabricated by adding
copper centrally or peripherally protected by a diffusion barrier
of tantalum in the fabrication stage and by soldering prereacted
conductors of rectangular shape between copper strips. For high
current applications transposed flat cables with copper strands
have been developed. The performance of these conductors in coils
has been successfully tested in magnets produced by the wind and
react method and by using prereacted conductors. Magnets with
central flux densities up to 15 T in a bore of 50 mm have been
fabricated.

REFERENCES

1. H. Hillmann and E. Springer, _Siemens Zeitschrift_ 49:739
      (1975).
2. H. Kuckuck, E. Springer, and G. Ziegler, _Cryogenics_ 16:350
      (1976).
3. H. Hillmann, H. Kuckuck, H. Pfister, G. Rupp, E. Springer,
      M. Wilhelm, K. Wohlleben, and G. Ziegler, _IEEE Trans._
      Magnetics MAG-13:792 (1977).
4. H. Kuckuck, G. Rupp, H. J. Weiße, M. Wilhelm, K. Wohlleben,
      H. Hillmann, and E. Springer, _Proc._ _6th_ _Int._ _Conf._ _Magnet_
      Technol. MT-6:996, Bratislava (1977).
5. H. Hillmann, H. Kuckuck, E. Springer, H. J. Weiße, M.
      Wilhelm, and K. Wohlleben, _IEEE_ _Trans._ _Magnetics_ MAG-15:
      205 (1979).
6. G. Rupp, E. Springer, and S. Roth, _Cryogenics_ 17:141 (1977).
7. H. Pfister and H. Hillmann, _World_ _Electrotechn._ _Congr._,
      Sect. _3B_, Moscow (1977).
8. I. Pfeiffer, E. Springer, _Ztschr._ _Metallkde._ 68:667 (1977).
9. G. Rupp, _IEEE_ _Trans._ _Magnetics_ MAG-13:1565 (1977).
10. H. Hillmann, _Proc._ _6th_ _Int._ _Conf._ _Magnet_ _Technol._ MT-6:972,
      Bratislava (1977).

WORK IN THE U.K. ON FILAMENTARY A15 CONDUCTOR DEVELOPMENT

J. A. Lee   C. A. Scott

Chemistry Division   Rutherford Laboratory
AERE Harwell   Chilton
Didcot, U.K.   Didcot, U.K.

INTRODUCTION

In the U.K. substantial work on Nb_3Sn conductor development began at Harwell in 1967, with a Department of Industry (DoI) funded programme to establish the technical and economic basis for the industrial production of Nb_3Sn tapes produced by the surface-diffusion (liquid reaction) route. Work on intrinsically stable filamentary Nb_3Sn composites, financed by the United Kingdom Atomic Energy Administration (UKAEA), began in 1969 with work on the tin-bronze/niobium composites. Other matrix materials, for example silver, were also examined. The work developed in close collaboration with the Rutherford Laboratory, with the emphasis on conductors suitable for pulsed field use in accelerators. By 1974 the Rutherford Laboratory had built a successful series of small magnets, using conductors incorporating high purity copper regions protected by diffusion barriers, both tantalum and phosphorus-poisoned niobium, and had established the principles of the wind-react technique for magnet construction.[1,2] In collaboration with a commercial bronze founder, C. C. Clifford, Birmingham, the production of 13.5 wt% Sn bronzes on a tonnage scale by continuous casting had also been established.

A larger 2.2 x 1.1 mm section conductor with 40,000 filaments was developed shortly afterwards and a magnet development programme at Rutherford High Energy Physics Laboratory (RHEL) led to the successful construction of a 450 mm long x 50 mm bore hexapole magnet.[3]

A parallel, but separate, programme of conductor development was carried out by Imperial Metal Industries (IMI) with emphasis on tantalum diffusion barrier development. The essential difference between the two approaches was the adoption by IMI of a hot extrusion first stage – a commerically more acceptable practice for large scale production. From 1976, under the auspices of the DoI, Harwell and IMI have collaborated on conductor development, primarily for the construction of high field research magnets, and two UK companies: Oxford Instruments and Thor Cryogenics, now market high field magnets using filamentary Nb_3Sn conductors. There has been no significant requirement for larger high-current conductors. In particular, there has been no UK programme to develop large coils for fusion reactor experiments from these materials, such as has stimulated developments elsewhere.

The present position on conductor development in the UK stems from a Science Research Council (SRC) proposal to unify the various technical and commercial interests, including those of the Universities, in a national programme of superconductor machine/-device development, with the SRC playing a coordinating role. Work on A15 superconductor development, within this programme was initiated at a number of universities in 1979. Broadly speaking, immediate commercial development is aimed at improvement in performance of already developed conductor configurations, and of better fabrication techniques to lower costs, with a supporting longer term technological development programme on new types of conductor for new applications but based on existing material. The universities' programmes are aimed at a better understanding of the behaviour of technologically interesting materials, new ways of preparing known materials, and of fabricating in filamentary form A15 conductors which have not so far proved amenable to a solid state reaction route technique.

Harwell's role was envisaged as intermediate between that of the universities and that of the commercial producers, carrying forward new developments to the production of trial lengths of conductor for construction of development magnets. However, at the present time, a forward programme has not been funded and Harwell involvement is limited.

PRESENT COMMERCIAL POSITION

IMI have successively developed and marketed three types of filamentary Nb3Sn conductors, all of which have been used in commercial magnets[4,5]:
1)   An all bronze conductor (AB)
2)   A conductor with a copper sheath protected by a tantalum diffusion barrier (CS). A development of this is a restacked version which gives a larger conductor with islands of bronze and niobium filaments in copper matrix (copper web design)

3)   The current standard conductor developed in collaboration
       with Harwell; a conductor with a central copper island (CI)
       protected by a diffusion barrier, typically with 2000-3000
       filaments of 5-10μm diameter at conductor sizes in the range
       0.5-0.7 mm (Fig.1).

A Harwell developed conductor with 4477 filaments, and
containing ∿15% high purity copper in six separate 'floret' units
(Fig.2) protected by phosphorus poisoned niobium barriers (Section
5), is also marketed by IMI and used in the development of the
Oxford Instruments NMR magnets and other commercial magnets. This
latter configuration offers shorter thermal/electrical diffusion
paths between filament and stabilising copper than for comparable
copper island conductors and hence improved stability. Studies on
the configurational distribution of pure copper in the composite
for quench protection and stability are discussed in Section 3.

The in-magnet performance of these conductors is discussed in
Section 6. The 'wind before reaction' technique has generally
been used and the conductors supplied with lapped or braided glass
insulation.

The fabrication routes adopted by commercial manufacturers
are generally similar. Two aspects in which we have differed from
general practice are perhaps worth further comment; the use of
13.5 wt% tin bronzes continuously cast on the tonnage scale
(Section 4), and the use of phosphorus-poisoned niobium diffusion
barriers (Section 5).

 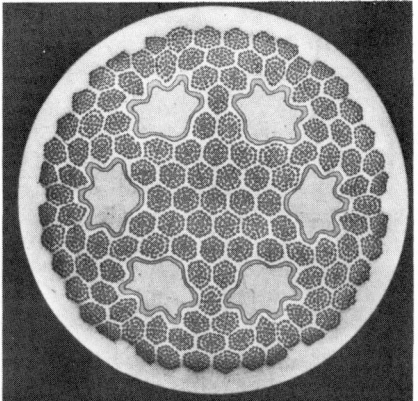

              Fig. 1   Fig. 2

FIG.1: Standard copper island conductor.   2000-3000 filaments,
        11% Cu in central island.

FIG.2: Harwell floret conductor 4477 filaments,
        15% Cu in florets.

STABILITY AND QUENCH PROTECTION

It is possible to make sufficiently small magnets from wire which contains no regions of pure copper.[4]   However special techniques have to be employed to protect such magnets when they quench. As a consequence the rate of current rise in the magnet is severely limited and one of the principal advantages of fila-mentary conductor is lost. Hence in general commercial conductors have to contain some copper just for quench protection. The amount of copper required for a given magnet design can be calculated using a computer simulation of the quench process.[7] For the laboratory-scale magnets which dominate the use of Nb_3Sn in the U.K. a combination of computer simulation and experience shows that 10-15% copper is adequate There is little point in lowering the proportion to 5%, say, because the resulting gain in overall current density would be small. As far as possible the use of more than 15% copper is avoided by employing alternative quench protection schemes (eg. subdivision of the winding and closely coupled shorted turns[8]).

Although the location of the copper within the composite is unimportant for quench protection it has a significant effect on stability.[9] Two sorts of stability are important. Firstly the self field stability of a large diameter wire is potentially a problem. (The limit on filament size imposed by the diffusion kinetics of the bronze process is always much smaller than that required for stability so filament stability can be ignored). Secondly "training" in magnets is believed to be related to the ability of a composite to return to its initial temperature after a temperature rise caused by a sudden local release of mechanical energy within the windings.[10] In this case there is a slight advantage in having the copper on the outside of the composite as in the copper sheath (CS) design if the disturbance originates in the impregnation material between the wires. The opposite applies if the disturbance originates within the conductor. However in practice training is not a problem in small laboratory-scale magnets and until larger magnets are common the choice between external or internal copper will be arbitrary.

As far as self field stability is concerned Wilson's analysis[9] shows that for the CS design the core can be regarded as a "macro-filament" and that at 10T a composite can be expected to be dynamically stable for diameters less than about 1.2mm. Extension of the macrofilament concept to the floret design suggests that as long as no filament is more than half this distance from a floret the composite should be stable. This is equivalent to allowing the central region of the composite to have a diameter of 1.2mm (see Fig.2): a less stringent restriction on the size than for the CS design. At higher fields larger sizes are allowed because the critical diameter is inversely proportional to the current

density. The location of the copper exactly in the centre of the
copper island conductor (Fig.1) makes it a special case. Here the
copper confers no dynamic stability on the composite because there
is no self field flux within the copper. So far this design has
only been used in magnets at diameters of about 0.6mm where it is
adiabatically stable, and no trouble has been experienced. Its
use at larger sizes should be approached cautiously.

CONTINUOUS CASTING OF HIGH-TIN BRONZE

     With demands for conductor at present limited, common
commercial practice has been to vacuum cast billets on a
relatively small scale from OHFC copper and 99.99 wt% pure tin.
For larger scale production casting in air is desirable. The
primary problem here is to control the level of phosphorus,
normally added in bronze casting as a de-oxidant and to improve
fluidity in casting. Levels in excess of $\sim$0.2 wt% P are commonly
used in commercial bronzes, but at this level formation of Nb_3Sn
in a filamentary composite can be completely suppressed, a thin
layer of a niobium phosphide forming at the niobium-bronze
interface. At lower phosphorus levels Nb_3Sn can form, but at a
slower rate than in pure bronzes and with a lower critical
temperature. The phosphorus content of the bronzes should be
$\sim$0.02-0.03 wt% or less[11] to produce Nb_3Sn with a good T_c.[12] This
level is adequate to keep the dissolved oxygen low, the solubility
of oxygen decreasing rapidly as the residual phosphorus content
increases, reaching an equilibrium minimum value of $\sim$0.0004%
oxygen and decreasing no further when the phosphorus content
exceeds 0.02% to 0.025%.

     Harwell, in collaboration with C.C. Clifford Ltd.,
Birmingham, developed tonnage scale continuous casting of 13.5 wt%
tin, low phosphorus bronzes at an early state.[13] Typically using
a Metatherm twin-strand machine with a nitrogen lanced 1.5 tonne
holding furnace and a 2 tonne casting furnace, but casting a
single strand through water-cooled graphite moulds, 88mm diameter
solid bar could be cast at a rate of 6cms/minute. Such bars, in
the as-cast state showed a fine (<0.5mm) chill cast grain
structure to a depth of $\sim$1.25cm, then fairly fine radial columnar
grains to the centre of cooling, where some equiaxed grains were
apparent. The 'eye' at the centre of cooling in this arrangement
would be offset $\sim$1cm from the bar centre, due to the asymmetry of
the die block. Porosity was very limited and on a micro scale,
and sections showed a few clean gas bubbles varying from 10 to
50μm diameter. Inter-dendritic δ-phase was present through the
body of the casting, but evenly dispersed and, in our experience,
on a much finer scale than with stakes of similar cross section
chill-cast into iron moulds, which showed large pools of δ-phase
and excessive porosity. A higher concentration of δ-phase,
associated with tin sweat was generally present at the surface to

a depth of ∿0.2mm, where the microhardness was measured as ∿185
VPH as against 160-162 VPH measured from 1mm below the surface to
the centre of the casting; the measured UTS varied from 295-350
MPa with elongation of 5 to 8%. Homogenisation in a reducing
atmosphere for 5-7 days at 650°C gave a uniform distribution of
fine, equiaxed grains across the section with the casting 'eye' no
longer apparent; the fine δ-phase within the body of the casting
dissolves, though residual δ-phase may remain in the outer surface
skin. The homogenisation anneal could be reduced to ∿1 day if the
casting were swaged, to give ∿20% reduction in area (RA) prior to
the anneal. Residual δ-phase and associated porosity at the
surface can lead to intergranular oxidation problems on subsequent
heat treatment, and transverse cracks developing on subsequent
drawing. Normal practice would then be to machine off the outer
surface layer to remove any surface oxide or flaws, and to
consolidate closing up internal porosity by a swaging-annealing
cycle with ∿40% RA between anneals after the first anneal. With
an interstage annealing cycle including a short heat treatment at
∿275°C, to reduce internal stresses prior to a recrystallisation
at 600 to 650°C, longitudinal stress-induced cracking could be
avoided.

    Analysis data for such a casting 29m long from samples taken
at regular intervals along the whole length (Table 1) shows good
compositon control. A similar quality control could be achieved
in casting tube 58mm OD x 38mm i.d.

    Casting on a similar scale, but with smaller capacity holding
and casting furnaces, Rotomead Ltd., Dundee have found it
necessary to use a pulsed withdrawal; for 88mm diameter solid
cast at 6.5 to 8cm/min a cycle 1.5 sec. pull, 2.0 sec. off, 0.5
sec. back, and with every two minutes a 15 sec. stop; for 58mm o.d.
tube, and a withdrawal rate 8 to 9.6 cm/sec, a straight pulling
sequence of 0.5 sec. on, 2.5 sec. off giving a net casting rate of
∿1cm/min.

TABLE 1. Composition of continuous casting

Sample No.	Cu wt %	Sn wt %	P wt %
1	85.90	14.06	0.03
2	86.14	13.81	0.03
3	86.21	13.75	0.03
4	86.62	13.31	0.03
5	86.19	13.72	0.03
6	86.83	13.11	0.02

DIFFUSION BARRIERS

In early conductors tantalum barriers were used to separate bronze and high-purity copper regions. This approach has been widely adopted in commercial fabrication and its advantages and disadvantages are generally known. Tantalum compatibility with copper and tin bronze is good; $TaSn_3$ can be formed, but at normal composite reaction temperatures the reaction rate is not significant and it creates no fabrication problems. Though the hardness of tantalum after the initial extrusion stage is high ($\sim$190–210 VPH) and interstage annealing temperature too low for softening, the work hardening rate is low and extensive deformation is possible before embrittlement occurs at hardnesses >270 VPH. Problems which can arise from a proclivity to shear fracture can be minimised by grain size control, and one approach, examined, in common with other manufacturers, has been to use a multiple layer, fine grain-size wrapped sheath.

Though tantalum bronzes can be fabricated successfully, niobium is less susceptible to failure. The hardness plateau at $\sim$125 VPH extends to elongations > x 10   Its use as a barrier keeps to a minimum the number of components in the composite with different working characteristics, and it has the advantage of reducing costs as compared to tantalum. Its primary disadvantage is that normally reaction will occur and a large diameter reacted barrier could exhibit instability. The Harwell group have developed niobium as a diffusion barrier, reaction at the barrier-bronze interface being suppressed by a local concentration of phosphorus. Ion implantation of the niobium has been exploited, but is likely to add significantly to the costs in commercial production. The simplest way is to sheath the niobium in a commercial phosphor bronze choosing the thickness of the sheath as a compromise between outward diffusion of phosphorus affecting adjacent filament performance and keeping the concentration of phosphorus adjacent to the barrier sufficient to inhibit reaction. With a commercial phosphor bronze containing $\sim$0.26 wt% phosphorus, thicknesses of 20–30μm can suppress reaction completely, for heat treatments of $\sim$20 hours at 750°C. With thinner sheaths some reaction occurs but the T_c of the Nb_3Sn formed at the barrier is reduced and with suitable configuration the reacted layer becomes normally conducting at operating fields. In a typical composite, Smathers and Larbalestier[12] have detected phosphorus at trace level in adjacent filaments, measuring the start of the superconducting transition as 17.2–17.8K, typical of bronze composites with sufficient tin to complete reaction,[14] and with a transition width of $\sim$1K for samples reacted at 700 and 800°C.

MAGNET PERFORMANCE

    Early work at the Rutherford Laboratory[1,3] established that filamentary Nb₃Sn composites could be made into useful magnets by winding unreacted conductor and then performing the reaction heat treatment on the complete magnet. In the UK this has now expanded into a small but thriving industry at Oxford Instruments, Thor Cryogenics and Cryogenic Consultants Ltd. All the commercially produced magnets are solenoids with outer sections wound from NbTi conductor and inner sections from Nb_3Sn conductor. Fig.3 shows the performance of a typical sample of copper island material used in many of these magnets.

    At present the largest use of conductor is in magnets for high resolution nuclear magnetic resonance (NMR) made by Oxford Instruments. The first   of the present type was a solenoid with a room temperature bore of 41mm made for Oxford University by a team from the Rutherford Laboratory and Oxford Instruments. It operates in the persistent mode at 11T, equivalent to proton

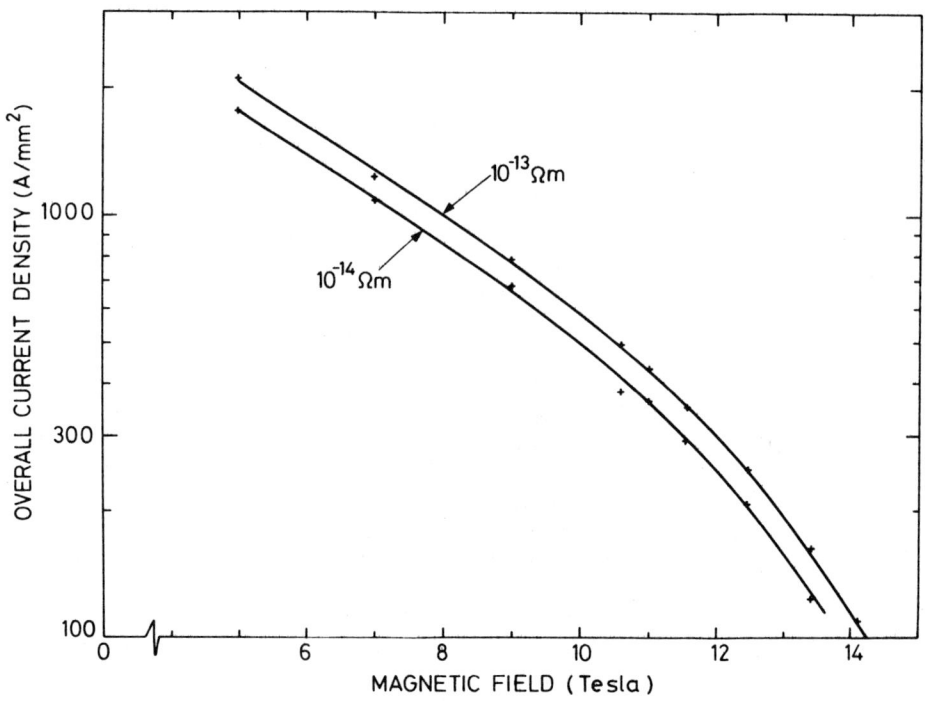

FIG. 3. Overall current density as a function of field for a typical copper island conductor, diameter 0.5mm, overall bronze:Nb ratio 3.5:1, heat treatment 72 hours at 700°C. Resistivity is defined over total cross-section.

resonance frequency of 470 MHz. Similar magnets are now being produced routinely with fields of 11.7T (equivalent to 500 MHz). An essential feature of these magnets is a special technique used to join the various pieces of wire in the magnet.   With this proprietary process it is possible to keep the total circuit resistance below $3 \times 10^{-9} \Omega$ and hence keep the decay rate of the current below 10^{-7}/hour. The technique is tolerant of small changes in conductor design; early magnets used the Harwell floret conductor, more recent magnets use both floret and CI conductor.

Other significant magnets include the series of split pairs and solenoids made by Thor Cryogenics,[4] particularly a 32mm bore solenoid which achieved 14T at 4.2K and exceeded 15T when cooled to 2K.[16]   This series includes magnets made from all bronze wire as well as from copper island and floret wire.

FORWARD PROGRAMME

Two other areas of work nearing completion are perhaps worth mentioning:

- The ac loss measurements reported earlier[17] have been extended and compared with the theory of Clem[18] confirming the conclusion arrived at earlier, that the surface barrier is the cause of the exceptional low losses at low amplitudes, and showing that the surface barrier is much reduced in the transverse field orientation.

- R Rawlings[19] has identified interstitial loops with a <100> Burgers vector in Nb_3Sn neutron irradiated at reactor ambient correlating their densities with integrated flux and changes in T_c. Some vacancy loops were also seen.

Forward long-term development programmes are concentrated at the Univerities, where three main programmes have been approved:

At Cambridge University a group under J.E. Evetts is working on the assessment and development of new copper matrix A15 multifilamentary composites. The work, at present in its early stages, will include studies of the effects of ternary additions, impurities and poisons on Nb_3Sn fabrication. A sophisticated ac technique has been developed which should allow property variations across the reacted layer to be measured.

At Oxford University programmes of work under J.W. Martin and M.J. Gorringe are aimed at a detailed characterisation of structural changes and internal stress developed during fabrication and heat treatment and on the micromechanisms of fracture, initially on filamentary composites, but in relation to latter studies on in-situ/continuous filament composites. Closely

related to these studies is proposed work at Harwell on the properties of close bundles of fine continuous filaments.

In a third programme at Manchester University, F. R. Sale
will study the effects of powder particle size, morphology and
composition on the superconducting properties of powder route
composites, including composites prepared by the ECN technique,[20]
based on an initial evaluation of different powder production
techniques.

A number of other programmes have been proposed but not yet
approved.

ACKNOWLEDGMENTS

It is a pleasure to acknowledge many useful discussions with
S. J. Warden at IMI Titanium and with our colleagues at Harwell
and Rutherford, particularly P. E. Madsen, D. Armstrong and M. N.
Wilson. We also wish to thank P. A. Hudson and H. Jones for much
help with measurements at the Clarendon Laboratory High Field
Facility.

REFERENCES

1.   D. C. Larbalestier, V. W. Edwards, J. A. Lee, C. A. Scott,
        and M. N. Wilson, IEEE Trans. on Magn. MAG-11:247 (1975).
2.   D. C. Larbalestier, P. E. Madsen, J. A. Lee, M. N. Wilson,
        and J. P. Charlesworth, IEEE Trans. Magn. MAG-11:555
        (1975).
3.   R. Q. Apsey, D. E. Baynham, and C. A. Scott, 6th Int. Conf.
        Magnet Tech. Bratislava MT-6:546 (1977).
4.   F. J. Brown, D. Phillips, F. D. Thornton, Proc. 7th Int.
        Cryo. Eng. Conf. ICEC-7:81 (1978).
5.   P. McDonald and W. Proctor, Proc. 8th Int. Cryo. Eng. Conf.
        ICEC-8, to be published.
6.   I. L. McDougall, Proc. 5th Int. Conf. Magnet Tech. MT-5:710
        (1975).
7.   M. N. Wilson, Computer Simulation of the Quenching of a
        Superconducting Magnet, RHEL/M151, Rutherford Laboratory,
        Oxfordshire, U.K. (1968).
8.   P. F. Smith, Rev. Sci. Instr. 34:368 (1963).
9.   M. N. Wilson, Proc. 5th Int. Conf. Magnet Tech. MT-5:615
        (1975).
10. M. N. Wilson, Superconducting Magnets, Oxford University
        Press, to be published.
11. P. E. Madsen, J. A. Lee and D. Armstrong, Brit. Patent
        1499507; J. A. Lee, P. E. Madsen, and R. F. Hills, Brit.
        Patent 11708/76.

12. D. B. Smathers and D. C. Larbalestier, <u>Int</u>. <u>Cryo</u> <u>Mat</u>. <u>Conf</u>.,
     Madison, WI (1979), to be published.
13. J. A. Lee, C. F. Old, and D. C. Larbalestier, <u>Colloques</u> <u>Int</u>.
     CNRS-42:87 (1974).
14. T. Luhman and M. Suenaga, <u>Appl</u>. <u>Phys</u>. <u>Letters</u> 27:61 (1976).
15. P. Styles, I. D. Campbell, D. I. Hoult, R. Porteous, R. E.
     Richards, and N. F. Soffe, <u>Proc</u>. <u>4th European Experimental</u>
     <u>NMR Conf</u>. (EENC), Grenoble/Autrans (1979), to be published.
16. D. Phillips, Thor Cryogenics, private communication.
17. J. P. Charlesworth and P. T. Sikora, <u>IEEE Trans</u>. <u>Magn</u>. MAG-15:
     260 (1979).
18. J. R. Clem, <u>J</u>. <u>Appl</u>. <u>Phys</u>. 50:3518 (1979).
19. R. Rawlings and C. Pitt, private communication.
20. J. D. Elen, C.A.M. van Beijnen, and C.A.M. van der Klein,
     <u>IEEE Trans</u>. <u>on Magnetics</u> MAG-13, 470 (1977).

DEVELOPMENT OF A-15 MULTIFILAMENTARY SUPERCONDUCTORS AT AIRCO

E. Gregory, E. Adam, W. Marancik,
P. Sanger and C. Spencer

Airco Superconductors
A Department of Airco, Inc.
600 Milik Street
Carteret, NJ 07008   U.S.A.

INTRODUCTION

In this paper an attempt is made to trace the commercial development of A-15 compound superconductors at Airco.

The work will be described under three separate headings – high energy physics, laboratory magnets and fusion conductors since the requirements demand somewhat different configurations in all three cases. This is not meant to imply, however, that the development work on A-15's has proceeded along separate lines for each application. The facts are quite the opposite and throughout a decade of development work a conscious effort has been made to develop a series of conductors which have a modular design ensuring flexibility in application.

HIGH ENERGY PHYSICS

The original development of multifilamentary A-15 conductors in the United States was carried out around the start of the last decade with the same use in mind as that which has sparked the development of fine multifilamentary NbTi some very few years before -- high energy physics accelerator magnets.[1,2] At Airco the development was started by investigating both the external diffusion approach and the bronze approach using several different configurations. The first work was with V_3Ga (Fig. 1) but this was abandoned because at that time the high cost of both the elements seemed to preclude the use of this A-15 compound for large scale applications.

It was known in the early stages of the development that it

47

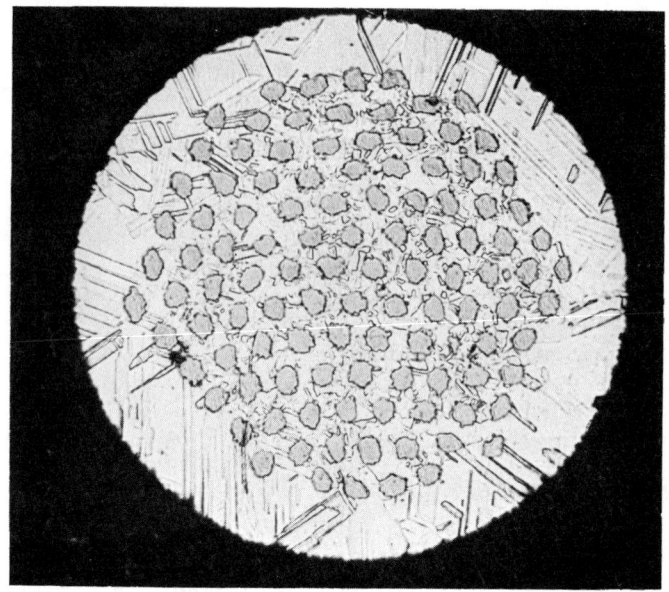

Fig. 1:   V$_3$Ga multifilamentary conductor consisting of 110 V
          filaments in a Ga bronze matrix 0.2 mm diameter

was desirable to have copper close to the A-15 compound in a commercial superconductor to provide stability and for this reason an
attempt was made to employ niobium tubes, either with copper in
their center (Fig. 2a) or surrounded by copper (Fig. 2b). The
concept was that the niobium would serve to form Nb$_3$Sn and also
act as a barrier layer. Problems were however encountered due to
the tube wall becoming nonuniform in thickness as the cross-section was reduced, and after heat treatment, the reaction layer
passed all the way through the niobium locally, so that the latter
no longer acted as a barrier and the stabilizing copper became
contaminated with tin. This tendency could be reduced by restric-
ting the amount of reaction, but the net effect of this was to
lower the overall critical current density (J$_c$) available. This
technique was therefore abandoned.

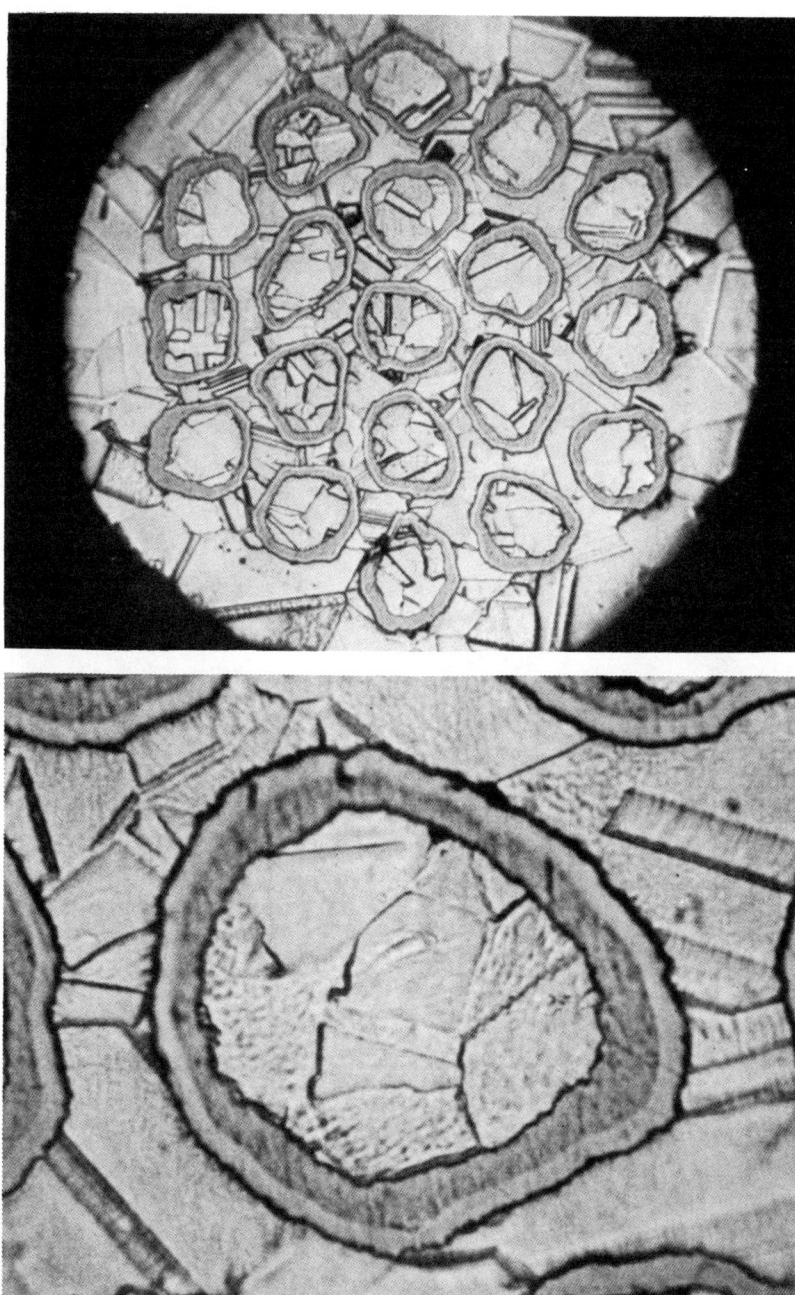

Fig. 2a: 0.3 mm diameter conductor consisting of 19 tubes of Nb
with copper cores in a bronze matrix.

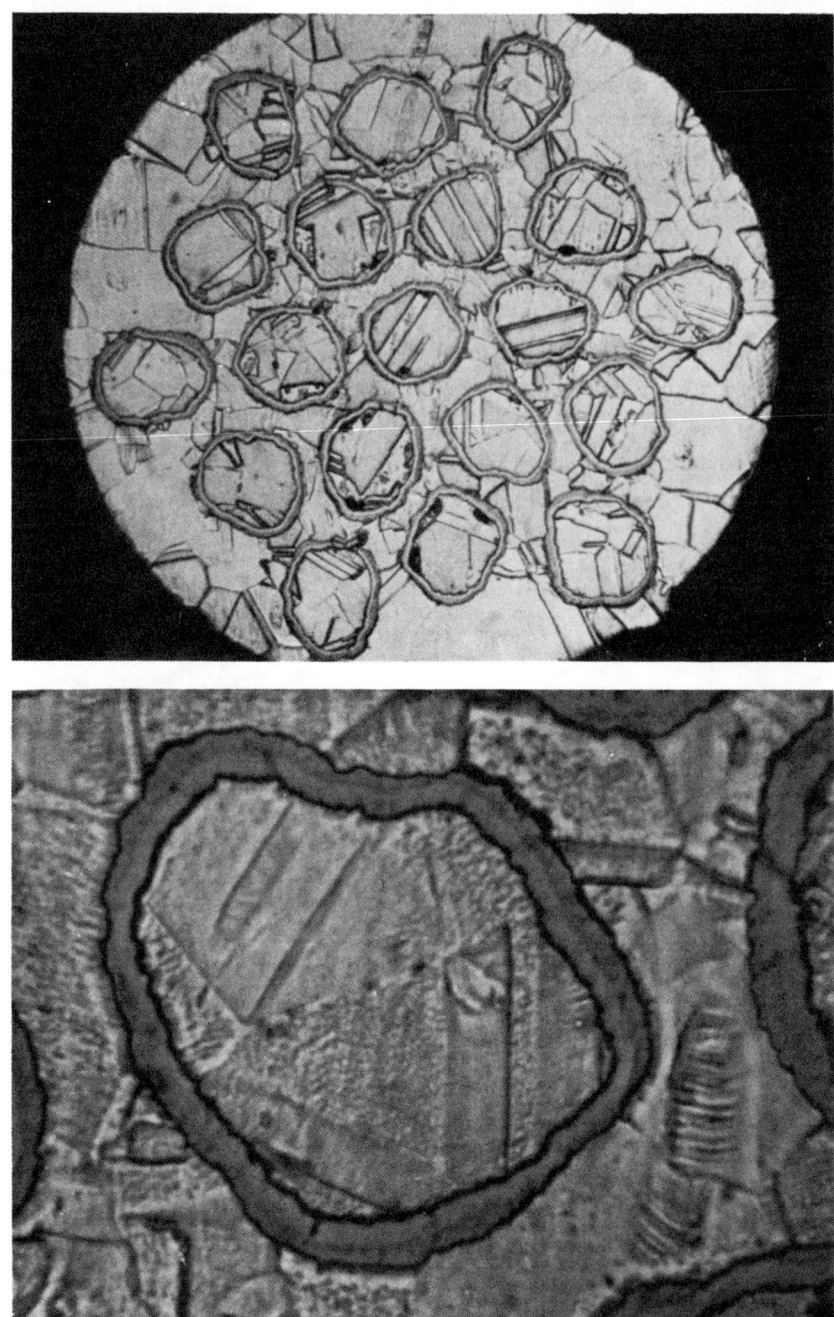

Fig. 2b: 0.3 mm diameter conductor consisting of 19 tubes of Nb
         with bronze cores and a copper matrix.

Effort was then concentrated on the use of niobium rods and since bronze tubes were not readily available at that time, a double extrusion technique was employed for the bronze approach. The first material developed commercially contained 342 filaments (19 x 18 - Fig. 3) and the first extrusion was made from a 10 wt % Sn bronze billet with 19 holes drilled in it. This technique enabled us to develop the 5 μm dia filaments necessary for suffi- cient reaction to give high J_c's. The results of this work were first published in 1974,[3] some considerable time after it had been completed. At about the same time, the results of some of Airco's even earlier work on V_3Ga[4,5] were published.

The next commercial material, supplied in kilogram quantities to the Brookhaven National Laboratory (BNL), is shown in Fig. 4a. This conductor contained 1045 (19 x 55) filaments, each 4 μm in diameter. It was supplied in two forms which were identical except that one was in a copper matrix for external diffusion and the second was in a bronze matrix. The third configuration (Fig. 4b) contained 361 filaments (19 x 19) and the groups of 19 were sur- rounded by thick tantalum barriers so that the copper could be

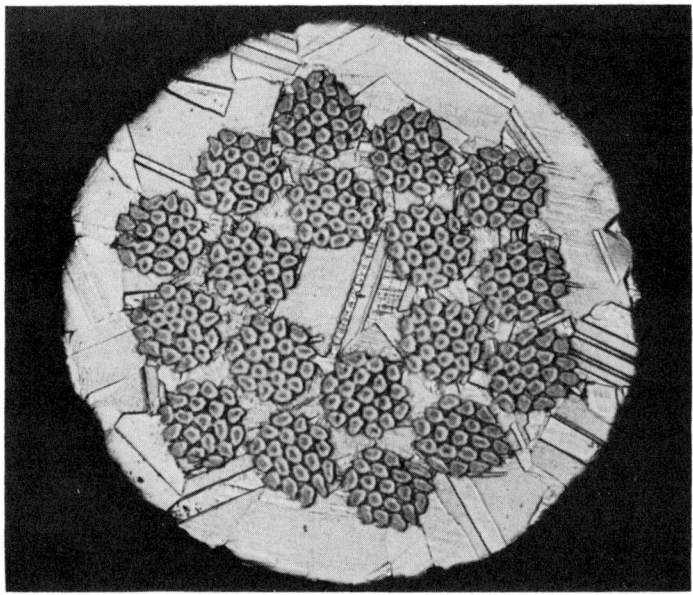

Fig. 3:   342 (19 x 18) filaments Nb_3Sn conductor 0.2 mm diameter 10 wt % Sn bronze matrix.

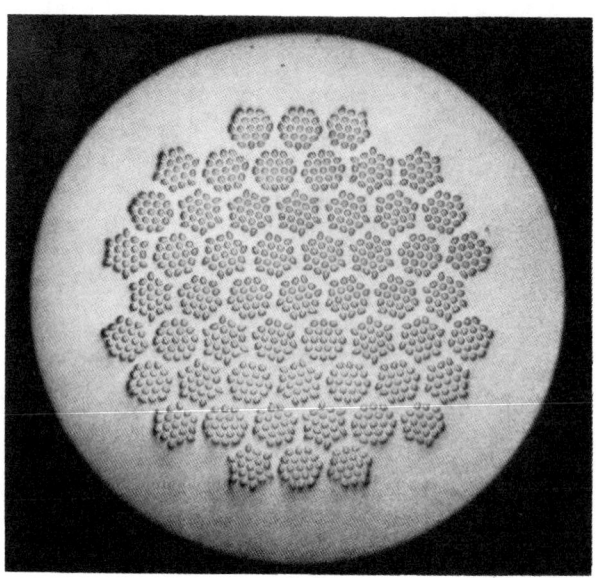

Fig. 4a: 1045 (19 x 55) filament Nb₃Sn conductor 0.3 mm diameter either copper matrix or 10 wt % Sn bronze matrix.

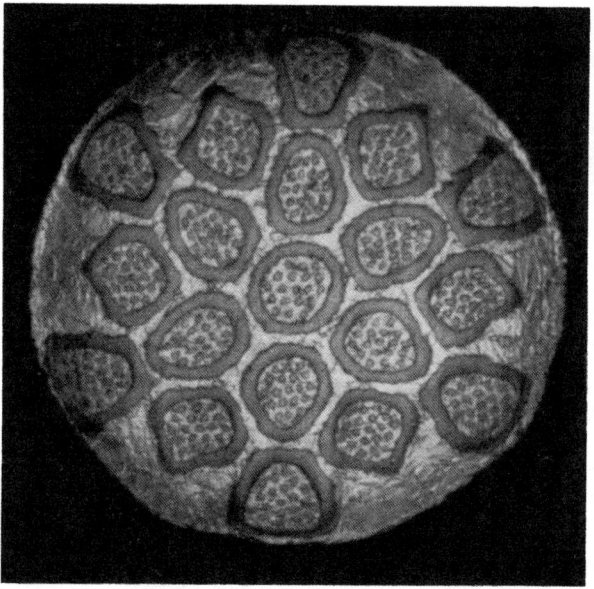

Fig. 4b: 361 (19 x 19) filament Nb₃Sn conductor with Ta barriers, copper matrix and 10 wt % Sn bronze cores.

incorporated to provide stability. This subsequently proved to be necessary from the results of magnet tests carried out at BNL.[6,7]

The next conductor supplied to BNL contained 1045 filaments in a 13 wt % Sn bronze matrix with a single tantalum barrier layer, still thick relative to present barriers, and stabilizing copper on the outside (Fig. 5). This material performed with considerable success in a prereacted braid used to make one meter long dipole magnets. The degradation of the conductor in the regions where the most severe bending takes place was compensated for by the use of a number of superconducting braids in parallel to reduce the local current density.[8]

Recent work on conductors for high field dipole applications has been directed towards the solution of the problem of the minimum bend radius. Sufficient in situ type Nb_3Sn is being externally tinned and then will be braided before reaction. The prereacted braid will be made into a small dipole at BNL.

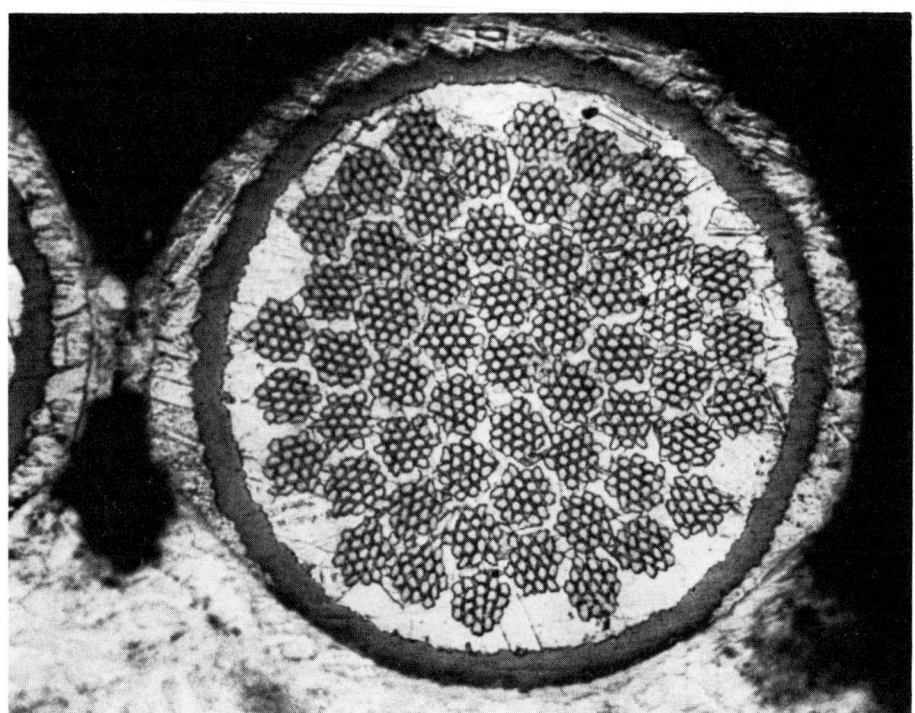

Fig. 5:   1045 (19 x 55) filament Nb_3Sn conductor 0.3 mm diameter with tantalum barrier stabilizing copper exterior and 13 wt % Sn matrix. Wire is in a BNL braid. (Photo courtesy of Brookhaven National Laboratory.)

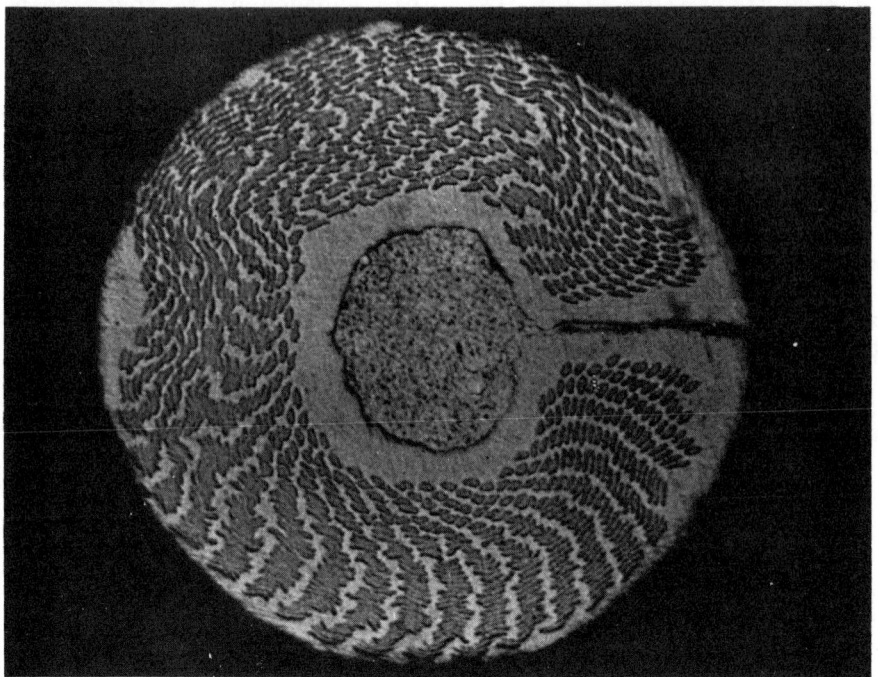

Fig. 6: Multifilamentary Cu matrix conductor made by wrapping a
         strip around a high Sn alloy core.

Airco has also developed a technique in cooperation with BNL
to wrap multifilamentary copper matrix conductor around a high
tin-copper alloy core and thus reduce some of the problems encountered by external tinning.[9,10] Figure 6 shows a conductor made
by such a technique. The same method could possibly employ in situ
material.

It is important to realize however that the in situ and
wrapping techniques mentioned immediately above are in their early
stages of development and the first dipoles have not yet been
made. The bronze wire was used for dipoles several years ago and
Airco has made several tonnes of similar material in recent years
and is presently engaged in large scale bronze process Nb_3Sn
wire manufacture for fusion applications as will be described
below.

LABORATORY MAGNETS

Airco did not build magnets in the early 1970's, and therefore
elected to supply potential customers with prereacted conductors.
A wire 0.76 mm in diameter with 3553 (19 x 187) filaments 5.0 μm
in diameter (Fig. 7) was developed and bend tests showed that it
was not possible to exceed 0.7% strain in the outer filaments
without degradation. The wire was thus limited by a minimum bend
diameter of approximately 100 mm. In order to decrease this, the
same wire was made into a rectangular conductor 0.2 mm x 2.1 mm
and this could be bent around a 25 mm dia. mandrel after reaction
without measurable degradation.[3]

All these bending strain data were obtained long before a
scientific explanation of the behavior was published.[11]   The
high aspect ratio of the above conductor led to manufacturing and
insulation problems and a conductor with a 3:1 aspect ratio (0.36
mm x 1.04 mm),[12] which was easier to manufacture and insulate,
was developed.

In order to preserve adequate ductility which was thought to
result from the remaining unreacted Nb core, only 75% of the wire
cross-section was converted to Nb_3Sn on wires which were prereacted, and thus peak current density was not achieved.

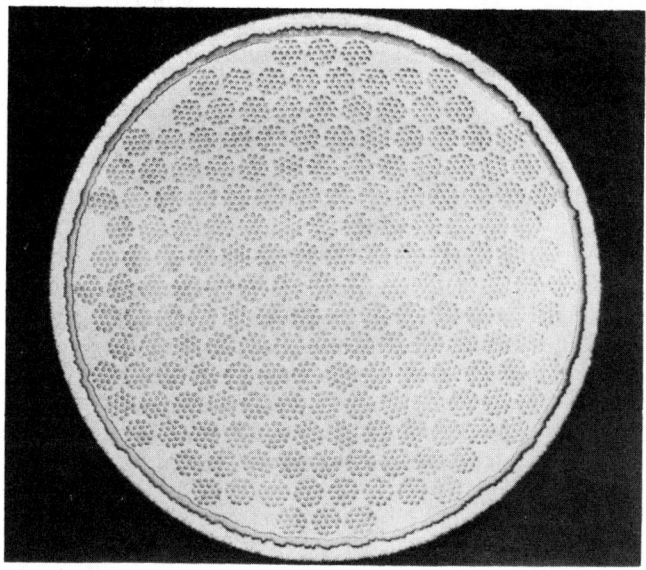

Fig. 7:   3553 (19 x 187) filament Nb_3Sn conductor 0.76 mm dia-
          meter with tantalum barrier layer, stabilizing copper and
          10 wt % Sn bronze matrix.

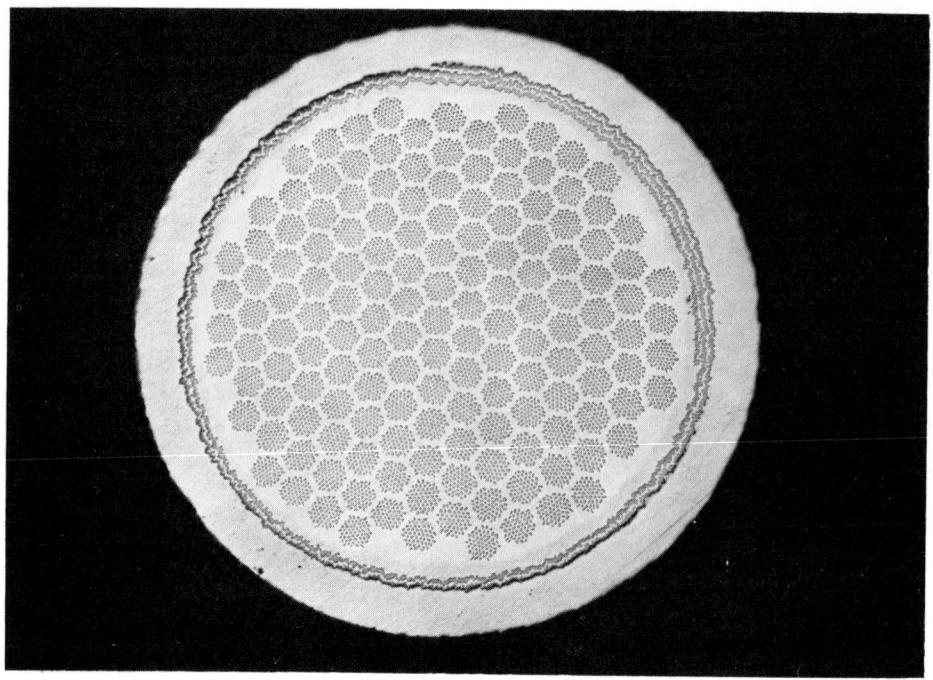

Fig. 8:   8305 (55 x 151) filament Nb$_3$Sn conductor with tantalum
          barrier layer, stabilizing copper and 13 wt % Sn bronze
          matrix used for small high field magnets.

     An insulation and impregnation procedure has recently been
developed at Airco which makes it possible to make small coils by
a wind and react technique. A 54 mm bore magnet for the National
Magnet Lab using a wire with 8305 (55 x 151) filaments (Fig. 8)
has been made recently. The magnet is in three sections, each
with differing wire diameter and amounts of copper stabilization.
A new manufacturing technique allows the stabilizer quantity to be
varied, but the conductor core is a standard configuration developed readily from the modular approach described below. The
three coils go into a background field of 9 T from a Bitter magnet
and the overall configuration is designed to achieve 15 T.

MAGNETIC FUSION

     In 1974, Lawrence Livermore Laboratory was charged with the
responsibility of developing reliable Nb$_3$Sn multifilamentary
superconductors and superconductor large scale production methods
for use in the construction of yin-yang magnets and for

other fusion needs.[13] The aim was first to develop a conductor
to carry 1 kA at 12 T, and then two conductors were required to
carry 3.5 kA and 10 kA at 12 T, respectively, for what was then
called FERF (Fusion Engineering Research Facility). Airco, because of its previous work for the high energy physics market, was
in an ideal position to develop such a conductors (Fig. 9). The
successful development was completed by 1975[14] using a modular
approach based on the 3553 (19 x 187) filament conductor unit with
a tantalum barrier layer,[12] similar to the wire developed for
high-energy physics and laboratory magnet applications (Fig. 7).

The 65,507 filament conductor made by a triple extrusion using
19 of the 3553 filament units was tested in a 270 mm bore coil at
Lawrence Livermore Laboratory in a 5 T background field. Flux
jumping in the coil was considerable due presumably to conductor
motion. The coil as a whole went to short sample at 9.94 T after
some training.[15] Parts of the conductor were strained to 0.6%.

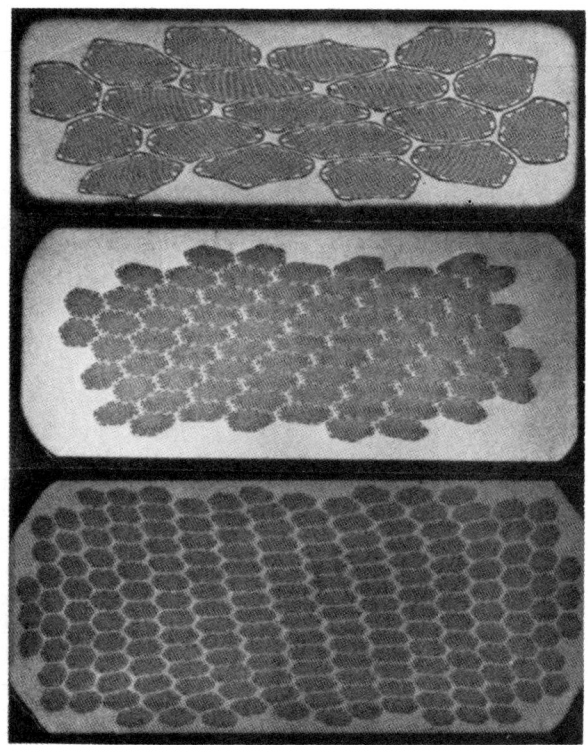

Fig. 9:   Three 10 wt % Sn bronze matrix conductors:
          a. 67,507 filaments (19 x 3553) 1.68 x 5.0 mm
          b. 259,369 filaments (73 x 3553) 3.9 x 9.3 mm
          c. 664,411 filaments (187 x 3553) 6.0 x 13.75 mm

It was therefore proved by mid-1975 that a high current 1 kA prereacted multifilamentary Nb_3Sn superconductor manufactured by the bronze method was a good candidate for the construction of high field magnets.

The conductor used in this magnet had, however, the following drawbacks for large scale fusion applications: (1) it had only been produced in 300 m lengths from a 100 mm diameter billet; (2) it had only been tested in coil form at 10 T; (3) it had only a short sample current of 1 kA at 12 T; (4) it was not cryostable and further stabilizer strips had to be added; (5) its cost was too high; (6) it was not ideal for pulsed applications. At the time, some of these shortcomings appeared trivial, but what was not appreciated were the problems involved in scale up, which turned out to be both technical and economic. It has taken five years to overcome them and to build a facility to supply Nb_3Sn conductors reliably in the quantities sufficient for the needs of the fusion program. The way that this has been done for LLL will be reported later in this paper, but to maintain the chronological sequence, the work for the Oak Ridge National Lab (ORNL) conductors will now be described.

In 1977 work was started on the LCP Project, the Airco-Westinghouse Large Coil being the only one to be made from a Nb_3Sn conductor. The design is a forced flow type which operates at 17.6 kA, 4 K and 8.4 T and has a square cross-section of 20.7 mm x 20.7 mm with rounded corners[16]   (Fig. 10). Cooling is provided by forcing supercritical helium at 4 K through the interstices of the cable. The basic strand is a 0.7 mm diameter wire containing 2869 (19 x 151) filaments surrounded by a tantalum barrier layer and 66 vol % of high conductivity copper (Fig. 11). The fabrication of the conductor from individual basic strands of wire can be divided into two parts: (1) cabling, and (2) jacketing and compaction.

The cable is made from 486 (6×3^4) of the above basic strands. This phase of the work is divided into five separate cabling operations.

The second part of the fabrication process is to continuously weld a sheath around the cable and to compact the conductor to final dimensions. Little development of the basic conductor was required for this work, but the main development problems lay in the cabling and sheathing area. As part of the LCP verification testing program on this type of partially compacted cable, a one-sixth scale model conductor was tested and shown to be highly tolerant of both bending strain and tensile strain as shown in Fig. 12.[17]

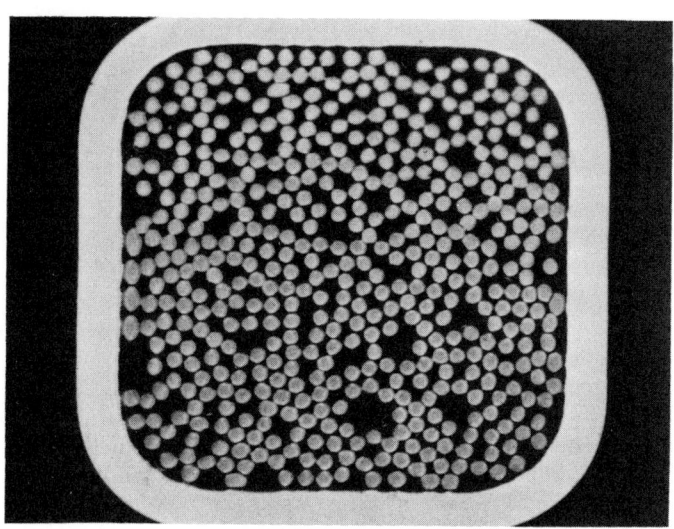

Fig. 10: Forced flow type conductor for the Large Coil Program
20.7 mm x 20.7 mm containing 486 (6 x 3^4) individual
strands.

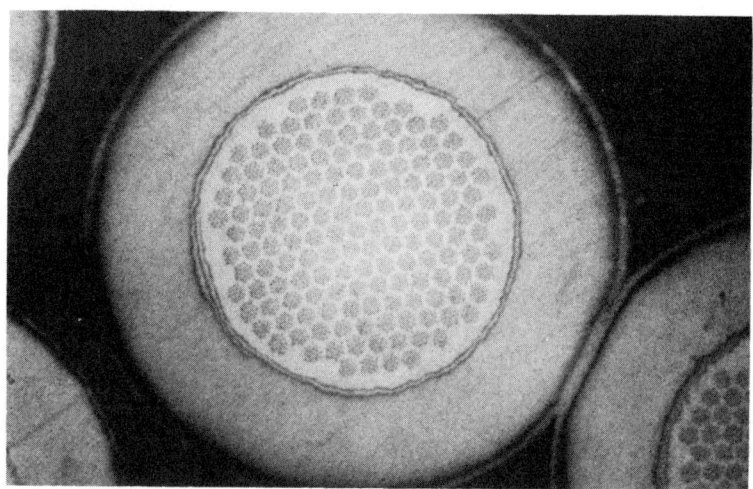

Fig. 11: Individual strand of LCP conductor 0.7 mm diameter containing 2869 (19 x 151) filaments with tantalum barrier
layer and 65% stabilizing copper and 13 wt % bronze
matrix.

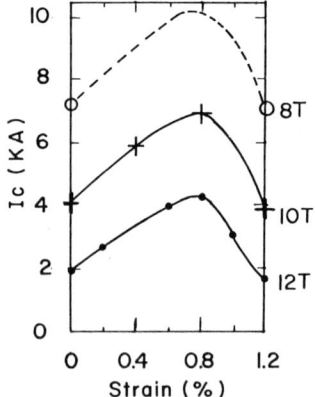

Fig. 12: Critical current of a one-sixth scale model of the LCP
forced flow conductor as a function of tensile strain.

It is however important to realize that for this one D coil 3 Mm
of 0.7 mm diameter wire is required which, although only onetenth
of the NbTi wire required for the Energy Doubler/Saver project at
Fermilab, necessitates a considerable scale up of Nb_3Sn wire
making capability. This scale up has started at Airco.

In 1977 also, the decision was made that the Lawrence Livermore Laboratory (LLL) would concentrate on building a high field
insert coil for their existing test magnet. The HFTF conductor
required for this insert was to carry 7.5kA at 12T and 4.2K, and
have added stabilizer strip. In an attempt to reduce costs, a
two-stage extrusion process was developed to produce a conductor
with 333,982 Nb_3Sn filaments.[18] This conductor was designed
to overcome the first five of the six objections held against the
1 kA conductor described earlier. It was still a monolith however

and therefore not ideal for use in pulsed applications. Although
this was not a drawback for the HFTF application, the core was
such that the conductor suffered from current transfer problems.
The requirement for a very large number of filaments and a long
continuous length of the monolith also led to fabrication problems[19]
and these in turn led to the development of a modification of
the core design which is shown in Fig. 13. It consists of a
"compacted monolith" made from an 18 (6 x 3) strand cable of 2.3
mm diameter wires, each containing 15,895 (187 x 85) filaments, a
tantalum barrier layer and external copper (Fig. 14). The final
filament size is approximately 5 μm and the copper content in the
core is 60 vol %. Two stabilizer strips, tinned on one side, are
added to this core to form the final conductor[20] (Fig. 15).
This conductor provides an answer to all the criticisms leveled
against the 1 kA conductor used in the test magnet in 1975. Airco
is now in the process of manufacturing 2 km of this conductor.
This includes 36 km of 2.3 mm diameter wire.

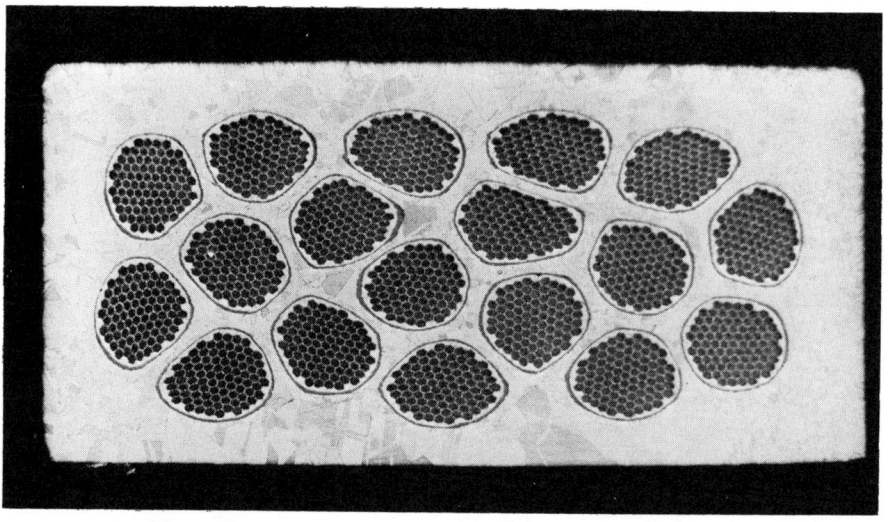

Fig. 13: Compacted monolith 5.4 mm x 11 mm made from 18 (6 x 3)
        strand cable.

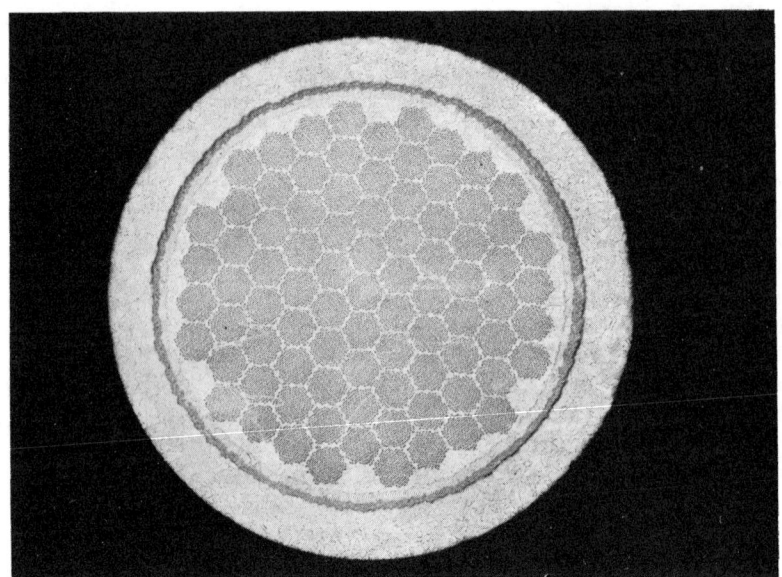

Fig. 14: Individual strand used in compacted monolith – 15,895
(187 x 85) filament Nb$_3$Sn conductor 2.3 mm diameter
with tantalum barrier, stabilizing copper and 13 wt % Sn
bronze matrix.

Fig. 15: Complete HFTF conductor with stabilizer strip added.

A critical current density (J_c) of 350 A/mm^2 at 12 T and
zero applied strain (based on the bronze, Ta, Nb, and Nb$_3$Sn area
of the conductor) was recently obtained on this "compacted monolith" containing 18 strands. The effect of applied tensile strain
on critical current for a conductor of the compacted monolith type
is shown in Fig. 16[20] compared with data previously reported[21]
on conventional monoliths fabricated by the triple extrusion
process. The J_c increases to about 118% of the zero strain at
0.5% strain and then begins to decrease as did the triply extruded
conductor.

The conductor described above is monolithic in appearance and
behavior, has none of the drawbacks of normal cables, is fully
transposed and designed to be readily scalable by simply varying
the number of wires cabled together. This provides another way in
which the basic unit can be adapted to different current and stability requirements.

A variation of this core could make an ideal component for
incorporation into a 12 T pool boiling conductor for an ETF and/or
INTOR design. Fig. 17 shows such a conductor which is proposed
for the Airco-General Dynamics-ORNL 12 T Model Coil. It is designed to carry 10 kA at 12 T and 4.2 K.

A modification of the LCP conductor is also proposed for the
Airco-Westinghouse-MIT 12 T Model Coil which will provide a 12 T
forced flow alternative for ETF and/or INTOR.

RECENT WORK

It should not be concluded from the above that all the activity at Airco on A-15 compounds is restricted to large scale manufacture of Nb$_3$Sn by one variation of the bronze process. Efforts are continually being made to improve the process and to
reduce manufacturing costs. Also, alternative and, hopefully,
cheaper methods of manufacture are under constant review. Nb$_3$Sn
multifilamentary wire of the LCP type (containing 60% copper) is
now sold in large quantities at less than twice the price of standard NbTi conductor of the same size. The problem is that so far
LCP has been the only large quantity requirement.

The high cost of Ta is causing us to consider ways of replacing it.

The lack of availability of bronze tubes led to the development of the double extrusion technique which has certain disadvantages one of which is that it is more expensive than the single
extrusion technique generally used for the manufacture of NbTi.
In recent months bronze tubing has been obtained on special order

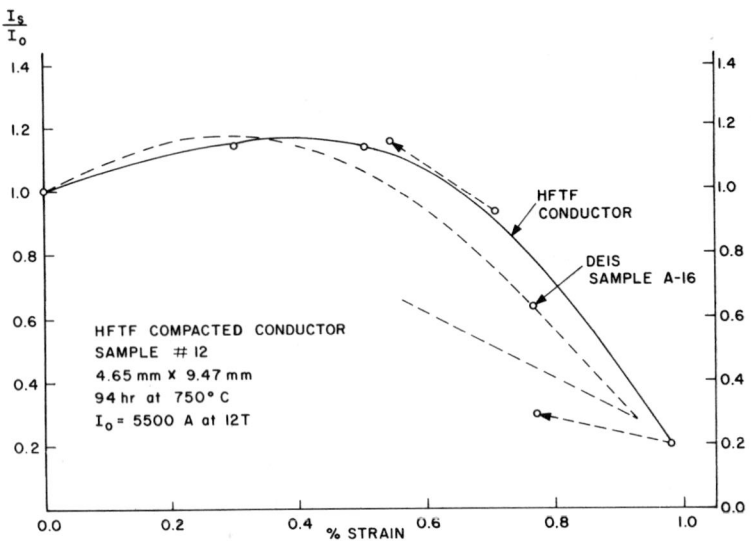

Fig. 16:   Critical current of an HFTF type compacted monolith as a
           function of strain compared with data previously obtained
           on conventional monolith fabricated by triple extrusion.

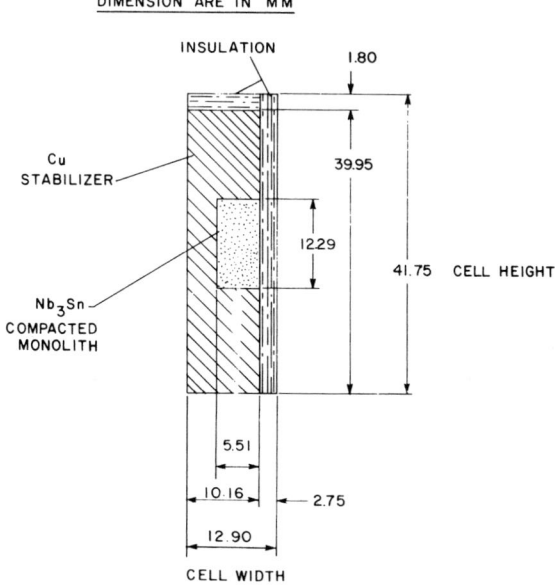

Fig. 17:   Schematic drawing of conductor proposed for the
           Airco–GD–ORNL 12 T Model Coil.

and a 2860 filament array in a 150 mm diameter can is presently
being assembled. The viability of this approach will depend on
the ultimate cost and availabiity of this tubing.

The in situ work and the technique of wrapping a material
consisting of a copper matrix containing Nb filaments around a
high Sn-Cu alloy core has been mentioned above.

Although the V_3Ga work was abandoned in the early 1970's,
work on the commercial fabrication of material made from vanadium - 8% gallium rods in a copper-gallium matrix has been started
in an effort to duplicate the work done on a laboratory scale by
the U.S. Naval Research Laboratory.[22] Considerable difficulty
has been experienced in obtaining the alloy rods in the degree of
purity necessary for fabrication by conventional methods into fine
multifilamentary wire. It is thought that the problem has now
been solved at least in part and suitable starting material is
expected to be delivered in the near future.

All the work described so far has been on multifilamentary
A-15 compounds but Airco has also developed a high rate magnetron
sputtering process for producing Nb_3Sn tape for transmission
lines, a variation on the bronze process.[23] A commercial continuous high rate magnetron sputtering line has been built and
operated.

Small quantities of Nb_3Ge tape with a T_c of 23 K have been
made in a small laboratory type high rate magnetron sputtering
device but no funding has been available to transfer the technique
to the commercial continuous machine.

SUMMARY

Five years ago, enough of the basic development work on
Nb_3Sn multifilamentary material made by the bronze process had
been completed to indicate that such a material would probably be
suitable for large scale applications. Airco has therefore concentrated on perfecting a commercial process which is suitable for
making A-15 compounds for large scale applications.

The process can produce, reproducibly and in long lengths,
materials in which:

1.   the Nb_3Sn can be in close proximity to copper;
2.   the amount of stabilizer can be varied;
3.   the current carrying capability can be varied while main-
      taining a reasonable overall J_c by use of a modular
      approach;
4.   the cost of manufacture is being constantly reduced by
      process improvements.

Since the filament diameter is controlled by a desire to achieve maximum conversion of Nb and Sn to Nb_3Sn and not by the physics of the application as it often is in NbTi conductors, the size is always small (less than 5 μm) and the number of filaments varies with the overall I_c required. In order to reduce the amount of material which must be placed in inventory to meet a range of I_c requirements, a modular approach has been adopted.

Three first extrusion units containing 19, 55 and 187 filaments respectively are available. Varying numbers of these first extrusions can be combined into second extrusions to give a series of small diameter conductor basic units to which varying amounts of copper can be added.

If high current conductors are required for operation at high fields, another degree of flexibility can be introduced by using a cabling, braiding or a "compacted monolith" approach. These techniques are frequently preferable to further extrusion for the production of large superconductor cores to which additional stabilizer can be added.

No significant effort towards optimizing the J_c in small billets by adding third elements, making ultrafine filaments and varying the bronze to Nb ratio in has been attempted recently at Airco, although some small amount of work has been done on varying heat treatments to optimize properties at various fields in both prereacted and "wound and reacted" conductors. Instead work has been concentrated on scaling up a standardized modular approach so that significant amounts of material, with properties markedly superior to those of available NbTi materials, can be produced reliably and reproducibly. Many problems have been encountered and overcome in accomplishing this scale up and Airco is now in a position to determine those process improvements reported in the literature which are most likely to be capable of scale up.

While the techniques have been developed and some of the equipment necessary for scale up purchased, only one half of the work required to make the HFTF conductor has been completed. The work on the LCP conductor, which involves eight times the amount of material used in HFTF, is just starting.

With the quantity production of the basic A-15 conductors now beginning, much of the equipment is in place and techniques developed for converting these wires into fully stabilized, fully transposed large conductors. It is in this area that a lot of the work in the last five years has gone. Cabling, braiding, embossing, plating, cleaning, wrapping and cladding lines have all been

built and are now operational. These lines are flexible and adapt-
able to different sizes and configurations of both NbTi and A-15
conductors.

Since the quantities of material, while becoming large by superconductor standards are very small by normal nonferrous melting and
fabrication standards (and likely to remain so until the turn of the
century), it has proved difficult to interest the companies engaged
in these activities in deep involvement in what still remains a highly
developmental activity with only long range payoff. To overcome the
problems of attempting to do one's development in someone else's production line, which is frequently engaged in routine work, Airco has
adopted the policy of purchasing and/or building as much as possible
of this ancillary equipment needed for the operations listed above.

All our equipment and personnel are now located in one 10,000 m^2
site in Carteret, New Jersey, and this plant is totally devoted to
the development and production of NbTi and Nb$_3$Sn superconductors and
coils. This facility, we believe, provides a national and international capability for the supply of the conductors needed for large
scale applications requiring large coils which will operate at 12 T,
(i.e. ETF, INTOR and even more advanced superconducting devices).

ACKNOWLEDGMENTS

The authors would like to acknowledge the continued support provided us by Airco, Inc. top management who have approved the funding
of much of the development work over the years.

We wish to acknowledge also the work of the many employees,
past and present, of Airco who have helped either directly or in-
directly in the development of the conductors described.

Without the continuing support of our principal customer, the
Department of Energy, both the headquarters and the various National
Laboratories with whom we have had development or material supply
contracts, all the work described would have been without value and
therefore not carried out.

REFERENCES

1. A. R. Kaufmann and J. J. Pickett, J. Appl. Phys. 42:58 (1971).
2. M. Suenaga and W. B. Sampson, Appl. Phys. Letters 20:443 (1972).
3. E. Gregory, W. G. Marancik, and F. T. Ormand, IEEE Trans. on
       Magnetics MAG-11:295 (1975).
4. Y. Iwasa, IEEE Trans. on Magnetics MAG-11:266 (1975).
5. P. R. Critchlow, E. Gregory, and W. Marancik, J. Appl. Phys.
       45:5027 (1974).

6. W. B. Sampson, M. Suenaga, and K. E. Robins, 5th Inter. Mag. Tech. Conf., Rome (1975).

7. W. B. Sampson, M. Suenaga, and S. Kiss, IEEE Trans. on Magnetics MAG-13:287 (1977).

8. W. B. Sampson, S. Kiss, K. E. Robins, and A. D. McInturff, IEEE Trans. on Magnetics MAG-15:117 (1979).

9. W. Marancik, E. Adam, E. Gregory, and M. Suenaga, ICMC Conf., Paper FA-10, Madison (1979).

10. M. Suenaga, W. Marancik, and E. Gregory, AIME Fall Meeting, Post Deadline Paper, Milwaukee (1979).

11. G. Rupp, IEEE Trans. on Magnetics MAG-13:1865 (1977).

12. E. Adam, E. Gregory, and F. T. Ormand, IEEE Trans. on Magnetics MAG-13:319 (1977).

13. E. J. Zuirys, 5th Inter. Mag. Tech. Conf., Rome (1975).

14. E. Gregory, W. G. Marancik, F. T. Ormand, and J. P. Zbasnick, 5th Inter. Mag. Tech. Conf., Rome (1975).

15. J. P. Zbasnick, R. L. Nelson, D. N. Cornish, and C. E. Taylor, Adv. in Cryo. Eng. 22:370 (1975).

16. P. A. Sanger, W. Marancik, E. Mayer, E. Adam, and E. Gregory, Proc. 7th Symp. on Eng. Problems of Fusion Research, Knoxville (1977).

17. P. A. Sanger, E. Ioriatti, E. Adam, and C. Heyne, Proc. 8th Symp. on Eng. Problems of Fusion Research, San Francisco (1979), to be published.

18. D. W. Deis, D. N. Cornish, D. G. Hirzel, and A. R. Rosdahl, Proc. 7th Symp. on Eng. Problems of Fusion Research, Knoxville (1977).

19. P. J. Singh, E. Adam, E. Gregory, W. Marancik, F. Ormand, M. Young, and C. Spencer, ICMC Conf., Madison (1979), to be published.

20. C. Spencer, E. Adam, E. Gregory, W. Marancik, P. Sanger, R. Scanlan, and D. Cornish, Proc. 8th Symp. on Eng. Problems of Fusion Research, San Francisco (1979), to be published.

21. D. W. Deis, D. N. Cornish, A. R. Rosdahl, and D. G. Horzel, Proc. 6th Int. Conf. on Mag. Tech., Brataslava (1978).

22. D. U. Gubser, T. L. Francavilla, D. G. Howe, R. A. Muessner, and F. T. Ormand, IEEE Trans. on Magnetics MAG-15:385 (1979).

23. E. Adam, P. Beischer, W. Marancik, and M. Young, IEEE Trans. on Magnetics MAG-13:425 (1977).

REVIEW OF SUPERCONDUCTOR ACTIVITIES AT IGC ON A-15 CONDUCTORS[*]

C. H. Rosner, B. A. Zeitlin, R. E. Schwall, M. S. Walker
and G. M. Ozeryansky

Intermagnetics General Corporation
P.O. Box 566
Guilderland, NY 12084

INTRODUCTION

In the past several years, IGC has developed a number of multi-filamentary A-15 conductors for specialized applications. Very high current density, flexible, low loss materials were fabricated using the more conventional internal bronze technique. In addition, a lower cost, versatile, high performance material is produced using the IGC external bronze technique. Experimental quantities of multifilament Nb-Ta-Sn have also been fabricated using the external bronze process.

In this paper we review the characteristics, performance and economics of these materials and their relationship to their predecessors, the G.E. cryostrand and tape processes.

The remarkable aspect of A-15 superconductor development activities remains the diversity and longevity of a number of metallurgical fabrication approaches. If one considers that Nb_3Sn is both the earliest high-field superconductor discovered, and that this particular compound continues to be the most attractive near-term candidate for future technological application, one wonders at the rate of progress made in the two decades since Bell Laboratories first measured the high field properties of a Nb_3Sn wire produced via a powder metallurgical approach.

Out of the various activities throughout the 1960's, several configurations of Nb_3Sn and V_3Ga evolved which led to the utilization of these materials in magnets producing fields in the range

of 10 tesla. However, the extreme strain sensitivity of these
materials limited their usefulness in wire form, and it became
quickly apparent that the mechanical properties were of paramount
importance. In addition, the relatively high reaction temperatures
(in the range of 900-1000°C) put great emphasis on finding compatible
structural and insulation materials that would allow formation of
specifically desired A-15 compounds without adding deleterious
impurities.

These problems were most readily addressed in thin tape-like
configurations and this led RCA and GE to pursue tape fabrication
based on their respective chemical vapor deposition and diffusion
processes.

The composite tape approach arose from a realization that three
key parameters had to be satisfied simultaneously in order to obtain
technologically useful superconductors. These parameters, as necessary today as then, are: highest critical current carrying capacity,
stability and mechanical strength. By laminating copper and stainless steel to the Nb_3Sn substrate, these parameters could be
achieved and combined selectively. The non-superconducting elements
could also be added after formation of the A-15 compound.

The tape approach was quite successful and even today provides
the best approach for constructing magnets producing fields over 12T.
Successful tape magnets operating up to 17.5T have been produced and
the production of 20T magnets is within the scope of present
technology.

From a physics standpoint, however, it was recognized even
earlier[1] that multifilament wire configurations would be necessary
to reduce ac losses. Several monofilament and multifilament wire
configurations, including the GE-"Cryostrand",were successfully
fabricated and used in the earliest 10T magnets.

Imaginative approaches pursued by others, notably the
"composite" process by Tachikawa and BNL, led to a realization that
lower temperatures for A-15 compound formation were possible. This
opened the possibility of incorporation of copper, bronze and other
materials, thereby extending the range of developmental possibilities.

During the 1970's, we at IGC have attempted to incorporate
these various considerations into the development of so-called
"bronze based" Nb_3Sn multifilamentary superconductors.

INTERNAL BRONZE PROCESS

Development of multifilament "Internal Bronze" process Nb_3Sn
at IGC has been guided by the requirements of the WPAFB Manufacturing

Technology Program.[2] The program has required the development of a
React And Wind (RAW) conductor suitable for bending about a 25.4 mm
diameter while carrying a current of from 200 to 500 Amperes at 7
tesla and 8 K [I_c(7T, 8K) $\cong$ I_c(10T, 4.2K)]. In addition, the
temperature rise of a rotor winding should be less than 5 K for a
ramp of 7 tesla at a rate of 10 T/sec. Lengths of from 3000 to
10,000 feet are also required.

These requirements have led to a series of designs which
focused on: 1) Filaments close to the neutral axis of the conductor.
            2) High current density achieved through 1 to 2μ
               filaments and a 13 wt % tin bronze.
            3) Cables and aspected configurations designed to
               lower the effective strain on the conductor.

These designs have been achieved through material, geometry
and processing choices that emphasized filament integrity. A 13
wt % bronze was chosen to provide as large a tin reservoir as
possible and enhance the kinetics of Nb3Sn formation. All materials
used were high purity wrought and annealed to give fine grain size.
The use of wrought materials also assures that impurities are uniformly distributed and that the filaments maintain their initial
uniform geometry. Tantalum diffusion barriers maintain the resistance ratio of the stabilizing copper. Fins, which were originally
introduced to reduce coupling loss, have been found to aid drawability. To minimize intermetallic formation during extrusion and
annealing, processing temperatures have been kept low. Annealing
temperatures and times are typically in the range of 425 to 475°C
for one to two hours.

The conductor produced in Phase I of this program consisted of
4453 filaments in an all-bronze matrix with a bronze to niobium
ratio of 2.4:1. 2μm diam filaments were achieved in the final wire
and resulted in current densities of 7 to 8 x 10^4 A/cm^2 at 10T
(4.2 K). 1μm diam filaments yielded current densities of 1 x 10^5
A/cm^2. At a 1% bend strain, a 20% decrease in critical current was
noted.[3] Figures 1 and 2 illustrate the conductor and its performance
characteristics.

The Phase I conductor basically met the program goals. Further
improvements such as extrusion clad copper and a more uniform filament array were then considered.

The Phase II conductor (as seen in Figure 3) is characterized
by a copper extrusion clad shell about the bronze Nb core. The
diffusion barrier is tantalum as are the radial fins. The design
utilizes 20,538 filaments which are 1μ at 0.4 mm wire diameter. The
volume fractions of the constituents are: Nb - 11.5%; Ta - 13.7%;
copper - 43.7% and bronze - 31.1%.

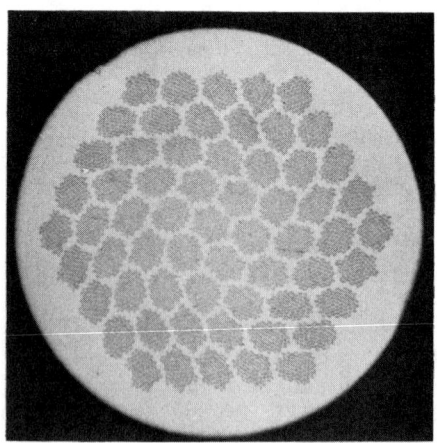

Fig. 1. Photomicrograph of IGC/WPAFB .25 mm diam bronze/Nb strand containing 4453 Nb filaments before reaction.

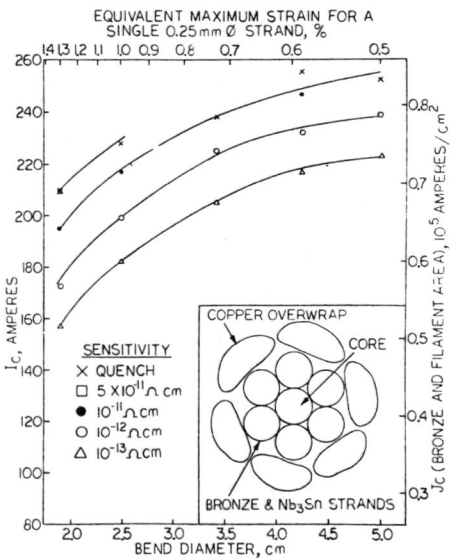

Fig. 2. Dependence of critical current upon bend diam for IGC/WPAFB cable. T = 4.2K, max. field = 10T, avg. field a. & t. Resistivities are averages over a 40 to 60 cm of the sample.

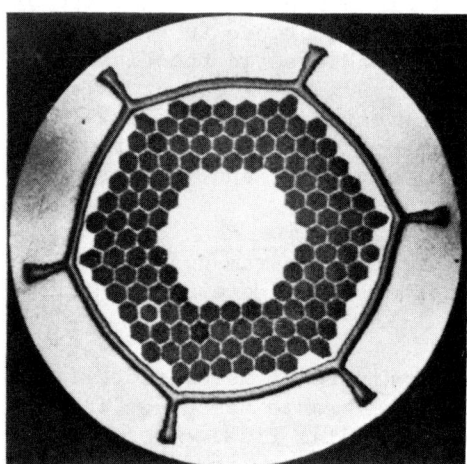

Fig. 3. Cross section of the Phase II composite at 5.8 mm diameter.

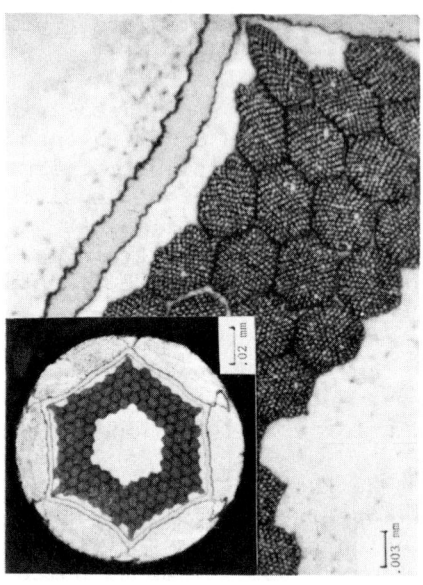

Fig. 4. 0.3µ diam filaments in Phase II .15 mm wire, unreacted.

The material processed smoothly in lengths of up to 4500 feet. At 0.635 mm wire diameter, breakage began to occur. Samples were drawn, however, to 0.15 mm diameter with 0.3μ diameter filaments as shown in Figure 4. Metallography and the sharp resistive onset of critical current measurements indicated that the filaments were continuous even at these small sizes.

The conductor at 0.4 mm wire diameter in cable form displayed a J_c at 10T, 4.2 K of between 1.3 and 1.4 x 10^5 A/cm^2 in the bronze and Nb core. A reduction in J_c of 30% with a 1.2% bend strain (calculated at the filaments) was observed. The V-I characteristics are shown in Figure 5. Further results and details are given in Reference 4.

We are currently manufacturing conductor of a very similar design to Phase II for rotating machinery applications. The material, shown in Figure 6 after the final extrusion, displays uniform filaments over the entire cross section.

EXTERNAL BRONZE PROCESS

It was recognized early by IGC and others that the addition of tin by diffusion into a copper niobium composite would be a more economical approach to the manufacture of MF Nb$_3$Sn than starting with a bronze matrix. IGC has independently developed a material suitable for use in small magnets and this development is now being augmented with external support.

The standard configuration consists of six 0.15 mm to 0.3 mm diameter stabilized strands cabled about a high strength core. This conductor provides a very high overall critical current density and an operating critical current in the range of about 100 amps, appropriate for most small magnets. A typical wire used in such a conductor is shown in Figure 7. This conductor provides J_c of 1 x 10^5 at 10T, 4.2 K in the bronze and filaments. It has a tantalum protected center containing 44% stabilizing copper.[5]

In the past 18 months, significant progress has been made in adapting this material for use in higher current cryostable con- ductors for use in large devices. This progress has followed two parallel paths. In the first, called the separated stabilizer approach, a goal is to produce somewhat larger diameter strands with no stabilizing copper and a very high overall current density. The high current conductor is then produced by cabling together a sufficient number of these strands to yield the required total current and adding copper stabilizer either by including pure copper wires in the cable or by soldering the completed cable into a copper monolithic stabilizer. Figure 8 illustrates a conductor suitable for the separate stabilizer approach.

In the second approach, called the internal stabilizer method, the aim is to increase the maximum diameter of strands, each of

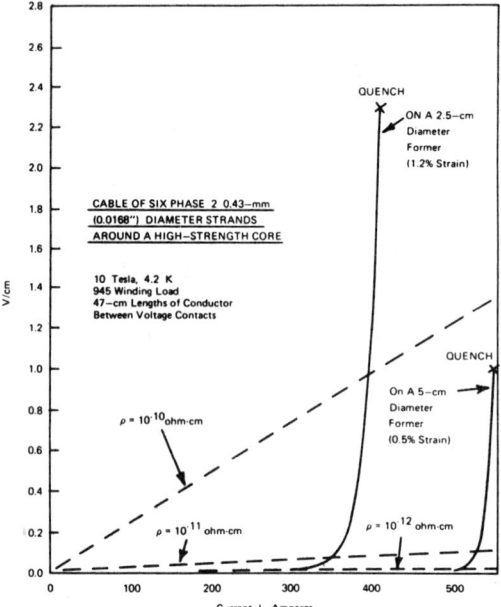

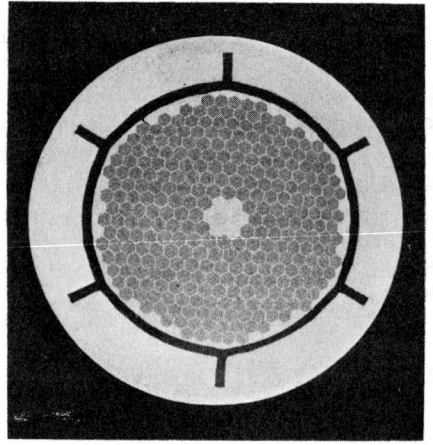

Fig. 5. Resistive onset for
Phase II cable as wound on a
5 cm and 2.5 cm former.

Fig. 6. Cross section of rotor
conductor after final extrusion
containing 24,480 niobium
filaments (not steady state).

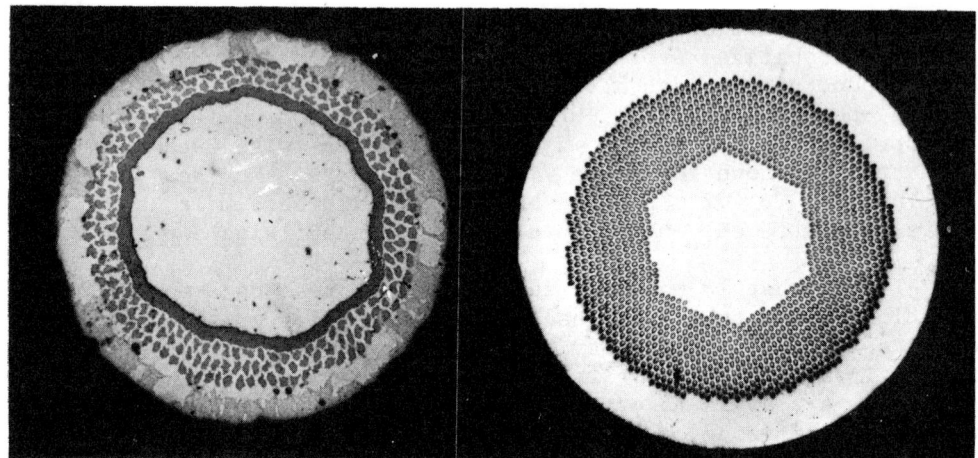

Fig. 7. 0.3 mm diameter
external bronze wire with a
tantalum protected copper core,
after homogenization, but
prior to final reaction.

Fig. 8. A copper niobium
composite prior to introduction
of the tin through external
diffusion.

which incorporates its own tantalum-protected copper stabilizer.
These strands are then combined through several stages of cabling
to again form the high current conductor. The latter configuration
is more suitable for conductor designs in which cooling is achieved
through the use of extended surface area and in forced flow designs.
In the relatively short course of this new effort, we have approximately doubled the critical current available in a single strand of
external bronze material and can now produce 0.30 mm diam solid
strands and 0.40 mm diam stabilized strands, each having a current
density in excess of 5×10^4 A/cm at 12T and 4.2 K over the filament
and bronze region. These sizes do not represent any fundamental
limitation in attainable conductor size, and IGC intends to continue
in the production of even larger strand sizes.

     Work has also progressed on the design of 10 - 15 KA conductors
using these strands. The bulk of such work has been funded by Oak
Ridge National Laboratory for use in the 12T model coil program,
and it is expected that one alternative for the 12T model coil will
be an external bronze conductor.

     One aspect of the external bronze process which is not uni-
versally appreciated, is the fact that the nature of the Nb_3Sn
formation itself allows for a higher critical current density in a
thicker reaction layer than is attainable in internal bronze process
material. This follows from the fact that the critical current in
the reaction layer is dependent on the rate of formation of the
Nb_3Sn. The faster the Nb_3Sn is formed, the higher the current
density in the reaction layer. The reaction kinetics themselves
are in turn dependent upon the tin concentration in the bronze
immediately adjacent to the filament. In the internal bronze
method, the tin concentration in the bronze is limited to approximately 13 wt % by the necessity of maintaining ductility in the
bronze during the wire drawing operation. In the absence of such
limitation, the external bronze process can use higher tin concentrations. These higher tin concentrations in turn maintain faster
reaction kinetics throughout the reaction, and hence, yield higher
overall critical current densities. These higher densities can be
seen in the J_c-H curve shown in Figure 9. In a recent series of
experiments performed in collaboration with Oak Ridge National
Laboratory, we have increased these current densities through the
incorporation of tantalum in the reaction layer. The fact that
tantalum increases the reaction layer critical current density in
the internal bronze process Nb_3Sn is well known, the phenomonen
having been discovered first by Livingston[6] and later confirmed in
more detail by Suenaga et al.[7] In the IGC/ORNL work, however, we
have confirmed that the increases attained through the addition of
tantalum are additive to those which are attained through the use
of high tin bronze. In the experiments, a number of small billets,
each incorporating approximately 75 filaments of NbTa alloy were
were extruded and drawn to wire using standard commercial wire drawing processes. The wire was then tin plated, diffused and reacted,

again using IGC's standard external bronze processing technology.
The resulting critical currents shown in Figure 9 are the highest
ever attained in any multifilament or A-15 superconductor above
10T.   Results are particularly significant in light of the relatively
large filament size (approximately 8.5 microns), since previous
very high current densities have always been attained in filaments
on the order of 1 to 3 microns in diameter.   Also notable is the
fact that these results are directly applicable to commercial
processing since no special procedures were employed in the
preparation of the test conductors.

    Another flexibility inherent in the external bronze process,
which is yet to be totally explored, is the possibility of adding
third elements to the reaction layer through the matrix material.
Third element additions such as gallium and aluminum have been explored by Dew-Hughes, Suenaga, Tachikawa[8-11] and others, and in
some cases have been found to substantially enhance the high field
critical current density.   In the internal bronze process, however,
these enhancements are generally limited by the fact that the third
element displaces tin in the bronze, hence slowing the reaction
kinetics and serving to decrease the critical current density.   In
the external bronze process, however, the third element addition
need not displace tin in the matrix and the possibility exists that
the critical current enhancement due to the third element addition
will not be offset by decrease in the reaction kinetics.

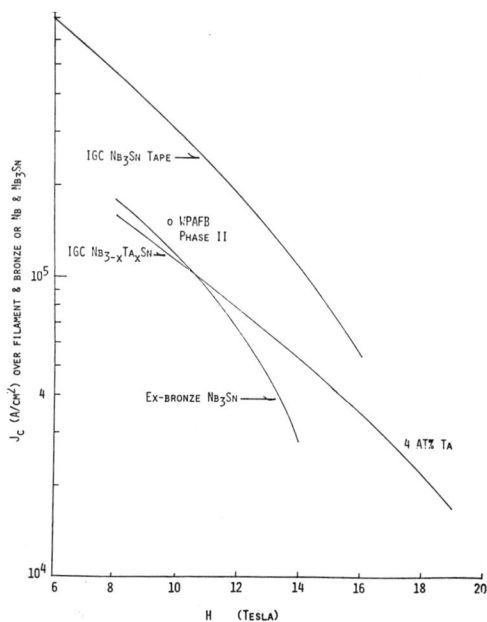

Fig. 9. J_c versus H for graph for IGC tape, internal bronze,
           external bronze and Nb 4AT% Ta external bronze.

One aspect of the external bronze technique which has received considerable attention recently is that of materials produced by the in-situ technique. In this process, the array of niobium filaments in a copper matrix is replaced by a dispersion of fine niobium fibers in copper. Once the material reaches wire size, however, further processing is identical to that of the normal IGC external bronze process. IGC has been actively engaged for a number of months in collaboration with various groups working in in-situ material, including those at Ames Laboratory at Iowa State University and at MIT. Results to date indicate that the copper niobium dispersions can be regarded as a homogenious material with tin diffusion coefficient somewhat lower than that of pure copper. It is expected that if the diffusion and homogenization heat treatment schedules take proper account of this reduced diffusivity, the material can be processed in a manner essentially identical to that of the multifilamentary external bronze material.

Comparison of External Bronze vs   Internal Bronze

When external bronze material is compared to internal bronze, the following is seen:

Advantages of External Bronze
1. Higher J_c can be obtained from coarser filaments as the tin reservoir is not limited to 13% by weight of the matrix.
2. The manufacturing costs are significantly less as described later.
3. Third elements can be readily incorporated in the matrix to enhance J_c or influence precompression strain without affecting fabricability.

Advantages of Internal Bronze
1. Bend strain tolerance is superior, as the filaments can be located close to the neutral axis.
2. The conductors can be insulated with film type insulation in the RAW configuration as the reacted conductor has smooth surfaces.
3. The maximum size of conductor is not limited by diffusion lengths.

Projected Economics

The costs of both methods can be broken down into materials and processing costs. Ignoring for the moment the question of yield, the costs of Nb and Ta are essentially the same for both processes. The cost of high purity wrought bronze is approximately five times that of the equivalent copper used in the external bronze process. This is offset only slightly by cost of the tin and chemicals used in plating.

   The fabrication costs of the internal bronze process are
significantly higher for several reasons:
   1.   The internal bronze process requires two extrusions to
produce filaments fine enough to compare in current density.
   2.   The yield of conductor in the internal bronze process is
significantly less. Each extrusion typically has a yield in the
range of 80%. Thus a double extrusion process starts with a 20%
yield disadvantage compared to a single extrusion.
   3.   Many anneals are required, typically every 50 to 60%
reduction in area. This precludes the use of modern high speed
machinery that can typically reduce external bronze 90% in area in
a single operation.

   These fabrication costs are offset to some extent by the
plating and diffusion operations used in ex-bronze. While the
plating of superconductive wire has not been accomplished on a
large scale, tin plated wire is produced in large quantities and
quite economically for the electronics industry. It appears reasonable to estimate that the total costs of plating (labor plus
materials) might equal the cost of the wrought bronze for an
equivalent internal bronze conductor.

   While the diffusion process for ex-bronze material may require
many hours, it is a relatively inexpensive operation since no
material handling is required and should have little effect on final
product cost.

CONCLUSIONS

   At IGC, the development of A-15 materials, mainly Nb_3Sn, has
focused upon metallurgical, configurational and the various processing approaches described in this paper. The objectives were
that this effort would yield understanding as well as technologically
useful conductors. The uncompromising nature of such intermetallic
compounds has also required careful attention to stress and strain
imposed limitations. Thus, in retrospect, slower than hoped for
progress has been made in fully developing the potential of Nb_3Sn
as a useful high field magnet material.

   In reviewing our work and assessing progress made elsewhere, we
consider the "external bronze" approach to be a generally attractive
processing avenue that has the potential for some intrinsic performance as well as economic advantages over the "internal bronze"
methods.

   Depending upon the ultimate application requirements, other
approaches such as "internal bronze" and in-situ processed conductors deserve attention as well. They have indeed specific
advantages on a case-by-case basis. For these reasons work at
IGC has encompassed, and exploration will continue on, a number of

process variations   leading towards useful bronze based Nb_3Sn conductors.

REFERENCES

1. C. P. Bean and J. C. Fisher, U.S. Patent #3214249.
2. B. A. Zeitlin, A. Petrovich, M. S. Walker, and J. R. Hughes, American Society of Metals, Manufacture of Superconducting Materials, Nov. 8-10, 1976, p. 27.
3. A. Petrovich, B. A. Zeitlin, J. M. Cutro, M. S. Walker, and C. H. Rosner, IEEE Trans. on Magnetics MAG-13:796 (1977).
4. M. S. Walker, J. M. Cutro, B. A. Zeitlin, G. M. Ozeryansky, R. E. Schwall, C. E. Oberly, J. C. Ho, and J. A. Woollam, IEEE Trans. on Magnetics MAG-15:80 (1979).
5. R. M. Scanlan and W. A. Fietz, IEEE Trans. on Magnetics MAG-11:287 (1975).
6. J. D. Livingston, IEEE Trans. on Magnetics MAG-14:611 (1978).
7. M. Suenaga, K. Aihara, K. Kaiho, and T. S. Luhman, to be published in Advances in Cryogenic Engineering 26 (1980).
8. David Dew-Hughes, Thomas S. Luhman, and Masaki Suenaga, Nuclear Technology 29:268 (1976).
9. D. Dew-Hughes and M. Suenaga, J. Appl. Phys. 49:357 (1978).
10. O. Horigami, Thomas Luhman, C. S. Pande, and M. Suenaga, Appl. Phys. Letters 28:738 (1976).
11. K. Tachikawa, this conference.
    *Part of this work has been supported by the Air Force Materials Laboratory under Air Force Contract F33615-75-C-5104 and by Oak Ridge National Laboratory under Contracts 22Y-21008C, 22Y-07871C and 22Y-07869C.

FILAMENTARY Nb₃Sn SUPERCONDUCTOR MANUFACTURED BY THE SOLID-LIQUID

DIFFUSION METHOD

S. Okuda, M. Nagata, M. Yokota, M. Watanabe* and
Y. Kimura*

Sumitomo Electric Industries, Ltd. R&D Group, 1-3,
Shimaya 1-chome, Konohana-ku, Osaka, 554 Japan
*Electrotechnical Laboratory, MITI, 1-4, Umesono
1-chome, Sakuramura Niiharigun, Ibaraki, 305 Japan

INTRODUCTION

There are a number of production processes for the fabrication
of multifilamentary Nb_3Sn superconductors. The most established
process is the bronze process which is carried out by reaction of
Nb cores and Sn in a bronze matrix.[1] As is well known, to obtain
a high critical current density throughout the conductor, it is
desirable that the Sn content in the bronze matrix be higher than
13 wt%, which is the solubility limit of Sn in Cu. If the Sn concentration is increased over this limit, break down of the wire at
drawing by work hardening of the bronze matrix is likely to occur.

One approach to prevent work hardening of matrices with high
Sn concentration is to make composites of pure Cu matrix with Sn
on the surface,[2] or in the matrix as a core.[3] Sn is diffused into
the Cu matrix during a preheating treatment before the final heat
treatment for formation of the Nb_3Sn. Another approach to obtain
multifilamentary Nb_3Sn wire is to use Nb tubes with bronze cores[4]
or with Cu and Sn alloy composites.[5] In this case, the Nb tubes
in the pure Cu matrix act as diffusion barriers against diffusion
of Sn into the Cu matrix.

From the viewpoint of the diffusion reaction, the formation
of Nb_3Sn by all of the production processes mentioned above use a
solid-solid diffusion reaction; that is, Nb and bronze are both
solid at around 700°C. Even in the internal bronze approach, where
the composites of Cu and Sn alloy are co-drawn to final size just

before the heat treatment,[3,5] the Sn concentration in the bronze
seems to be not over 50 wt% at final heat treatment.

We developed multifilamentary Nb_3Sn superconductors manufactured by the solid-liquid diffusion method, which uses Sn-rich Cu
alloy cores in Nb tubes. In this paper, we describe the dependence
of critical current on the Sn concentration, and the results of
experiments to obtain samples of Sn-7 wt% Cu with a high critical
current.

EXPERIMENTAL PROCEDURE

Manufacturing Process of Nb_3Sn

Figure 1 shows the cross-section of our Nb_3Sn superconductors.
These samples were produced by the bundle drawing technique. The
Sn-7 wt% Cu alloy core rod drawn from a billet cast in an induction
furnace was first inserted into an Nb tube, drawn to 2-3 mm diameter, and then cut into a number of segments of the same length.
These segments were bundled in a Cu tube which functions as a stabilizer after formation of the Nb_3Sn superconductor. The composite,
about 30 mm in outer diameter, was then drawn to 1 $\sim$2 mm diameter
and finally heat treated for formation of the Nb_3Sn layer between
the inner side of the Nb tubes and the Sn-Cu cores.

The composite of soft Sn-7 wt% Cu alloy and work-hardened Nb
is difficult to co-draw with uniform deformation. We, however,
developed the superconductor by making the most suitable arrangement of the soft and hard materials as shown in Fig. 1. Sample
parameters are summarized in the Table 1.

To vary the Sn concentration in the Sn-Cu alloy core, we used
a core consisting of seven wires, some of pure Sn and the rest of
pure Cu. For example, if the core of seven wires consists of two
Cu wires and five Sn wires, the Sn concentration after the preheating treatment of 36 h at several temperatures becomes 75 wt%Sn.
In the case of 93 and 98 wt% Sn, wires of Sn-rich Cu alloys were
used. Eight types of samples from 20 wt% Sn to 98 wt% Sn were
made for experiment I.

Measurement of Critical Current

To avoid scatter in the results, a 37-core multifilament wire
was prepared for the experiment. Critical current was measured
at 7 T, 4.2 K. Samples 7 cm long were bent into U-shape, and the
critical current was determined at the point where a voltage of
3 μV is generated across 2 cm at the center of the sample.

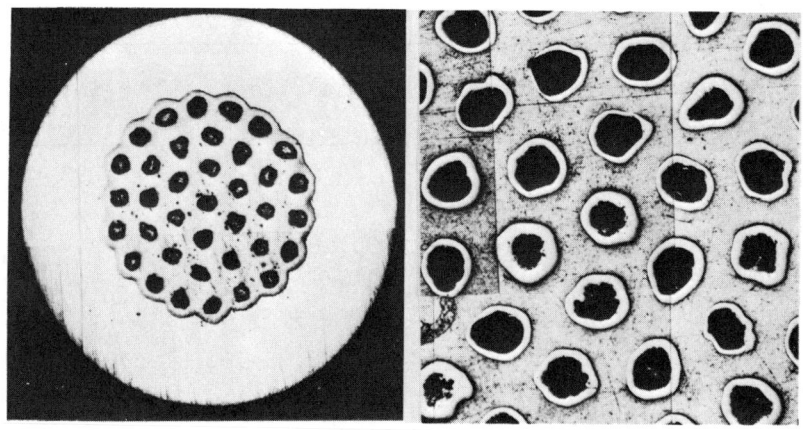

(a)   (b)   250 µm

Fig. 1.   Cross sections of Nb₃Sn superconductors.
        (a) 1.29 mm diameter wire with 37 Sn—Cu cores (Sample III).
        (b) Nb₃Sn tubes in Nb matrix. Boundary between Nb tubes is
            not visible.

Measurement was conducted on three U—shaped samples which were heat treated under the same conditions, and the data were averaged.

Measurement of Nb₃Sn Layer Thickness

Thickness of the Nb₃Sn layer was measured by a micrometer by viewing a cross section through a microscope. All of the samples were mechanically polished and chemically etched. As is well known, the measurement of thickness of Nb₃Sn layer is very difficult because of the irregular shape of the boundary. Therefore the calculated J_c of Nb₃Sn layer is uncertain to within a factor of two.

Table 1. Experimental Conditions

Sample Parameters	Experiment No.		
	I	II	III
Purpose of Experiment	Sn Content	Size Effect	Optimization
Tin Content (wt%)	20,33,48,62, 75,88,93,98	93	93
SnCu Core Diameter (μm)	45	30,38,49, 61,77	47
Cu : Nb : Sn–Cu	21 : 7 : 1	11 : 3.5 : 1	14 : 5 : 1
Heat Treatment Condition	690°C, 5– 100 h, 610°C–810°C, 50 h	690°C, 5–100 h 610°C–810°C, 50 h	610°C–810°C, 5–100 h
No. of Filaments	37	19	37
Wire Diameter (mm)	1.29	0.5 – 1.29	1.29

EXPERIMENTAL RESULTS AND DISCUSSION

Sn Concentration Dependence

Although data[5] already exist showing that the solid-liquid reaction had inferior properties, we investigated Sn concentration dependence again. Figure 2 shows the results of critical-current measurements. The diameter of the samples was 1.29 mm, and core diameter before heat treatment was about 45 μm. All samples in this figure were heat treated at 690°C. If Sn is richer than 40 wt%, the liquid phase will occur in Cu-Sn alloy system at 690°C.[6]

From this figure, it is evident that alloy cores with high tin concentration have high critical-current values. Another property to be noted is the rapid increase of critical current es-

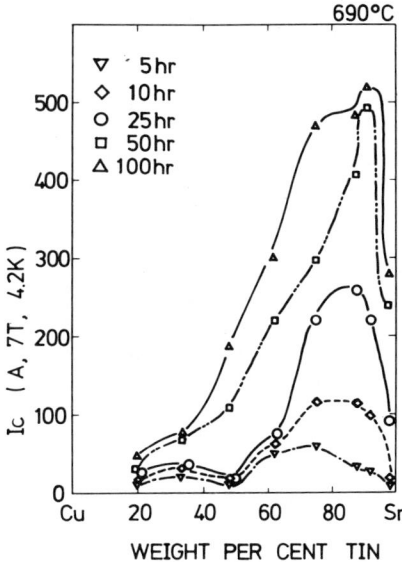

Fig. 2.   Sn concentration dependence of critical current.

pecially in the high tin samples, if the heat treatment time is
longer.   And in the case of 88 and 93 wt% Sn, the critical current seems to become saturated if the heat-treatment time is 50 h
or more.   In the case of heat treatments of 50 and 100 h, the
critical current shows a peak at 93 wt% Sn; that is, Sn–7 wt% Cu.

Since the wire diameter and Cu ratio are the same for each
sample, the overall current densities excluding Cu stabilizer
shows the same profile as in Fig. 2.   The superior properties of
the solid–liquid reaction are due to the high rate of diffusion in
the liquid and to the high tin content.

Effect of Core Size

In the case of the Cu-rich bronze process, it is essential to
decrease the size of Nb filaments to about 3 μm in order to increase overall current density.   In our solid–liquid diffusion
process, which uses the Sn–7 wt% Cu alloy cores, the thick Nb_3Sn
layer is formed in a heat treatment time shorter than 100 h, so it
is not necessary to drastically decrease the size of cores.   But
it is necessary that the core size not be too large in order to
obtain adiabatic stability.

Figure 3 shows the overall current density change versus the
heat-treatment time. It takes more time to increase current density
in the case of larger cores. But all of the curves coincide at the
same value when the heat treatment exceeds 50 h. So, overall J_c
tends to have no relation to the size in the case of lengthy treatment and appears to level off, as is shown more clearly in Fig. 4.
At high temperatures above 690°C, there is no size dependency, and
the overall current density is constant. This means that the reaction has gone as far as possible. But at the temperature below
690°C, the influence of size can be seen. Therefore, in the case
of the samples of 40 ∿50 μm core, it would be best to choose 690°C
as the heat-treatment temperature. As is shown in Fig. 3, the
formation of Nb3Sn is slow in the case of large core diameter and
short heat treatment. So we must choose the condition where the
formation of Nb3Sn is saturated to obtain high current density and
reproducible results.

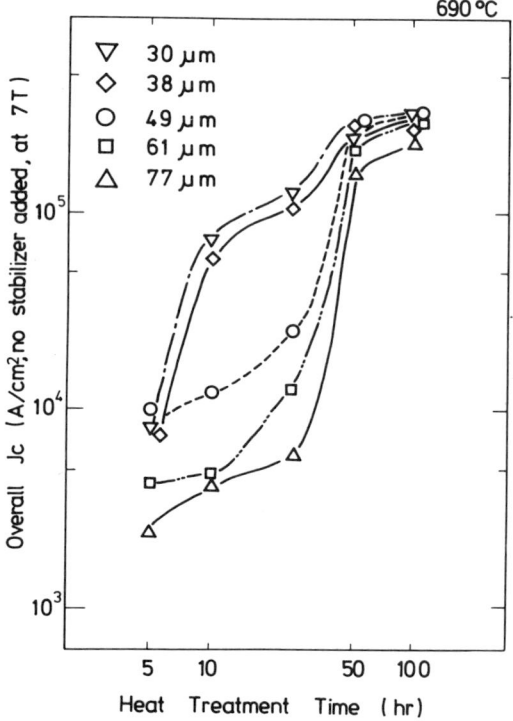

Fig. 3.   Overall critical-current density vs reaction time for
          samples with several core diameters.

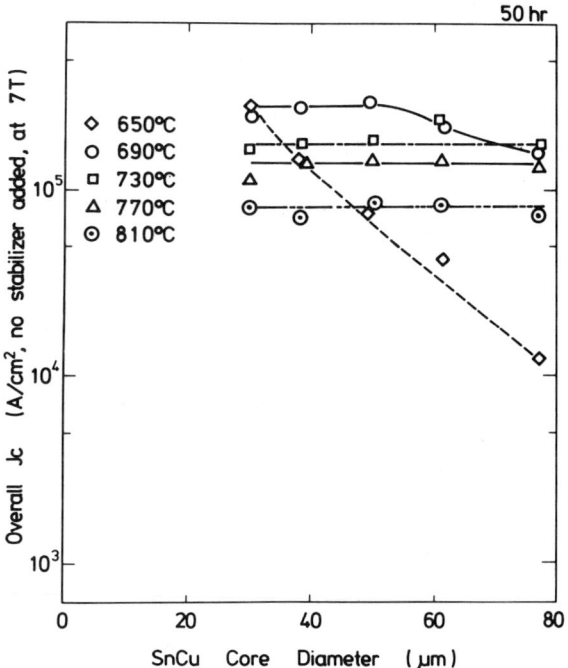

Fig. 4. Overall current density as a function of core diameter for samples heated at various temperatures.

If we assume that the Nb_3Sn layer thickness t is proportional to the Sn quantity of the core at the saturated condition, t is expected to have the following relation to the core diameter, d.

$$t \propto \frac{\text{tin quantity}}{\text{boundary length}} \propto \frac{\pi(d/2)^2}{\pi d} = \frac{d}{4} \quad .$$

The experimental result coincided with this expectation, and t/d was $\simeq 0.28$.

From these experiments investigating the size dependency, it becomes clear that the high overall current density cannot be obtained by short reaction times for large-core-diameter samples. We think this is the main reason why a large difference exists between other researchers' results and our results for the Sn concentration dependency.

Optimization of Critical-Current Density in Nb₃Sn

From the two results of Sn concentration and core size dependency, it can be seen that Sn-7 wt% alloy core has the best overall current density. Figure 5 shows the critical-current change
as a function of heat-treatment time and temperature. High critical currents are obtainable by heat treatments at 690°C and
50∿100 h. Heat treatments at higher temperatures than 730°C do
not give a high critical current, whereas the low-temperature
(650°C) heat treatments increase the critical current as the time
becomes longer. Figure 6 shows the thickness change with heat-
treatment temperature and time. Surprisingly, the thicknesses of
Nb₃Sn formed at more than 730°C and longer than 50 h are almost
twice that formed at lower than 690°C and 50 h. These thick layers
do not carry the high current, as shown in Fig. 5. This may be
attributed to low critical-current density of Nb₃Sn layers. We
suppose that there is a drastic change in diffusion reaction for
formation of the Nb₃Sn layer at around 710°C. One possible explanation of this change is that the measured Nb₃Sn layer actually
is two layers, Nb₃Sn and other compound which will occur at higher
temperature than 710°C.

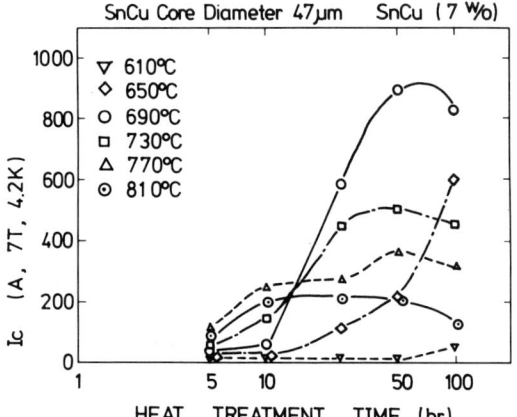

Fig. 5.   Critical current change vs heat treatment time at
          various temperatures.

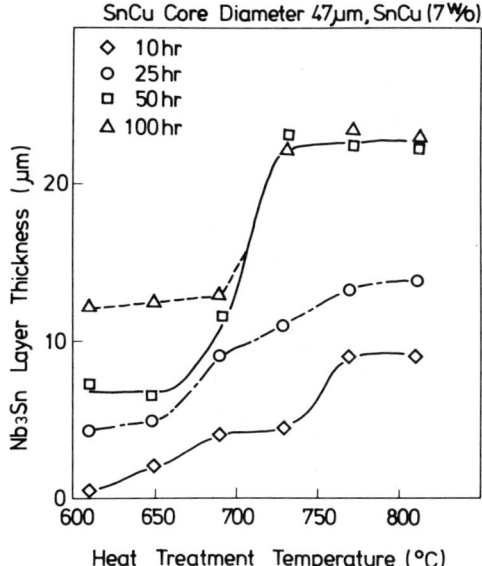

Fig. 6.   Nb₃Sn layer thickness heat treated at several tempera-
          tures.

        From the data of Figs. 5 and 6 the J_c of Nb₃Sn layer is about
2×10^6 A/cm² at 7 T, and is the same for a heat treatment of 25
∿100 h at 690°C. This value seems to be extremely high compared
to other reports.[7,8] As the critical-current density in Nb₃Sn was
computed for the maximum obtainable volume of Nb₃Sn, Nb₃Sn is
thought to have better properties than the usual bronze processed
materials.

        In the case of heat treatment at 690°C, the critical-current
increase with time (Fig. 5) is caused by increasing layer thickness,
whereas the increase of layer thickness and critical-current density in Nb₃Sn both bring an increase of critical current at 650°C.
So the heat-treatment temperature should not exceed 700°C and
preferably should be 650 ∿ 690°C for times longer than 100 h. One
approach to shorten the heat treatment time is to decrease the core
diameter. Then the saturation will occur at lower temperatures and
shorter times.

CONCLUSION

A new manufacturing process to produce Nb_3Sn utilizing the internal bronze approach was developed. From the extensive survey of Sn concentration in bronze, Sn-rich bronze, especially 93 wt% Sn gave the highest critical current. Solid-liquid diffusion reaction produces higher current densities and thicker Nb_3Sn layers than solid-solid reaction does. To obtain the optimized critical-current density in the solid-liquid reaction, it is essential to heat treat at saturated conditions which strongly depend on the Sn concentration and core diameter. That is, the heat treatment should be longer than 50 h at 690°C in the case of about 50-μm core diameter. The peak around the optimum condition is very sharp, and the best value obtained for overall current density and for critical-current density of the Nb_3Sn layer is 3×10^5 A/cm^2 or 2×10^6 A/cm^2 at 7 T, respectively. We believe that this process is a very promising method, and now we are in the initial stage of producing a high-field magnet.[9]

ACKNOWLEDGMENTS

We wish to thank Dr. Y. Aiyama, Dr. Y. Akiyama, and Dr. T. Yasui, for their continuous encouragement of this development. The authors are grateful to Mr. M. Kawashima, Mr. Y. Hosoda, and Mr. H. Takei for technical discussions.

REFERENCES

1. K. Tachikawa, ICEC 3 at Berlin, 339 (1970).
    A. R. Kaufmann and J. J. Pickett, J. Appl. Phys. 42:58 (1971).
2. M. Suenaga and W. B. Sampson, Appl. Phys. Lett. 20:443 (1972).
3. Y. Hashimoto, K. Yoshizaki, and M. Tanaka, ICEC 5, M4:332 (1974).
4. R. M. Scanlan, D. N. Cornish, J. P. Zbasnik, R. W. Hoard, J. Wong, and R. Randall, ICMC at Madison (1979).
5. S. Murase, M. Koizumi, D. Horigami, H. Shiraki, Y. Koike, E. Suzuki, M. Ichihara, F. Nakane, and N. Aoki, IEEE Trans. Magnetics MAG-15:83 (1979).
6. M. Hansen,"Constitution of Binary Alloys," McGraw-Hill, 634 (1958).
7. D. C. Larbalestier, IEEE Trans. Magnetics MAG-15:209 (1979).
8. C. A. M. van Beijnen and J. D. Elen, IEEE Trans. Magnetics MAG-15:87 (1979).
9. Y. Aiyama, Workshop on High-Field Superconducting Materials for Fusion at Tokyo, paper D-1, March 13-15, 1980.

MULTIFILAMENTARY Nb_3Sn BY AN IMPROVED EXTERNAL DIFFUSION METHOD

S. F. Cogan, D. S. Holmes, J. D. Klein, and R. M. Rose

Massachusetts Institute of Technology
Department of Materials Science and Engineering
77 Massachusetts Avenue
Cambridge, MA 02139

INTRODUCTION

The external diffusion technique permits easy mechanical fabrication of Nb_3Sn-based superconducting multifilamentary composites due to the lower work-hardening rate of Cu compared to Cu-Sn bronze. However, Kirkendall porosity can appear to a disastrous extent in external diffusion composites rather than the lesser degree apparent in internal diffusion products.[1] Figure 1 shows a cross section of the extremely fragile material resulting from the application of the external diffusion technique to multifilamentary composites with no precautions taken to suppress Kirkendall porosity. Figure 2 shows the vastly improved cross section resulting from the incorporation of a solution preanneal prior to electroplating in the external diffusion method. This approach was developed in our laboratory[2] upon recognition of the heterogeneous nucleation mechanism in the observed Kirkendall porosity. The heterogeneously nucleated voids precipitate from the temporary supersaturation of lattice vacancies caused by the differing diffusion mobilities of copper and tin in the copper matrix. The nucleation substrates are mainly particles of oxide, tarnish, or other foreign matter adherent to the surfaces of the original composite components or included in the copper itself. When these heterogeneous nucleation sites are dissolved or dispersed by the solution preanneal, the catalytic potency of the composite (at least for void nucleation) is reduced. Composites produced by preannealing before applying the external diffusion technique are totally free (or nearly so) of Kirkendall porosity and display good mechanical and electrical properties.

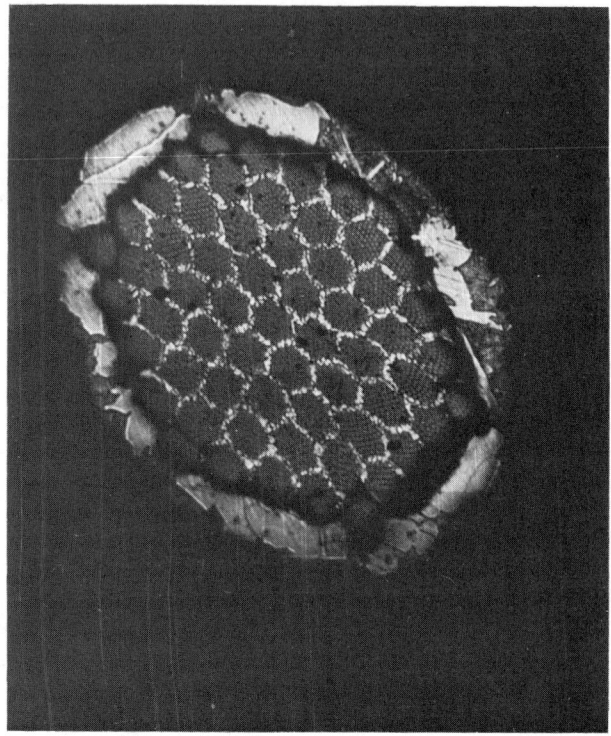

Fig. 1.   Cross section at 525X magnification of a multifilamentary composite with tin plating and diffusion annealing
directly after wire drawing.

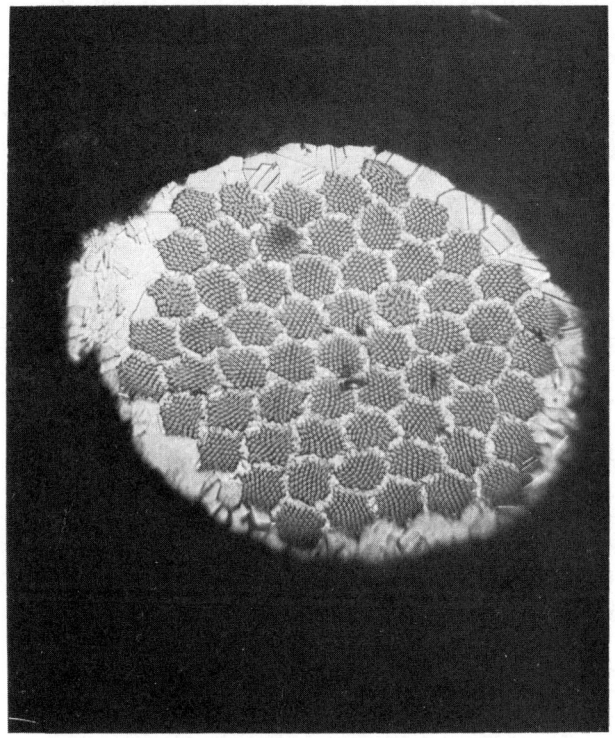

Fig. 2. Cross section at 525X magnification of a multifilamentary composite produced by the external diffusion
technique but with a solution anneal at 650°C for 16
hours before tin plating.

FABRICATION

      Small (1-inch diameter) extrusion billets were produced by
drawing Cu tubes onto Nb rod (through round and then hexagonal
dies) before assembling the hexagonal composite rods in groups inside Cu jackets. Electron-beam welding of nose and heel pieces
preceded room temperature extrusion of the billets. After drawing
the resulting rod through hexagonal dies, the bundling process was
repeated. The final volume fractions of the 0.010-inch (0.25 mm.)
wires were controlled by varying the makeup of the billets. Nominal
Nb fibre sizes at the final diameter were 12.5μ , 1.25μ , and 0.13μ.
The final wire was cleaned, preannealed for 16 hours at 650°C,
electroplated with Sn, and then annealed to diffuse and react the
Sn. The diffusion and reaction heat treatments were performed in
a stepwise manner; typically from an initial few hours at 200°C,
through as many as 25 hours at 500°C, to a final reaction anneal
between 600 and 750°C.

MEASUREMENTS

      The four-point method was used for J_c (critical current density)
with a criterion of 0.5μ V/cm. The mechanical properties (which
have been reported previously[4] were measured using the apparatus
shown in Figure 3. This design has been used to assess the mechanical and electrical properties of Cu-Sn alloys at 4.2 K[5] and is
described in detail in that work. The cruciform section has low
torsional rigidity and a highly reproducible torque-deflection curve,
eliminating one persistent source of mechanical irreproducibility
at low temperatures, i.e., the friction associated with bearings
or pivots. Strain is accurately measurable only up to ∿2.5% with
the present device.

RESULTS AND DISCUSSION

      To explore the possibilities of lower-temperature reaction
annealing in the composites with finer fibres, Nb_3Sn layer thickness was measured (by scanning electron microscopy) as a function
of annealing time at 600°C. The measured layer thicknesses as a
function of time at 600°C are plotted on Figure 4, along with the
700 and 800° data of Farrell et al.[6] For critical current density
the best reaction anneals seemed to be 100-200 hours at 650°C for
the composites with 1.3 μ Nb fibres. Figure 5 summarizes our best
results thus far. By way of comparison, Figure 6 shows specimens
with other reaction heat treatments, with poorer results. The
optimum reaction times for the 0.13 μ composites are not consistent
with the growth data, as the latter show that less than one hour
at 600°C should be more than adequate to convert the Nb to Nb_3Sn.
We may be encountering additional diffusional impedances due to
the close spacing of the fibres compared to the wire diameter; or

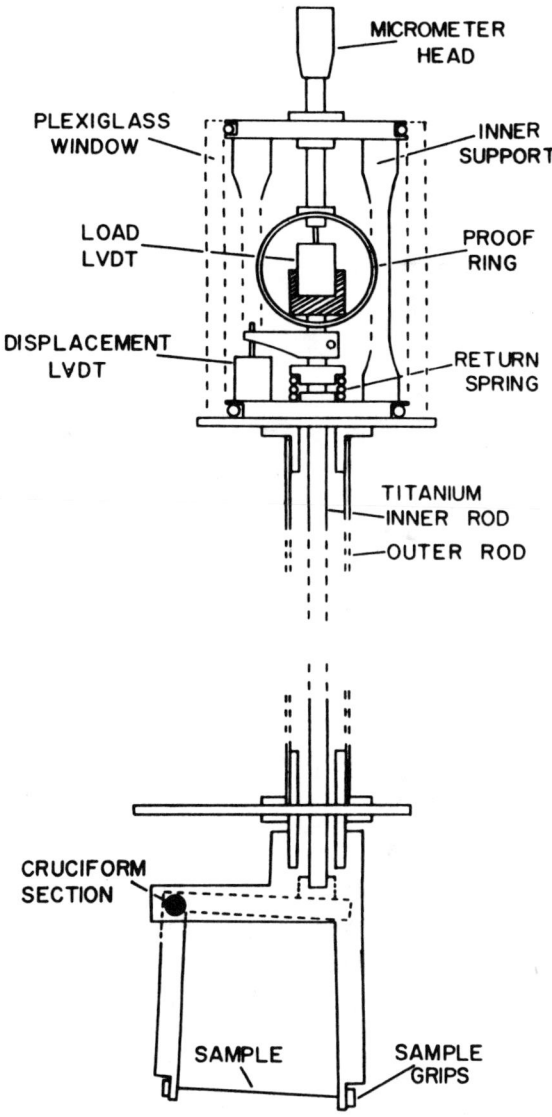

Fig. 3.   Apparatus for simultaneous measurement of J_c, stress and
          strain in high transverse magnetic field.   By permission
          of Cryogenics.[5]

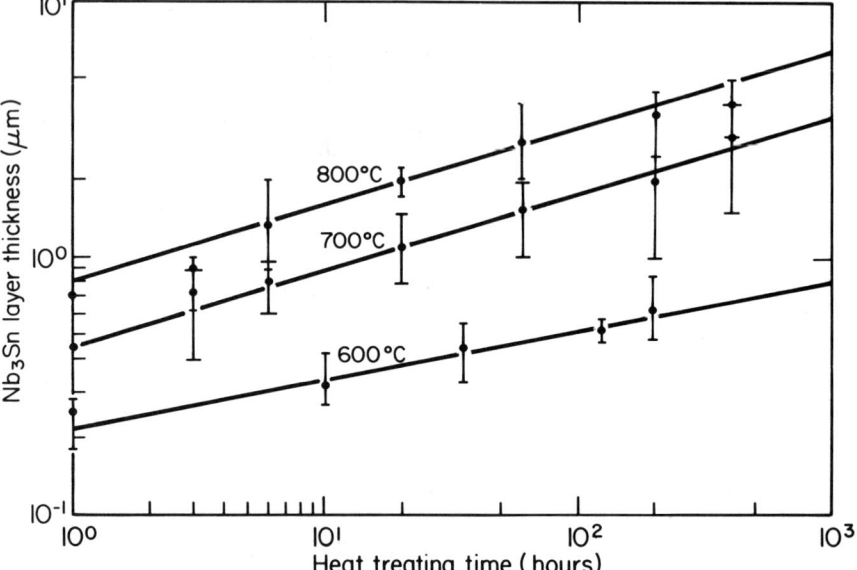

Fig. 4. Nb$_3$Sn layer thickness as a function of annealing time at 600°C, together with the data of Farrell et al.[6] for 700° and 800°C.

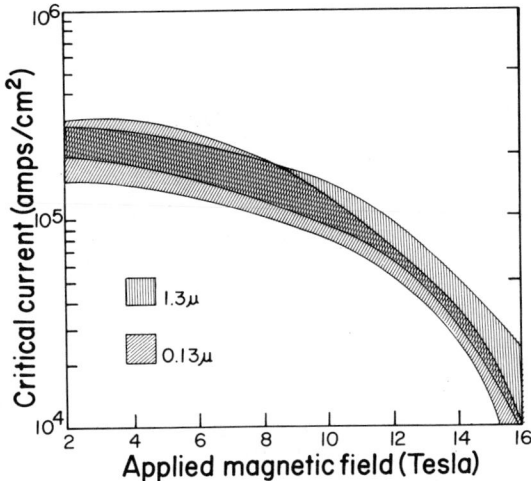

Fig. 5. J$_c$ vs H at 4.2 K for the best heat treatments, on 25%
Nb composites. For the 1.3 μ fibre size, five specimens
are summarized. For the 0.13 μ fibre size, four specimens are summarized.

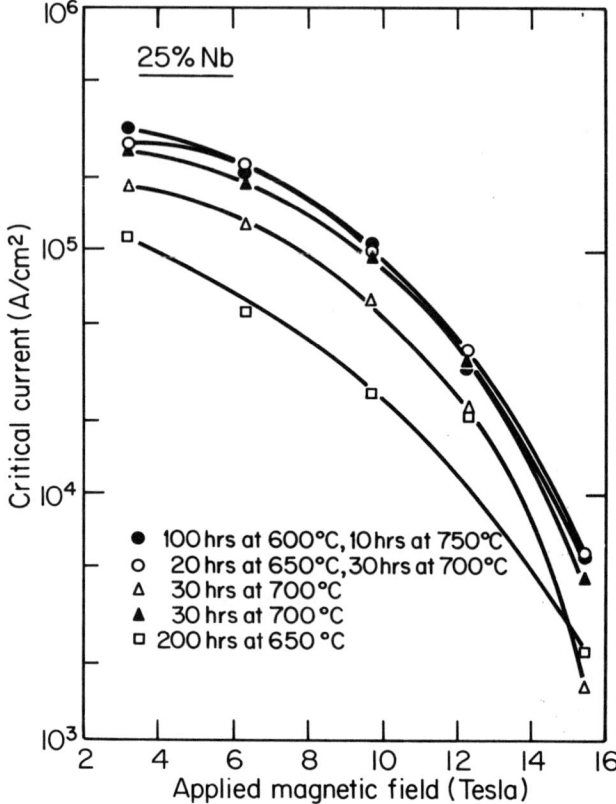

Fig. 6. J_c vs H at 4.2 K for 25% Nb composite with 0.13 fibres
and with nonoptimal heat treatments of various kinds.

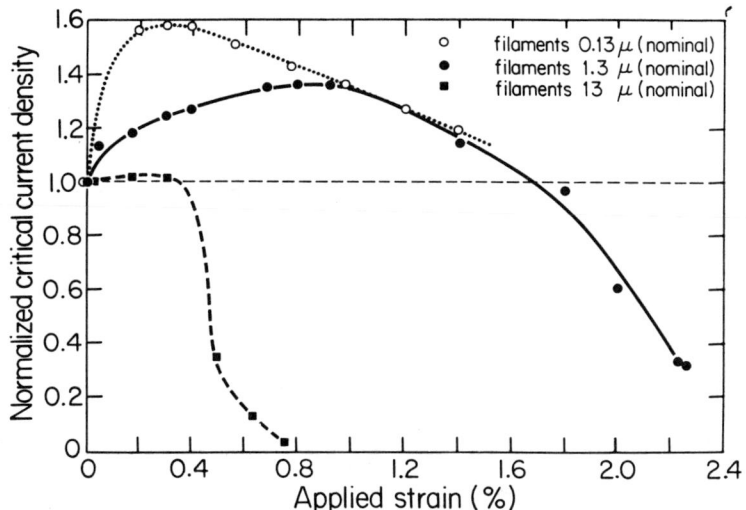

Fig. 7. J_c vs   strain for three fibre sizes in composites which
showed prestress effects, at 8 T and 4.2 K. J_c measured
with load applied.   (Courtesy of J. of Appl. Phys.[4])

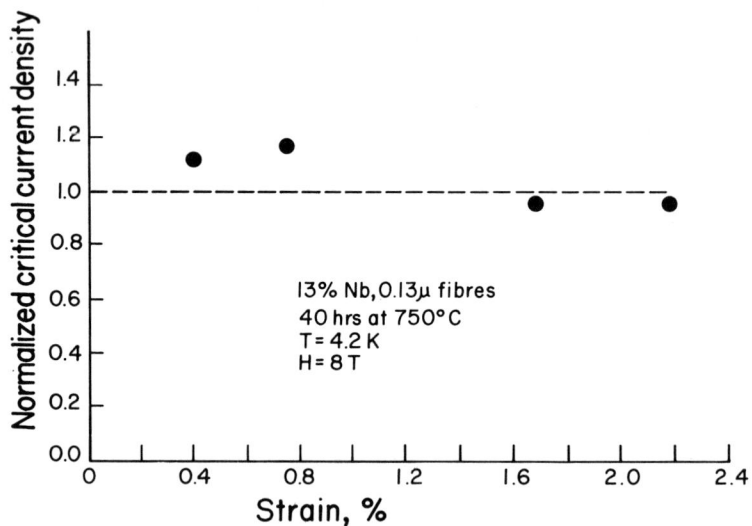

Fig. 8. J_c vs strain at 8 T and 4.2 K, J_c measured with load removed.

the multiplicity of interfaces. Some of the composites with 0.13μ fibres had higher J_c's at low fields and lower J_c's at higher fields than the composites with $1.3~\mu$ fibres. This tendency may be seen in Figure 5, and implies the usual tradeoff of pinning strength and T_c (and therefore H_c) in A15 materials. Also of interest (but not presented in the figures) is the J_c data for lower volume fractions. Some, notably 12% Nb with $1.3~\mu$ fibres, had J_c above 10^4 amps/cm^2 at 15.4 T, a performance almost as good as the 25% Nb composites, the best of which had J_c of 2×10^4 amps/cm^2 at 15.4 T, $3-4 \times 10^4$ amps/cm^2 at 14 T and 10^5 amps/cm^2 at 12 T applied field (e.g., see Fig. 5).

Figure 7 is taken from our earlier measurements on J_c as a function of stress and strain.[4] The finer sizes, 0.13μ and 1.3μ are capable of sustaining strains up to 1.6-1.8% under load without deterioration of J_c below the zero-strain value. If the load is released before J_c is measured, higher strains can be tolerated before J_c deteriorates.

Figure 8 was taken from a nonoptimal specimen which showed remarkable strain tolerance, where strain was measured then load removed and J_c measured. As reported,[4] the stresses were measured simultaneously with the strain; typically the stresses rose to the 500-700 MPA range as the strain was increased.

CONCLUSIONS

With measures taken for the suppression of Kirkendall porosity, the external diffusion method appears to be a practical means for the manufacture of superconducting multifilamentary composites with superior properties, particularly J_c and tolerance to strain. However, it is also apparent from the kinetics and processing data that the properties of this class of material have yet to be fully optimized.

REFERENCES

1. D. S. Easton and D. M. Kroeger, IEEE Trans. Magn. MAG-15:178 (1979).
2. S. F. Cogan, D. S. Holmes, and R. M. Rose, Appl. Phys. Letters 35:557 (1979).
3. A. Smigelskas and E. Kirkendall, Trans. AIME 171:130 (1947).
4. S. F. Cogan, D. S. Holmes, and R. M. Rose, to appear in J. Appl. Phys., August 1980.
5. S. F. Cogan and R. M. Rose, to appear in Cryogenics.
6. H. H. Farrell, G. H. Gilmer, and M. Suenaga, J. Appl. Phys. 45:4025 (1974).

EFFECTS OF FORMATION TEMPERATURE ON THE SUPERCONDUCTING

PROPERTIES OF V_3Ga WIRES

D. G. Howe and T. L. Francavilla

U.S. Naval Research Laboratory
Washington, DC 20375

INTRODUCTION

Multifilamentary V_3Ga wire has a promising potential for use in several advanced systems applications. These include such high-field systems as electrical motors and generators, plasma confinement magnets for thermonuclear fusion, and magnetohydrodynamics channel magnets.

At NRL, we have, through a modification of the "bronze process", developed V_3Ga wires that exhibit exceptionally high critical current densities (J_c).[1,2] The superior superconducting properties of these wires can give the designers of high field systems increased options for particular applications. Physical size could be reduced, refrigeration requirements eased and a greater safety margin built in for increased reliability.

V_3Ga conductors are available today with greater high field capabilities than any other material. Work in progress indicates that improvement over the entire magnetic field range is possible. If present efforts bear fruit, then realization of the intrinsic properties of V_3Ga will result in a conductor with great practical utility.[2,3]

SAMPLE PREPARATION AND MEASUREMENTS

The V_3Ga composite wires used in our studies were prepared from high purity metals. The V–Ga alloys were arc melted in helium. Compacts of V (99.9%+) and Ga (99.99%+) of 160 gram mass were melted and solidified on the water cooled copper hearth of the arc furnace. The homogeneity of the alloy was insured by inverting

103

the cast button and remelting it several times. Cylindrical castings were produced by melting the button on top of a split copper
mold whereby the molten alloy ran into the mold cavity forming a
12.7 mm diameter rod. A cast V-Ga rod produced in this manner is
shown in Fig. 1. The Cu-Ga alloys were prepared from high purity
metals (99.999% Cu and 99.99%+ Ga) by induction melting in a high
purity Al_2O_3 crucible under a partial pressure of argon. These
alloys were cast as 31.8 mm diameter rods in a ceramic-coated iron
chill mold. A Cu-Ga alloy rod casting produced in this manner is
shown in Fig. 2.

The headers were removed from the alloy castings and the rods
were cleaned prior to homogenization heat treatments. These heat
treatments were used to minimize solidification segregation effects.
The V-Ga and Cu-Ga castings were swaged to final size prior to
machining and assembly as a composite rod. The composite assembly
of V-Ga rods in the Cu-Ga matrix was capped with a vented Cu-Ga
end plug, evacuated under high vacuum and sealed by an electron
beam weld (see Fig. 3). This procedure removes entrapped air from
the cavity, which otherwise would interfere with the V_3Ga growth
by reacting to form compounds that would act as barriers to the Ga
diffusion from the Cu-Ga to the V-Ga interface. The composite
assembly was reduced in diameter to 0.81 mm (.032 inch) diameter
wire by swaging and wire drawing. Intermediate vacuum anneals
were employed after each 37% reduction in area. The annealing
temperatures were 500°C (1 h) for the first and second anneals and
550°C (1 h) for the third anneal. This reduction and annealing
schedule was repeated until the final wire size was reached. Sections of these wires were encapsulated in silica ampoules under
high vacuum and subsequently reacted isothermally to form V_3Ga as
an interfacial layer around each of the filaments.

The V_3Ga cross-sectional areas were determined using an electronic digital planimeter to measure the peripheries of the filaments in an enlarged photomicrograph (530X magnification). These
values along with the V_3Ga layer thickness, previously determined
from growth kinetics studies, were used in calculating the total
cross-sectional area of V_3Ga superconductor in the wire. The J_c
value was obtained by dividing the critical current I_c by the
total cross-sectional area of the V_3Ga reaction layer. Details
of the superconducting property measurements have been described
elsewhere.[2]

RESULTS AND DISCUSSION

The growth rate and superconducting properties of V_3Ga are
very dependent on the Ga contents of the matrix and filaments in
the composite wire. The use of V-Ga alloys, rather than unalloyed
vanadium filaments in combination with a given Cu-Ga bronze matrix

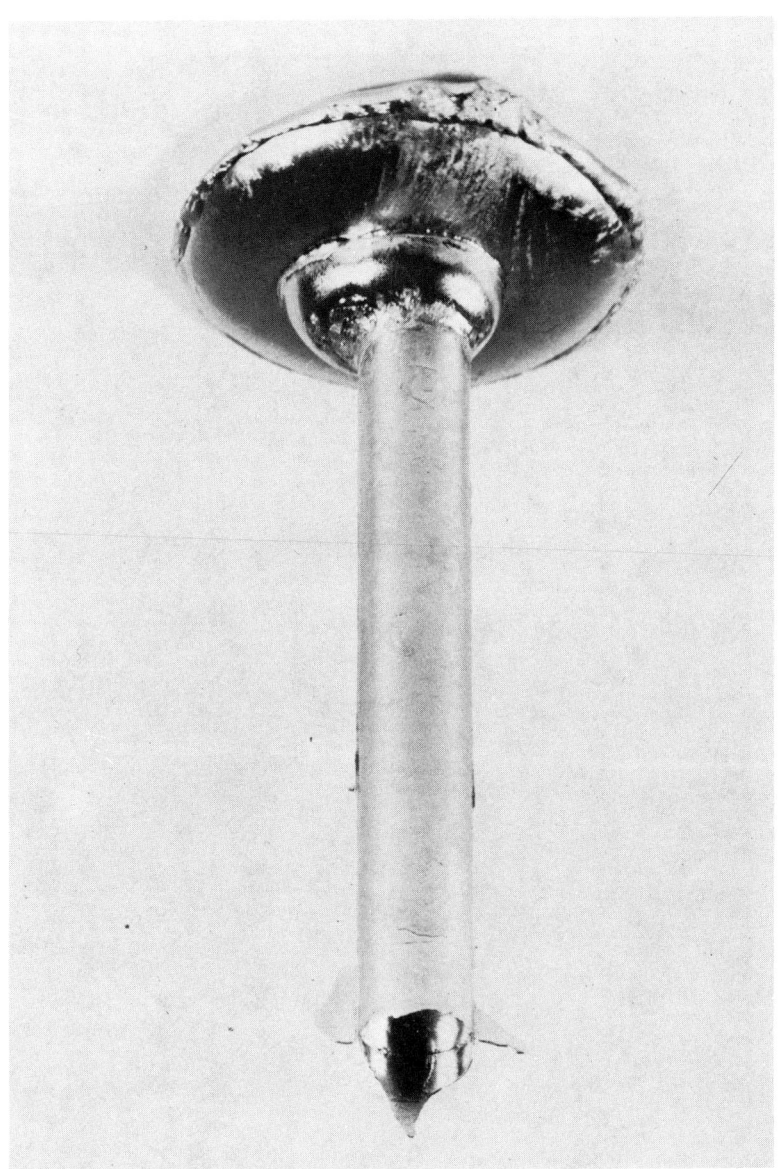

Fig. 1.   A V-8 at.% Ga (12.7 mm diameter) filament rod casting.

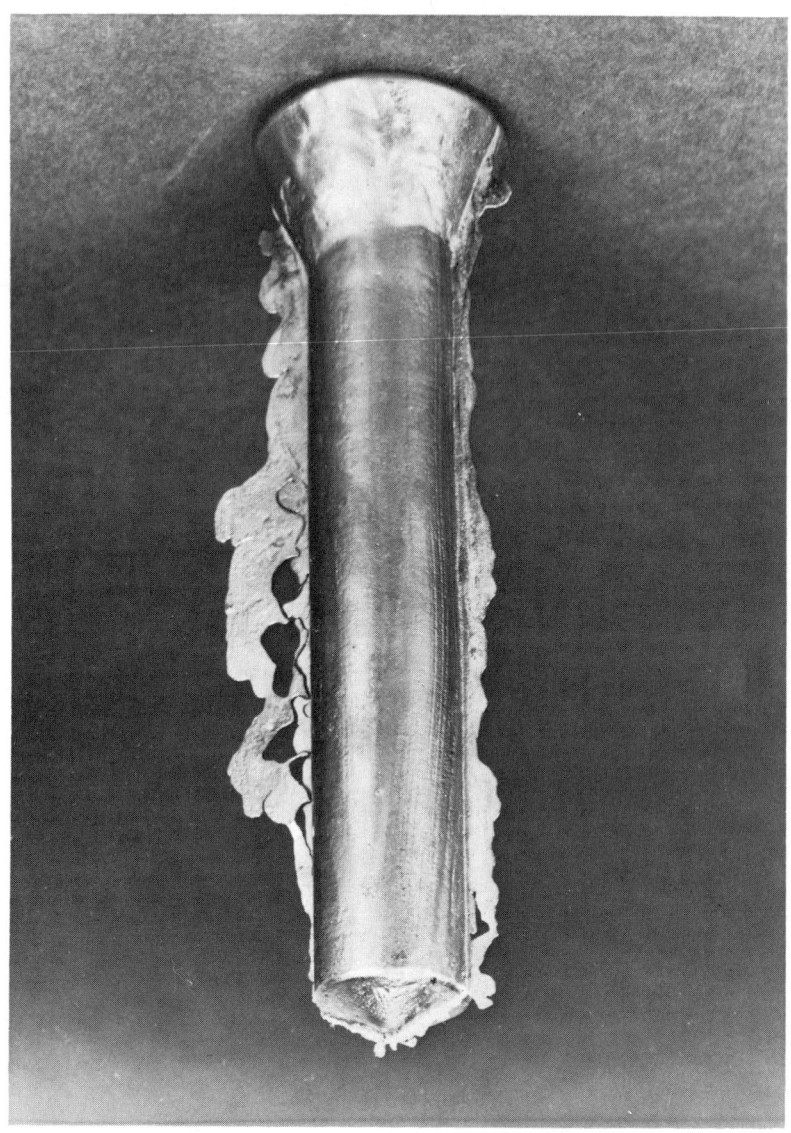

Fig. 2.   A Cu-17.5 at.% Ga (31.8 mm diameter) bronze matrix rod
           casting.

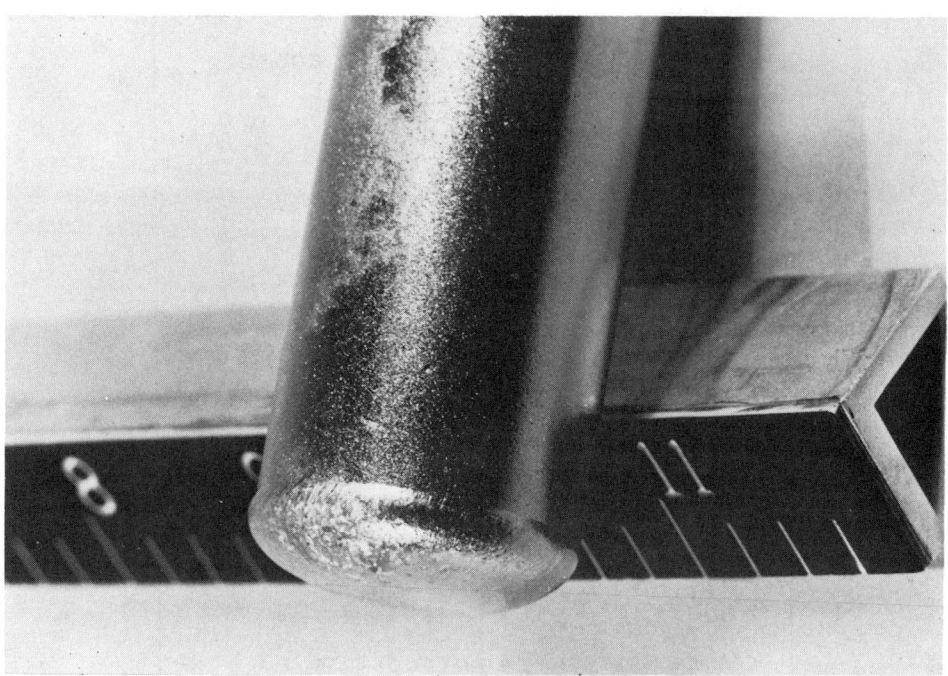

Fig. 3.   The V-8 at.% Ga/Cu-17.5 at.% Ga composite assembly show-
          ing the electron beam weld seal.

composition, results in much faster V_3Ga growth and higher J_c
values.   In the bronze process, Ga diffuses from the Cu-Ga matrix
to the V filament surface and reacts to form V_3Ga.   The V_3Ga layer
grows from the filament-matrix interface to the V filaments.   As
the V_3Ga interfacial layer grows, some Ga diffuses through this
layer and is retained in the unreacted V filament.   The reaction
rate may slow down with thicker V_3Ga layers due to Ga depletion
from the matrix.   Consequently, a longer reaction time is required
to achieve a desired V_3Ga thickness resulting in larger A15 grain
sizes and hence lower J_c values.

When V-Ga alloys are used in place of unalloyed V filaments
less Ga is required to diffuse from the bronze matrix to the
V-Ga interface in order to grow a given V_3Ga layer thickness.
Thus increased growth rates are obtained and lower V_3Ga formation
temperatures can be used.

This is illustrated in Fig. 4 (also shown in Table 1).   The
highest V_3Ga growth rates and J_c properties in our wires were
obtained by increasing the Ga contents of both the filament and
the matrix alloys to near their solubility limits.   For example,

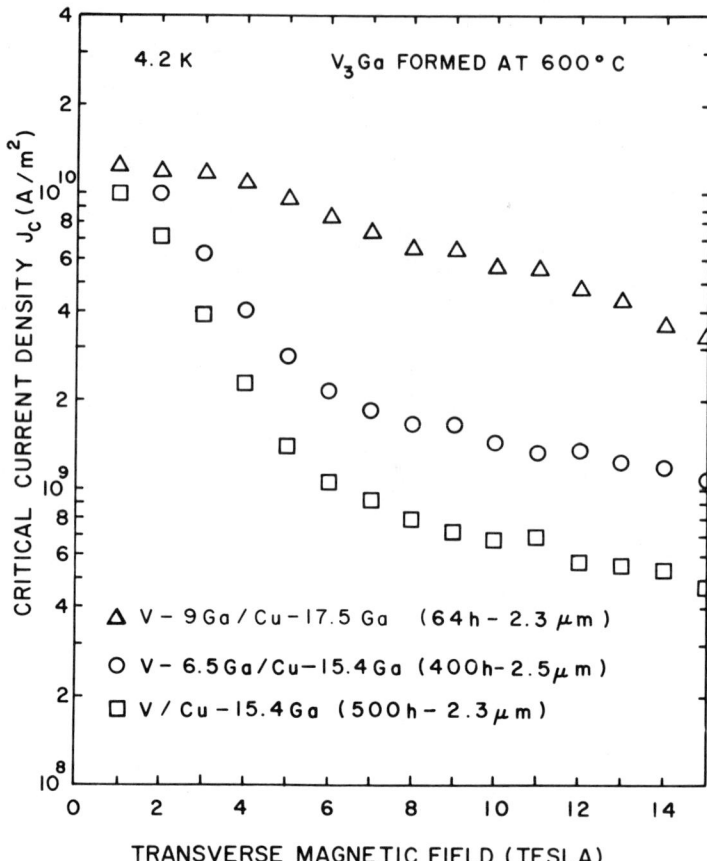

Fig. 4., The effects of alloy composition on the critical current
density versus transverse magnetic field properties of
V_3Ga formed at 600°C.

the V-9.2 at.% Ga filament/Cu-17.5 at.% Ga matrix alloy (9/17.5
alloy) combination resulted in an eightfold increase in V_3Ga
growth rate over the unalloyed vanadium/Cu-15.4 at.% Ga composite
wire (0/15.4 alloy). The J_c properties as well exhibited a seven-
fold improvement in the high Ga content composite wire as shown in
Table 1. The critical temperature (T_c) of the V_3Ga wire also increased from 14.3 to 14.6 K. It is felt that the modest increase
in the V_3Ga critical temperature (shown in Table 1) in higher Ga
content composite wires is due to the probability that the V_3Ga
formed is closer to stoichiometric composition.

Table 1.   Effect of Alloy Composition on Properties of
           V_3Ga Formed at 600°C

Alloy*	0/15.4	6.5/15.4	9/17.5
Reaction Time (h)	500	400	64
V_3Ga Layer (μm)	2.3	2.5	2.3
Critical Temp. (K)**	14.3	14.4	14.6
$J_c(10^9$ A/m^2) 6T	1.2	2.2	8.4
$J_c(10^9$ A/m^2) 10T	0.8	1.4	5.7
$J_c(10^9$ A/m^2) 13T	0.6	1.2	4.4

*
Alloy designation for these 0.81 mm diameter wires indicate the
gallium contents of the vanadium-gallium filaments and copper-
gallium matrix in atomic percent, respectively.
**
Critical temperatures shown here were measured using the ac
susceptibility method.[1]

The critical field-temperature dependence of two 30 filament
(8/17.5 alloy) composite wires reacted at low V_3Ga formation temperatures are shown in Fig. 5.   One sample was reacted at 480°C
for 600 h and the other at 556°C for 116 h.   These heat treatments
formed V_3Ga layers of 0.25 microns and 0.8 microns, respectively.
The critical field at 11 K was 12.3 tesla.   The critical temperature (resistance method) for the 556°C-116 h sample shown in Fig.
5 is 15.3 K.   (This method of measuring T_c gives midpoint values
that are about 0.5 K higher than obtained when the ac susceptibility method is used.)

The I_c versus temperature profiles at 6, 10, and 13 tesla
for a 30 filament (8/17.5 alloy) composite wire reacted at 480°C
for 865 h is shown in Fig. 6.   The V_3Ga layer thickness was 0.3
microns.   The J_c properties of this sample are compared to two
other high J_c samples reacted at low temperatures in Fig. 7.   The
samples reacted at 480°C and 500°C are both 30 filament 8/17.5
alloys whereas the 550°C sample was a monofilament 9/17.5 alloy
(see Table 2).

The faster V_3Ga growth rates exhibited in our high Ga content
composite wires allow the use of lower formation temperatures than
previously possible.   Lower V_3Ga formation temperatures improve
materials compatibility of composite conductor and insulator when
you wind a coil and react to form the superconducting compound
(wind and react method).   Lower V_3Ga formation temperature results
in a finer grain size A15 layer and thereby more grain boundary
area in a given V_3Ga layer thickness.   Since grain boundaries are
the primary flux pinning sites in A15 superconductors,[5-8] it

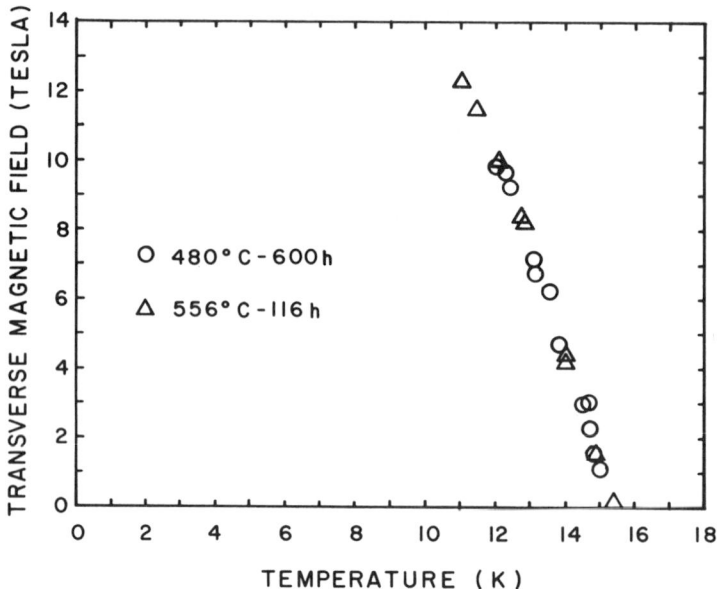

Fig. 5.   The transverse magnetic field versus temperature relation-
          ship for two 30 filament V_3Ga wires reacted at low forma-
          tion temperatures.

follows that the critical currents should be greater in materials
with the highest concentration of flux-pinning sites. Our experimental results indicate agreement with this.[1,2,9,10]

     It is desirable to increase the number of filaments in a
superconducting wire not only for stability but also to allow complete or nearly complete reaction of the filaments to V_3Ga in a
reasonable time. This would allow for the optimization of overall
J_c properties by the high J_c's obtained in thin V_3Ga layers.[2] In
order to do this you need very ductile filament rods that can be
processed in the Cu-Ga bronze matrix down to 1.0 to 2.0 micron
diameter filaments. In addition, it would be helpful to increase
the growth rates even further to decrease the A15 grain size. The
addition of third element additions to both the matrix and the
filaments is one approach that is being considered to achieve both
objectives. For example, Tachikawa[11,12] has shown that V_3Ga growth
rates can be improved by adding Al or Mg to the Cu-Ga matrix. We
have attempted to improve our best J_c values by using either a
V-8 at.% Ga or V-8 at.% Ga-0.5 at.% Ti filament in a Cu-14 at.% Ga-
4 at.% Al matrix. Our results[10] show that at the higher reaction
temperatures there is an increase in J_c probably due to grain refinement contribution due to the third element addition. The J_c

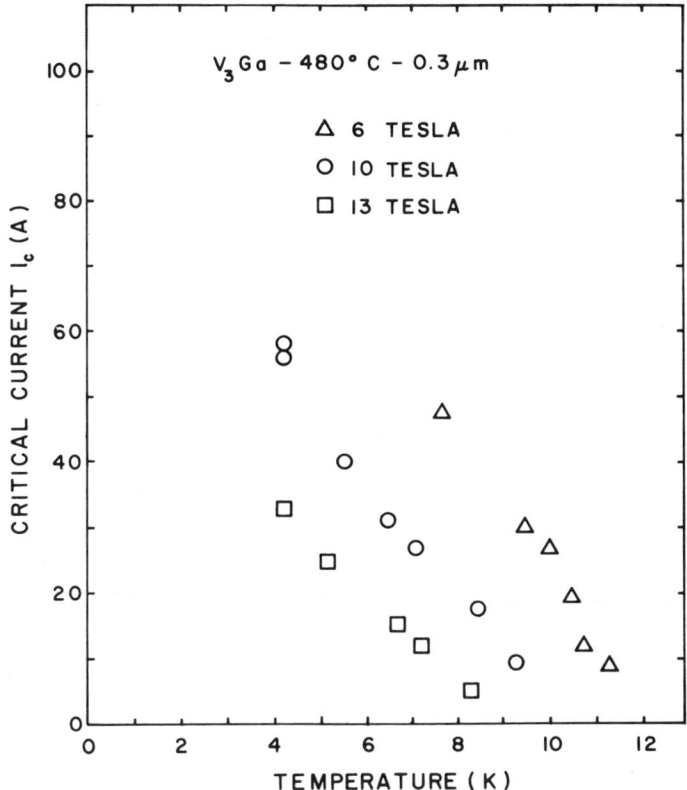

Fig. 6. Critical current versus temperature profiles at 6, 10, and 13 tesla for 0.3 μm V₃Ga layer formed at 480°C.

values we obtained in these alloys with third element additions were still lower than our best J_c values obtained in the 8/17.5 alloy reacted at low formation temperatures.[2] It is possible that an optimum alloy combination will be found with a third element addition where a large increase in J_c will be observed but as of now this has not been demonstrated.

In order to further improve the ductility of V-Ga alloy rods we have added third element additions of either 0.2 at.% cerium or 0.2 at.% yttrium and obtained softer alloy rods. The binary V-8 at.% Ga alloy microhardness was 170 kilograms/mm² without the third element additions. The addition of 0.2 at.% cerium or 0.2 at.% yttrium to the V-8 at.% Ga alloy resulted in lowering the microhardness of the alloys to 160 and 158 kilograms/mm², respectively. The effects of these third element additions on the superconducting properties of V₃Ga wires will be reported at a later date.

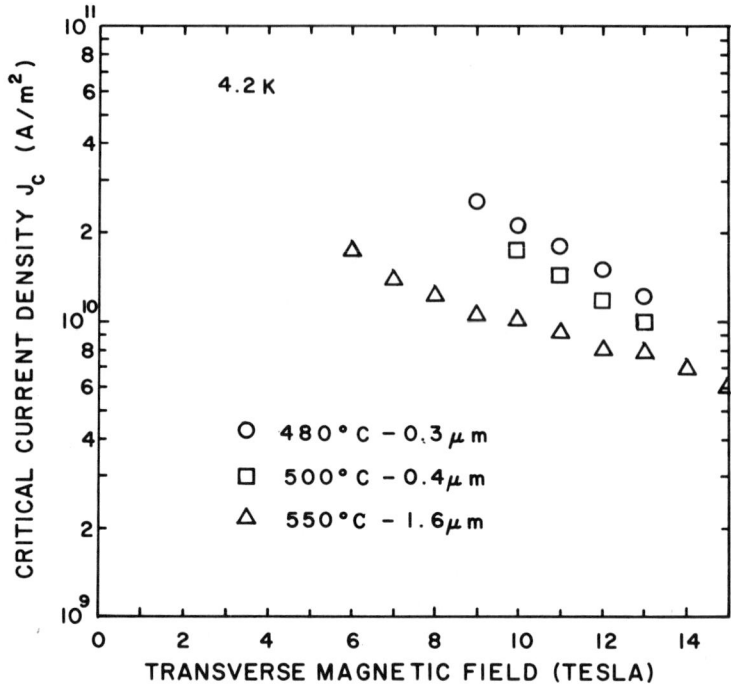

Fig. 7.   Critical current density (4.2 K) versus temperature
          magnetic field comparison of our highest J_c samples
          of V_3Ga reacted at low formation temperatures.

As a first step in the technology transfer of our in-house
developed high J_c (8/17.5 alloy) wire, we are scaling up to long-
length industrially produced multifilament V3Ga wires. Our commercial contract is with AIRCO, Inc., and the conductor design calls
for approximately 2800 filaments in a 0.5 mm diameter wire. This
should result in a higher overall J_c in a practical superconducting
wire with a greater filament to matrix ratio than the NRL wires.

Table 2.   Superconducting Properties of V_3Ga Wires at 4.2 K

Alloy*	8/17.5	8/17.5	6/17.5	9/17.5	9/17.5
No. of Filaments	30	30	19	1	1
Heat Treatment Temp. (°C)	480	500	556	550	600
Time (h)	865	475	122	400	64
V_3Ga Layer Thickness (μm)	0.3	0.4	0.8	1.6	2.3
H = 10 T					
I_c (amperes)	57.8	64.0	54.0	21.6	17.7
J_c (10^9 A/m^2)	21.3	17.0	10.6	10.1	5.7
J_c X H(10^{10}AT/m^2)	21.3	17.0	10.6	10.1	5.7
H = 13 T					
I_c (amperes (A))	33.0	35.7	36.0	16.8	13.6
J_c (10^9 A/m^2)	12.2	9.6	7.1	7.9	4.4
J_c X H(10^{10}AT/m^2)	15.9	12.5	9.2	10.3	5.7

*Alloy designation for these 0.81 mm diameter wires indicates the gallium contents of the vanadium-gallium filaments and copper-gallium matrix in at.%, respectively.

ACKNOWLEDGMENTS

The authors wish to thank C. R. Forsht for his valuable assistance during the metallurgical processing of our V3Ga wires and C. D. Carpenter for the silica ampoule encapsulations.

REFERENCES

1.   D. G. Howe, T. L. Francavilla, and D. U. Gubser, IEEE Trans. on Magnetics MAG-13:815 (1977).
2.   D. G. Howe and T. L. Francavilla, Report of NRL Progress (June 1978).
3.   D. U. Gubser, T. L. Francavilla, D. G. Howe, R. A. Meussner, and F. T. Ormand, IEEE Trans. on Magnetics MAG-15:385 (1979).
4.   D. G. Howe and L. S. Wienman, Proc. of 5th Intl. Cryogenic Eng. Conf. ICEC-5:326, Kyoto, Japan (1974).
5.   E. Nembach and K. Tachikawa, J. Less Common Metals 19:359 (1969).

6. M. Suenaga, T. S. Luhman, and W. B. Sampson, J. Appl. Phys.
    45:4049 (1974).
7. M. Suenaga, O. Horigami, and T. S. Luhman, Appl. Phys.
    Letters 25:624 (1974).
8. R. M. Scanlan, W. A. Fietz, and E. F. Koch, J. Appl. Phys.
    46:2244 (1975).
9. T. L. Francavilla and D. G. Howe, Cryogenics 19:41 (1979).
10. D. G. Howe and T. L. Francavilla, Advances in Cryogenic Eng.
    26 (1980), publication pending.
11. K. Tachikawa, K. Itoh, and Y. Tanaka, IEEE Trans. on Magnetics
    MAG-11:240 (1975).
12. K. Tachikawa, Y. Tanaka, Y. Yoshida, T. Asano, and Y. Iwasa,
    IEEE Trans. on Magnetics MAG-15:391 (1979).

DEVELOPMENT OF MULTIFILAMENTARY COMPOUND SUPERCONDUCTORS

Y. Furuto,[*] Y. Tanaka,[*] S. Meguro,[*] T. Suzuki,[**] and
I. Inoue[**]

[*]Central Research Laboratory
The Furukawa Electric Co., Ltd.
Tokyo, Japan
[**]Nikko Laboratory
The Furukawa Electric Co., Ltd.
Nikko, Japan

INTRODUCTION

Various types of multifilamentary compound superconductors
have been developed consistently for the past ten years at the
Furukawa Electric Company.

In 1972 V_3Ga[1] and Nb_3Sn multifilamentary stranded wires having
current-carrying capacity of 45-65 Amps at 10 tesla were commercialized for the first time in the world. A 10-tesla magnet was
wound from the wire and delivered to National Research Institute
for Metals (NRIM[2]). The wires have outstanding properties such as
high current density, high stability and excellent windability due
to its two-dimensional flexibility. These are indispensable for
high-field, high-homogeneity magnet. However, the current-carrying
capacity of the wires was at most 100 Amps at 10 tesla, making it
unsuitable for use in the windings of medium and large magnets.

In the following several years, there were increased demands
for larger compound superconductors for synchronous motor, syn-
chrotron, medium-high-field magnets and the like. Thus, medium-
capacity compound superconductors have been developed having current-carrying capacity of 100-1000 Amps at 10 tesla.[3] One example
is a multifilamentary V_3Ga tape for a 12.5-tesla magnet for NRIM.
Another one is Nb_3Sn 12-strand compacted cable of 780 Amps in critical current at 7 tesla[4] delivered to Fuji Electric Company for the
field windings of a synchronous motor.

115

Recently, fusion research has given further stronger motive force for the development of large-capacity conductors. Large Nb_3Sn superconductors of 3000-10,000 Amps in critical current at 10 tesla are under research and development.

This article deals with the development of medium- and large-capacity compound superconductors, and describes designing, fabrication, superconducting and mechanical properties, and coil performances using these conductors.

MEDIUM-CAPACITY CONDUCTOR

Design

The extreme brittleness of V_3Ga, Nb_3Sn, and other A15 intermetallic compound superconductors is well known, and degradation in their properties normally occurs at tensile strains of 0.2-0.4%. The authors have already carried out a dynamic evaluation[5] of the flexibility for various configurations of the compound superconductors. Tape-shaped conductors are flexible only along one axis, while stranded or compacted-stranded types are practically flexible in two-dimensional bending.

In order to withstand the tensile stress produced in handling during conductor fabrication and coiling as well as thermal stress and electromagnetic forces, some reinforcement is necessary for compound superconductors. A tension member of tungsten or stainless steel should be arranged at the center portion or in symmetrical geometry for the balance of force. Allowable tensile stress is designed to be 7-15 kg/mm^2.

Stabilization and protection against conductor burn-out for medium-capacity conductors are more severe compared to small conductors. The adding of 25-70% copper in cross-sectional area of the conductor is required depending on the conditions of cooling and current decay after coil quenching.

Fabrication

Multifilamentary V_3Ga tape-shaped conductor and Nb_3Sn compacted cable were prepared using bronze technique in the following sequence:

(1) Casting and homogenizing of 10-20 wt% Ga (or 5-14 wt% Sn) bronze
(2) Assembly of bronze/61-2400 core V(Nb) into billets
(3) Hot-working of billets into rods
(4) Rod drawing and twisting

    (5)   Wire forming into tape-shaped or compacted stranded con-
ductor

    (6)   Reaction at 600-700°C for 10-100 h

    (7)   Insulation

    (8)   Testing

The specifications of compound superconductors are shown in Table 1. Figures 1 and 2 show the cross-sections of the multifilamentary V_3Ga tape and Nb_3Sn compacted cable, respectively.

Critical Current

Figure 3 shows $H-I_c$ characteristics for the CS12 and various tape-shaped conductors. The critical current criterion is here taken to be the value when effective resistivity in the supercon-ductor reached 10^{-12} cm. The critical-current density is $5-9 \times 10^5 A/cm^2$ at 7 tesla for Nb_3Sn, $5-8 \times 10^5$ A/cm^2 at 7 tesla for V_3Ga. As for the critical current anisotropy of multifilamentary compound supercon-ductors, the I_c measured was less by as much as 5-20% with the mag-netic field parallel to the flat surface than in the perpendicular direction. This value is remarkably improved compared to that of conventional single-cored conductor.

Mechanical Properties

Figure 4 shows typical stress-strain curves for compound super-conductors. It can be seen that for reinforced conductors (1S or

Table 1. Specification for Compound Superconductor

Spec. \ form \ code	Stranded 1S	Tape B2	Tape B3	Tape C2	Compacted-stranded CS121	Compacted-stranded CS122
No. of filament	55×6	4,921	2,367	990	505×12	505×12
Filament dia. (μm)	10	5.5	5	12	5	6
Twist pitch (mm)	20	34	50	30	7	7
Stabilizer (%)	15	24	25	39	40	68
Reinforcement	W	Nb	Nb	Nb	Nb, SUS	Nb, SUS
No. of strand	6	—	—	—	12	12
Stranding pitch (mm)	6.5	—	—	—	22	22
Insulation	Yes	Yes	Yes	Yes	Yes	Yes
Overall size (mm²)	0.375	0.2×5.8	0.16×5	0.2×6.4	1.2×2.0	2.1×3.6

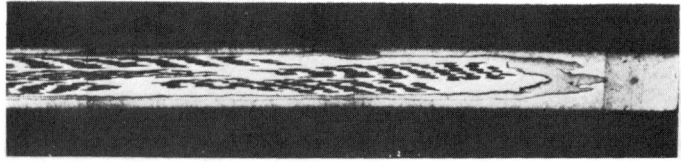

Fig. 1.   Cross-section of multifilamentary V₃Ga tape.

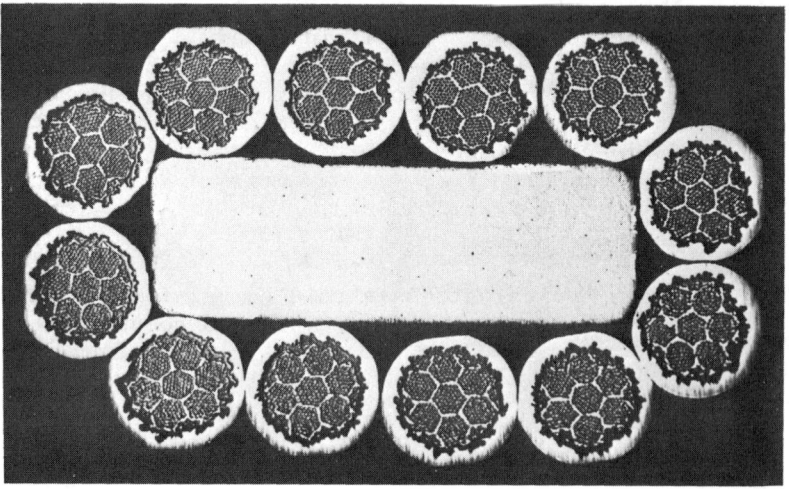

Fig. 2.   Cross-section of CS12 conductor.

CS12) the mechanical strength is larger than that for tape-shaped conductor which contains no tension member except a Nb tube for a diffusion barrier and residual Nb filaments. Figure 5 indicates the relationship between the normalized critical current (I_c/I_{co}) and actual maximum bending strain of compound superconductors. In 1S or CS12 conductor, tungsten or stainless steel exerts a marked reinforcing effect on the current degradation at tensile strains of 0.5% or more. However, even tape-shaped conductors exhibit no degradation in critical current until a maximum bending strain of 0.8% if the critical-current criterion is changed from 0.3 μV/cm (6.5×10^{-12} Ωcm) to 1 μV/cm (2.1×10^{-11} Ωcm).

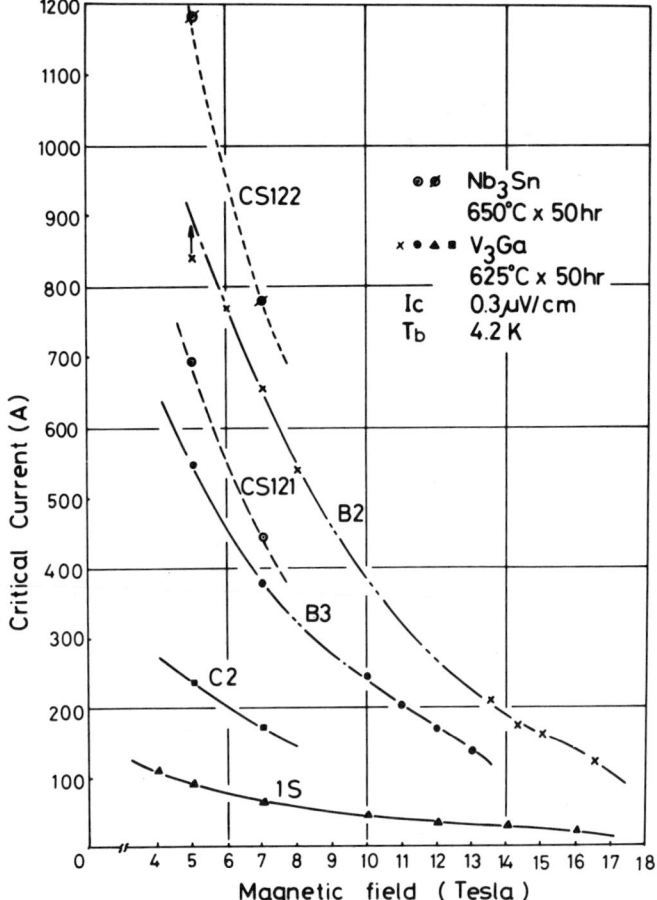

Fig. 3. H-I_c characteristics.

Coil Performance

    The specifications for typical coils wound from compound
superconductors are shown in Table 2. Coil 1 is the first one
wound from multifilamentary V3Ga superconductor in the world. A
10-tesla hybrid magnet was composed of coil 1 and NbTi alloy wire-
wound coil and delivered to NRIM. The magnet showed little degradation in quenching current at changing rates of up to about
2.5 tesla/sec.

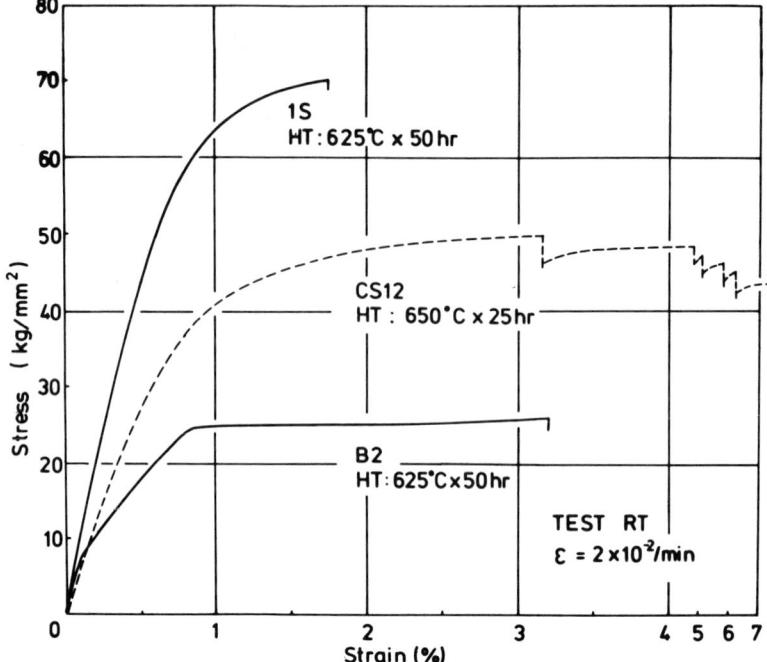

Fig. 4.   Stress-strain curves for compound superconductors.

Table 2.   Specifications for Coils Wound from Compound Superconductors

Spec. Coil	coil 1	coil 2	coil 4	coil 5
Conductor code	1S	C2	CS121	B3
Compound	V_3Ga	V_3Ga	Nb_3Sn	V_3Ga
Clear bore (mm)	30	70	64	30
Magnet OD. (mm)	72.8	103.8	100	82
Magnet length (mm)	99.7	15.3	60	11
Coil	solenoid	pancake	solenoid	pancake
Min. bending dia.(mm)	34.1	80	74.7	38
No. of turn	11,196	50×2	174	115×2
Field constant (T/A)				
center	1.24×10^3	1.36×10^3	2.19×10^3	4.95×10^3
max. in coil		3.35×10^3	3.07×10^3	7.97×10^3

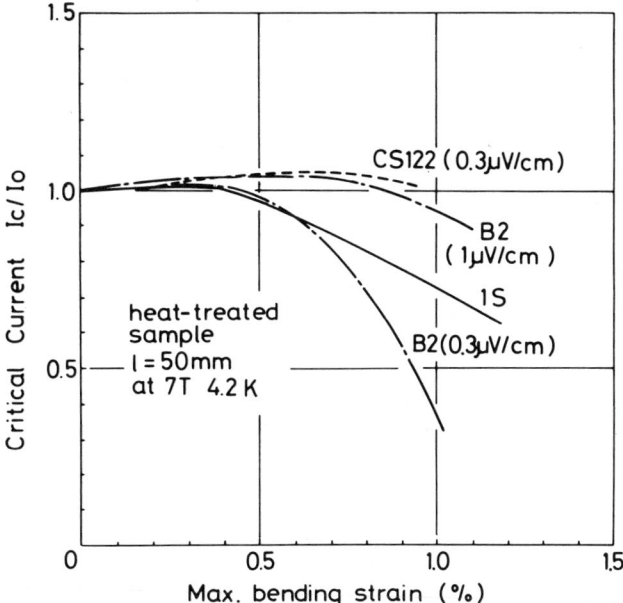

Fig. 5.   Current degradation due to bending strain.

Figure 6 shows the excitation characteristics of coil 2 and coil 5. Both coils exhibited no quench at the short-sample critical current defined by the resistive electric field of 0.3 μV/cm (10^{-12} Ωcm) and were extremely stable against normal excitations. A 12.5-tesla magnet was made of inner 10-double pancakes similar to coil 5 and an outer NbTi alloy wire-wound coil and was delivered to NRIM this spring. The magnet still remained superconducting at excitations of up to 13 tesla.

Coil 4 was made by Fuji Electric Co. for preliminary testing for the windings of 30-MVA synchronous condenser and test results were satisfactory. The actual coils are now under construction.

LARGE-CAPACITY CONDUCTOR

Design

Trial-manufacturing has been performed of a large-capacity

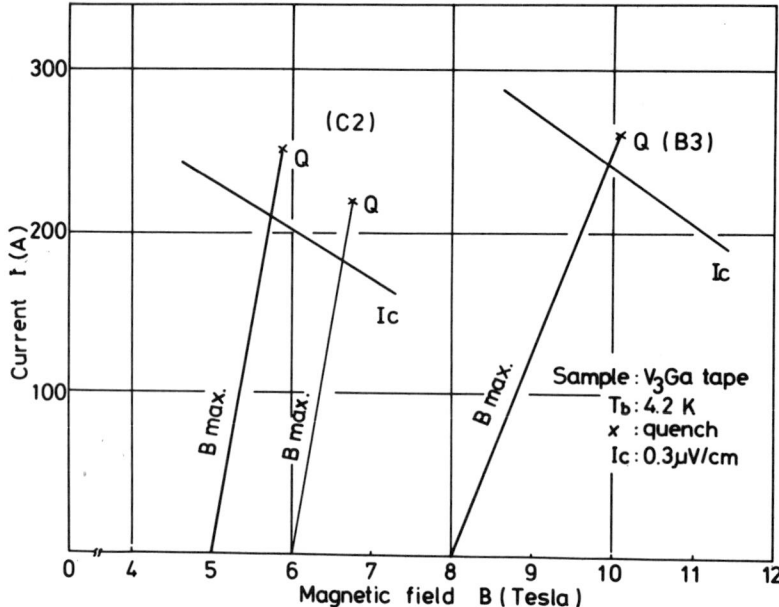

Fig. 6. Excitation characteristics of V3Ga pancake coils.

Nb_3Sn conductor for fusion reactor applications. The design thoughts adopted are as follows:

    (1) Pool-boiling cooling is adopted for cooling coils. The conductor is designed upon the criterion of full stabilization.

    (2) Hard copper is used as both the stabilizer and tension member.

    (3) An assembled conductor structure is chosen wherein superconducting cable is soldered into a stabilizing copper housing.

    (4) The surface of the copper stabilizer is grooved in order to meet the high heat flux of 0.5 W/cm^2.

The specifications for the conductor are shown in Table 3.

Fabrication[7]

    The conductor consists of two parts: superconducting compacted cable and copper housings. The superconducting cable is made of 11 strands which are multifilamentary Nb_3Sn composite wire with a diameter of 1.5 mm and copper sheath of over 15% in cross-

Table 3. Specification for Large Nb₃Sn Conductor

Item	Specification	Mesured Values
Basic Parameters		
Conductor size (mm²)	17×10.8	17×10.8
Operation current (A)	6000	
Critical current at 10T (A)	≥7200	7720
Yield strength at R.T. (Kg/mm²)	> 20	22.2
Minimum bending radius (mm)	≤300	300
Recovery heat flux (W/cm²)	≥ 0.5	0.6
Superconducting Strand		
Strand diameter (mm)	1.5	1.5
Filament diameter (μm)	~ 5	4.8
Number of filaments	20000	20750
Copper sheath (%)	≥ 15	17.5
Barrier	Nb	Nb
Twisting pitch (mm)	~ 30	31
Overall Jc without Cu at 10T (A/cm²)	4×10^4	4.4×10^4
Superconducting Cable		
Cable size (mm²)	8.3×2.8	8.3 × 2.8
Number of strands	11	11
Stranding pitch (mm)	~120	117
Stabilizing Copper		
Access groove size (mm²)	8.8×5.2	8.8×5.2
Cross section area (mm²)	152	152
Copper grade	OFHC Copper ½ Hardness	
Surface treatment	Chemical treatment (FC-2)	
Surface work	3 small grooves (1W×1.2H) per side	

section area. Nb₃Sn composite wire was produced through the usual process as mentioned in Section 2.2.

The copper housing was manufactured by extrusion and drawing from an OFHC copper billet. In order to increase the cooling perimeter, both sides of copper housing were provided with several longitudinal grooves. Moreover, the cooling surface with grooves was coated with a thin film of copper oxide by means of a newly developed chemical treatment (FC-2[8]) to raise heat-transfer characteristics.

The superconducting cable, copper housing and copper covering were incorporated into the assembled conductor by a soldering process using Pb-50% Sn solder. Figures 7 and 8 show the cross-sectional views of Nb3Sn composite wire and the finished conductor, respectively.

Test Results

The various properties were measured of the conductor. The major test results are summarized in Table 3.

Critical current. Since the critical current could not be directly measured owing to the limitations in our test facility, superconducting strands were taken as specimens from an end of the manufactured conductor. The overall critical current of the

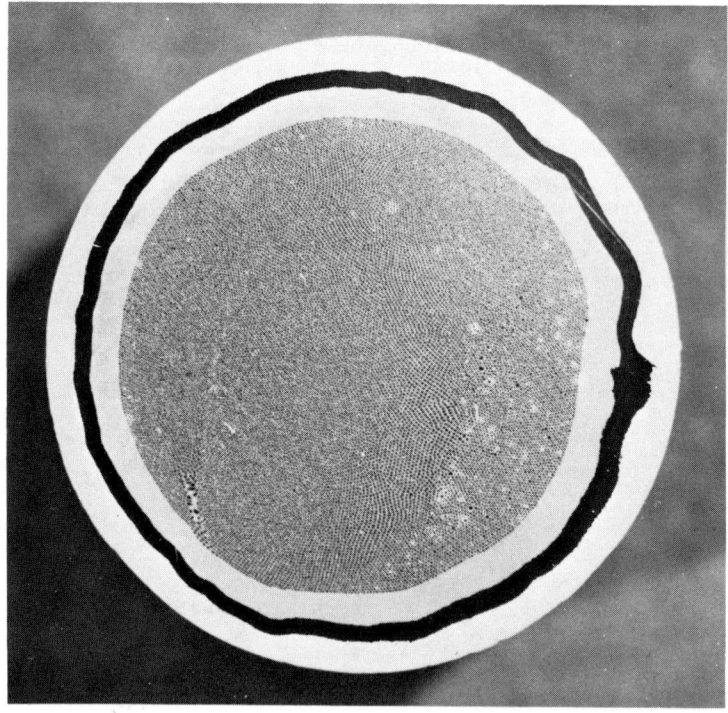

Fig. 7.   Cross-sectional view of Nb3Sn composite wire.

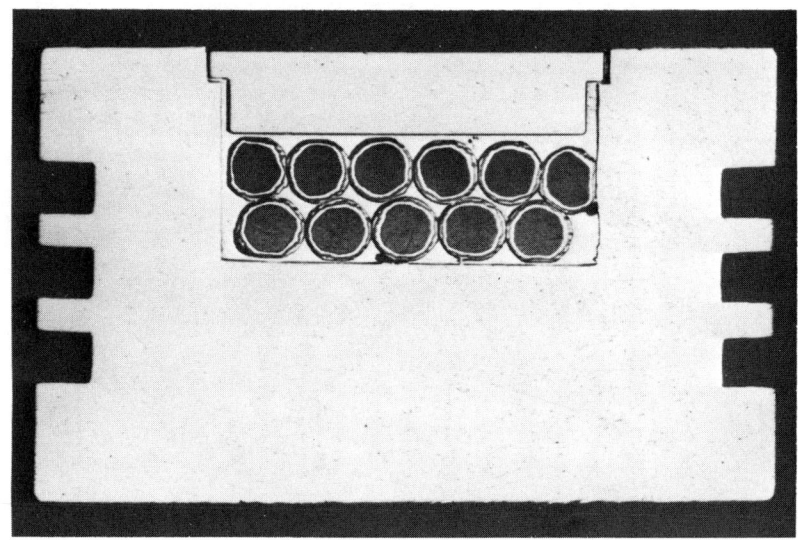

Fig. 8.   Cross-section of finished conductor.

conductor was estimated by multiplying the mean value of strands
by the number of strands.   Figure 9 shows the test results of critical current.   The conductor was confirmed as good enough to meet
the current specification.

     Strain effect.   Figure 10 shows the relationship between the
normalized critical current (I_c/I_{co}) and the actual bending strain
of compound superconductor as indicated in the figure.   It can be
seen qualitatively that I_c/I_{co} monotonically decreases with bending strain (compressive).   However, the degradation is neglegibly
small until the strain reaches 0.45% which corresponds to a bending radius of 300 mm.

     Resistivity and tensile strength.   Electrical resistance mea-
surements were made of the copper housing cold-worked for various
reduction ratios in liquid helium and transverse magnetic field.
The measurement results are shown in Fig. 11 and Table 4.   Designed
reduction ratio was decided to be about 6%.

     Heat-transfer characteristics.   Figure 12 shows the heat-trans-
fer characteristics of the conductor.   The heat flux is large
enough to meet the heat-flux requirement.

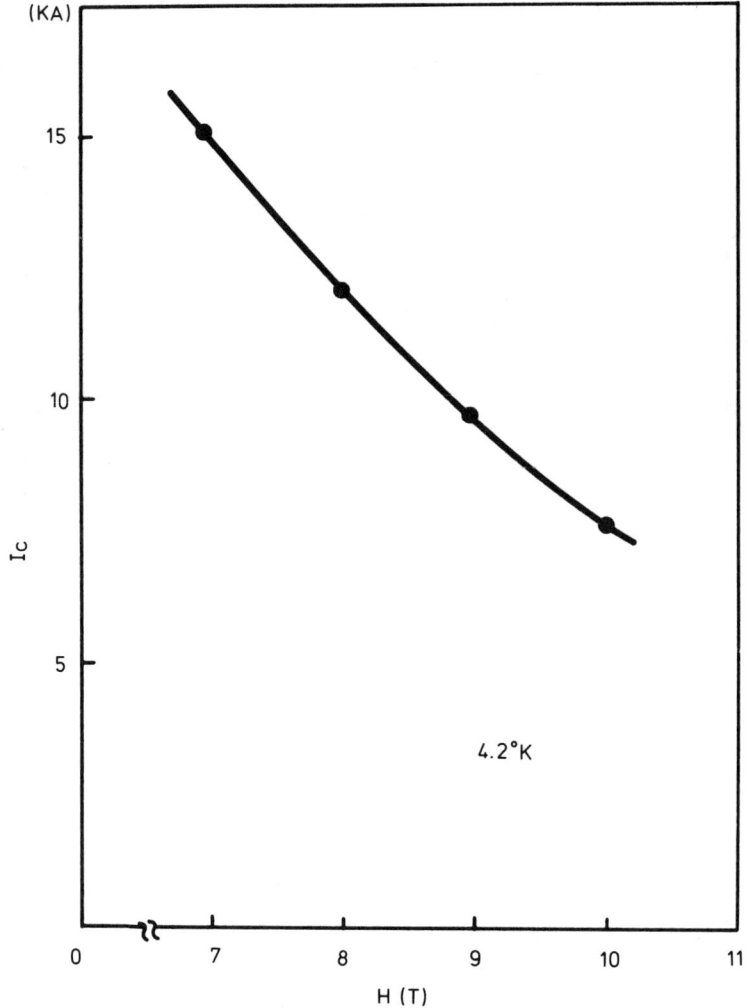

Fig. 9.   H–I$_c$ curve of large Nb$_3$Sn conductor.

ACKNOWLEDGMENTS

     The authors' thanks are due to M. Ikeda, K. Oishi, M. Ban,
H. Hamaoka and H. Terai of the Furukawa Electric Co. for the measurements and the preparation of specimens and coils.

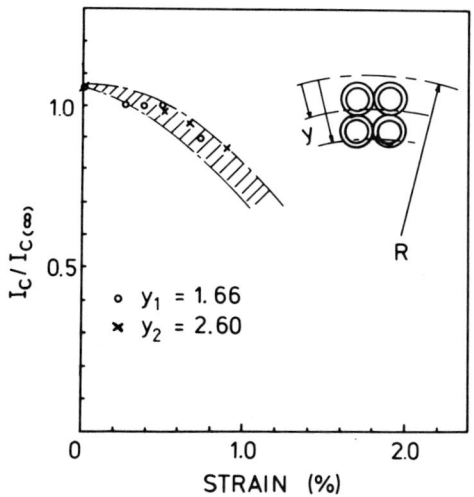

Fig. 10. Relationship between I_c/I_{co} and actual strain.

Table 4. Effect of Cold Reduction on the Mechanical Strength and Resistance Ratio of Copper

Cold reduction (%)	4	6	9	15
Ultimate tensile strength σn (Kg/mm^2)	24.8	25.8	27.3	29
Yield strength $\sigma_{0.2}$ (Kg/mm^2)	19.7	22.2	24.4	26.3
Residual resistance ratio γ	220	180	122	105

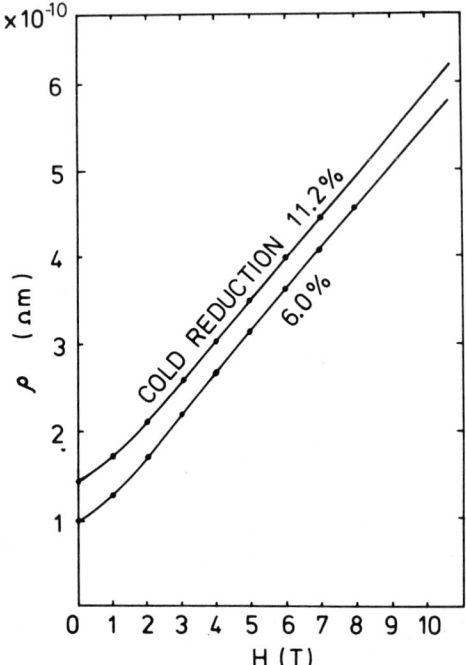

Fig. 11. Magneto-resistance of hard copper.

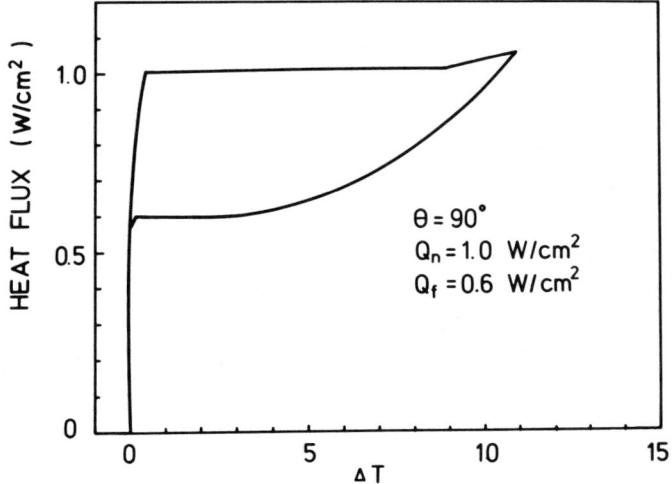

Fig. 12. Heat-transfer characteristics.

REFERENCES

1. Y. Furuto et al., Appl. Phys. Letters 24:34 (1974).
2. M. Ikeda et al., Proc. MT-5:715 (1975).
3. Y. Tanaka et al., Proc. ICEC-6:381 (1976).
4. Y. Tanaka et al., 17th Convention of Cryogenic Association
    for Japan (1977).
5. Y. Tanaka et al., Cryogenics 17:233 (1977).
6. S. Nose et al., US-Japan Workshop, "High Field Superconducting
    Materials for Fusion" (1980).
7. I. Inoue et al., 23rd Convention of Cryogenic Association of
    Japan A3-16 (1980).
8. S. Meguro et al., 8th Symposium on Engineering Problems of
    Fusion Research (1979).

FLUX PINNING IN BRONZE-PROCESSED Nb_3Sn WIRES*

M. Suenaga and D. O. Welch

Division of Metallurgy and Materials Science
Brookhaven National Laboratory
Upton, New York 11973

INTRODUCTION

With the increasing importance of multifilamentary Nb_3Sn conductors for technological uses such as for the production of very high magnetic fields in fusion magnets, means of improving the superconducting critical current density J_c at very high magnetic fields (H > 10 tesla) have been sought intensively, and it has been found that metallurgical factors such as heat treatment conditions,[1] alloying additions,[2] and mechanical strains[3] can strongly influence the critical current density. The correlation of changes in J_c with such metallurgical variations in the Nb_3Sn wires has been facilitated by the use of scaling laws for magnetic flux pinning in hard superconductors, and the scaling law developed by Kramer[4] has been used frequently.[5] We have found in the course of our investigations of the properties of monofilamentary Nb_3Sn wires produced by the "bronze process" that the magnetic field dependence of J_c at high fields can qualitatively be characterized well by Kramer's scaling law. However, when a detailed comparison of the scaling law and available experimental results was made, we found serious inconsistencies in the values of the parameters which appear in the scaling equation. In this article, we will point out those instances where the equation appears to work well and other cases where the use of the equation leads to unrealistic results.

*Work performed under the auspices of the U.S. Department of Energy.

THE SCALING LAW

The magnetic flux-pinning strength, F_p, in the <u>high magnetic field</u> regime is given by Kramer as[4]

$$F_p = |\vec{J}_c(H) \times \vec{H}| = K_s h^{1/2}(1-h)^2 \tag{1}$$

where $h = H/H_{c2}$ and $K_s = 0.56H_{c2}^{5/2}\kappa_1^{-2}(1-a_o\sqrt{\rho})^{-2}$, and κ_1 is the Ginsburg-Landau constant ($\kappa_1 = H_{c2}/\sqrt{2}H_c$), a_o is the flux lattice spacing ($a_o \simeq (\phi_o/H)^{1/2}$) and ρ is the density of the flux-pinning sites. This equation was derived assuming that flux pinning could be described by two regimes: at low fields flux motion occurs primarily by unpinning, whereas at high fields flux motion occurs by synchronous shear of the flux line lattice around pins too strong to be broken. Furthermore, the equation describes the dynamic pinning force produced by a series of pinning planes, each of which consists of a series of line pins, and which lie parallel to the Lorentz force.

In the past, the scaling law has often been tested by plotting experimental data as (F_p/F_{pmax}) vs h or F_p vs $h^{1/2}(1-h)^2$. A difficulty associated with such plots is the determination of a consistent value of the upper critical field H_{c2}, needed to calculate the reduced magnetic field h appearing in Eq. (1). Also, when both axes are normalized it is not easy to make a critical comparison.

For studying the high field behavior of J_c or F_p, it is found that another form of Eq. (1) is more convenient and useful.[6,7] Simple algebraic manipulation of Eq. (1) yields:

$$J_c^{1/2}H^{1/4}(1-a_o\sqrt{\rho}) = 0.7\kappa_1^{-1}(H_{c2}-H) \quad . \tag{2}$$

Now, since the right-hand side of Eq. (2) is linear in H, the equation can be used to determine H_{c2} by plotting the left-hand side against H and extrapolating. In most cases, at sufficiently high field $a_o\sqrt{\rho} \ll 1$, and $J_c^{1/2}H^{1/4}$ is linear in H over a reasonably wide range of H; and thus H_{c2} is obtained simply by linear extrapolation and the Ginzburg-Landau parameter κ_1 is obtained from the slope of the plot without adjustable parameters. In the following section, the applicability of the scaling law, in the form of Eq. (2), is examined for bronze processed Nb_3Sn wires with regard to: 1) the linear dependence of $J_c^{1/2}H^{1/4}$ on H, and 2) the values of κ_1 determined from the slope.

EXPERIMENTAL RESULTS AND DISCUSSION

In order to examine the applicability of the scaling law, various experimental results for critical current densities in the bronze-processed wires are used. The details of fabrication for the wires and tapes are discussed elsewhere.[1] All of the measurements for J_c were made at the National Magnet Laboratory using a 19 T and a 23 T Bitter coil and at 4.2 K. The criteria for J_c was ~1 μV/2 cm in most cases.

First, J_c data for a composite monofilamentary wire with a matrix-to-core ratio of ~15 and with a heat treatment of 725°C for 120 h are plotted in Fig. 1 as suggested by Eq. (2), illustrating

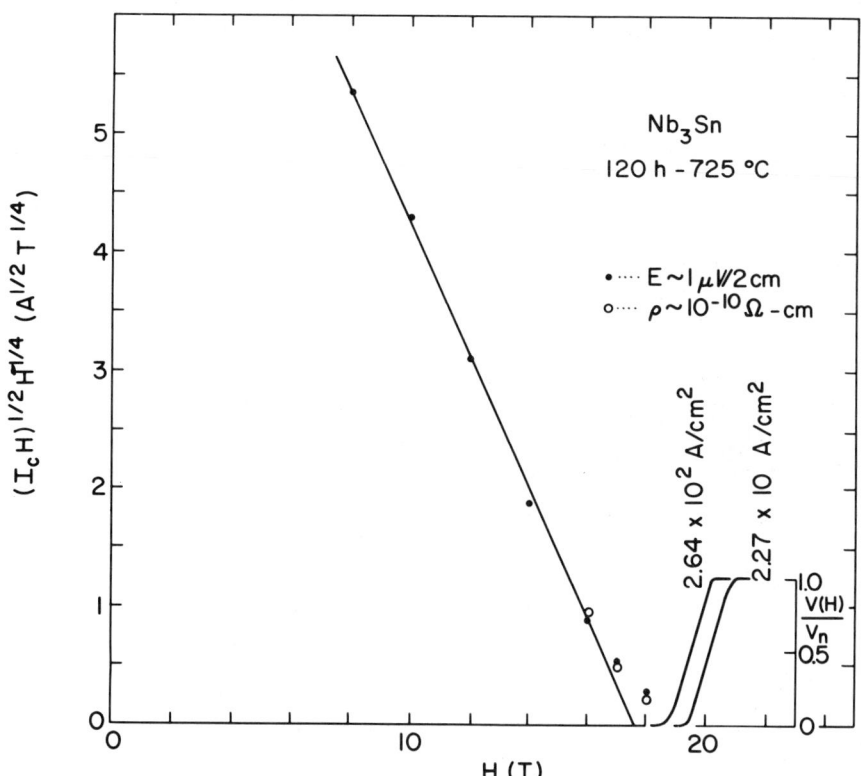

Fig. 1.   A plot for $J_c^{1/2} H^{1/4}$ vs H for a monofilamentary Nb$_3$Sn wire which was heat treated at 725°C for 120 h indicating the extrapolated H_{c2} and the low current density H_{c2}.

its use in analyzing the magnetic field dependence of the critical
current density. Here it was assumed that $a_o\sqrt{\rho} \ll 1$ in calculating the left-hand side of Eq. (2), and this assumption seems to be
justified since the plot is a straight line with H over a wide
range in magnetic field ($\sim$8 to $\sim$16 T). This implies that the distance between the pinning sites in this wire is considerably larger than the flux lattice spacing, a_o. Therefore a measure of the
upper critical field H_{c2} can be obtained by the linear extrapolation of the straight segment of the plot. This value of H_{c2} is
not the magnetic field where superconductivity vanished totally
from the wire, as may be seen in the figure. The experimental
data deviate from a straight line near the critical field. In
fact, superconductivity persists to considerably higher values
($\sim$2 T) than the H_{c2} as determined by linear extrapolation. This is
probably due to inhomogeneities in the Nb_3Sn such as a composition
variation across the Nb_3Sn layer or a variation in the strains due
to the matrix.[3] It is also possible that the bulk of the Nb_3Sn
still carries superconducting currents beyond the extrapolated
H_{c2}, and that the dependence of J_c on H does not follow Eq. (2)
in that region, i.e., the scaling law fails here. Although the
values of H_{c2} obtained by extrapolation will differ from those
determined with other criteria, such as the midpoint of the transition from the normal to the superconducting state, H_{c2} as determined by this method will be consistent with Eq. (2) for the
purpose of examination of the validity of the scaling law. Thus,
in the following discussion, the values of H_{c2} quoted were all determined by the linear extrapolation method.

   In the majority of cases, our measurements on monofilamentary
Nb_3Sn wires show that J_c varies with magnetic field as described
by Eq. (2) with $a_o\sqrt{\rho} \ll 1$, and thus yield linear plots of $J_c^{1/2}H^{1/4}$
versus H (except very near H_{c2}) as shown in Fig. 1. However in
several instances, such simple quasi-linear plots were not obtained, and we believe such deviations fall into two categories.

   In the first category, a plot of $J_c^{1/2}H^{1/4}$ vs H yields a plot
with concave-up curvature. We believe that such curvature results
from the erroneous assumption that $a_o\sqrt{\rho} \ll 1$. This behavior is
illustrated in Figs. 2 and 3 which show data for a "bronze-
processed" Nb_3Sn wire which was electron irradiated during a 500°C
heat treatment[8] and for an "in-situ processed" wire, heat treated
at 550°C for 6 days,[9] respectively. As shown in the figures, in
both cases the data can be made to produce a wide region of linearity in the plot if the $(1-a_o\sqrt{\rho})$ term is included. Also in
both cases, the selection of $(10^3\text{ Å})^{-1}$ for the value of $\sqrt{\rho}$ resulted in a linear dependence on H. This value was found to give a
better straight line fit than is obtained with $(500\text{ Å})^{-1}$ or
$(2000\text{ Å})^{-1}$. It is interesting to note that in these wires the
grain size of Nb_3Sn is $\sim$400Å[9] and yet the best fit for the

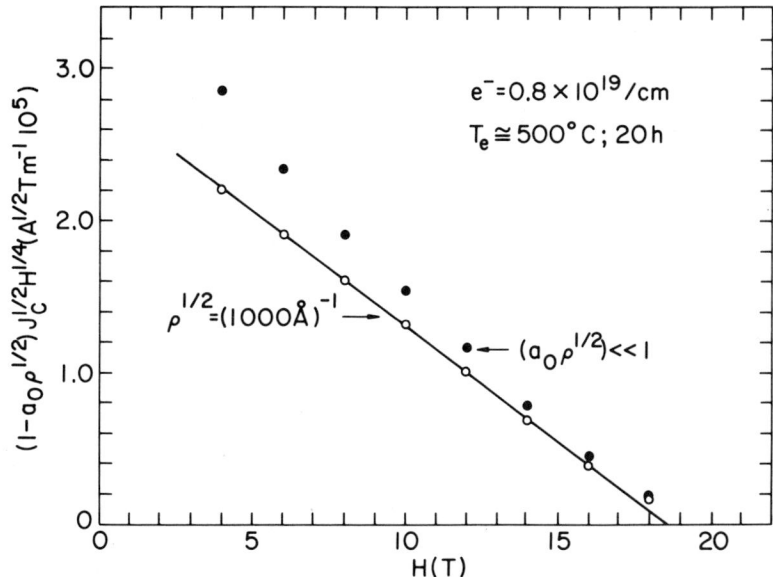

Fig. 2. A plot for $J_c^{1/2}H^{1/4}(1-a_o\sqrt{\rho})$ vs H for a Nb$_3$Sn wire which
was electron irradiated during the heat treatment at
500°C for 24 h. This plot shows that the inclusion of
$(1-a_o\sqrt{\rho})$ with $\sqrt{\rho}$ = $(10^3$ Å$)^{-1}$ is necessary to make the
plot linear with H.

straight line was obtained when $1/\sqrt{\rho}$ was approximately twice as
large as the grain size. This may imply that only about half of
the grain boundaries are effective in pinning the flux lines.
(It is well established that grain boundaries are the primary
flux pinning sites in Nb$_3$Sn.[2])

    The second category of results in which plotting $J_c^{1/2}H^{1/4}$
versus H does not yield a simple linear plot is illustrated by
the behavior of bronze-processed Nb$_3$Sn wires which were made with
a Ga-containing matrix (Cu-Sn-Ga).[10] Representative examples of
$J^{1/2}H^{1/4}$ vs H plots for these wires are shown in Fig. 4. In this
figure, data from Sekine and Tachikawa[11] are also included. It
is speculated that the deviation from linearity for these wires
is due to the paramagnetic limit on H_{c2},[10] as in V$_3$Ga, in which

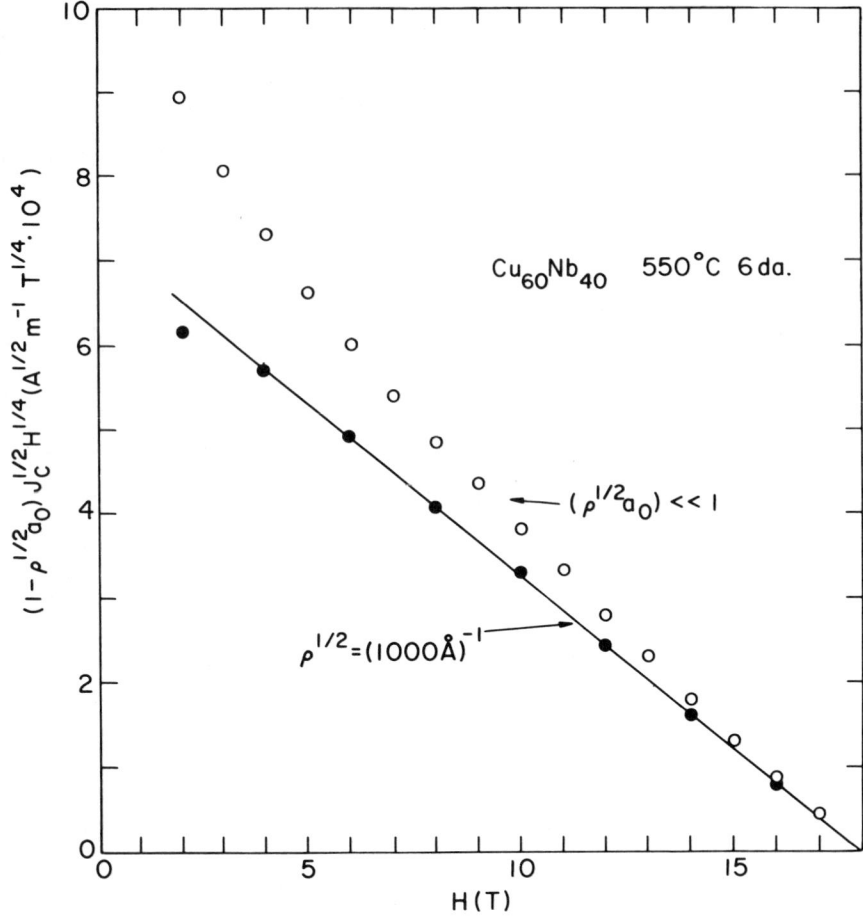

Fig. 3.   A plot similar to Fig. 2 for a Nb_3Sn wire fabricated by
the in-situ process [Ref. 9].

case, the conditions for which Kramer derived the scaling law do
not apply.

The examples cited above, Figs. 2 and 3, appear to indicate
that the Kramer scaling law adequately describes the effect on J_c
of variations in the grain size of Nb_3Sn.   However an experiment
on the anistropy of J_c casts doubt on this conclusion.   Recently,
Tanaka et al.[12] reported the observation of very large differences

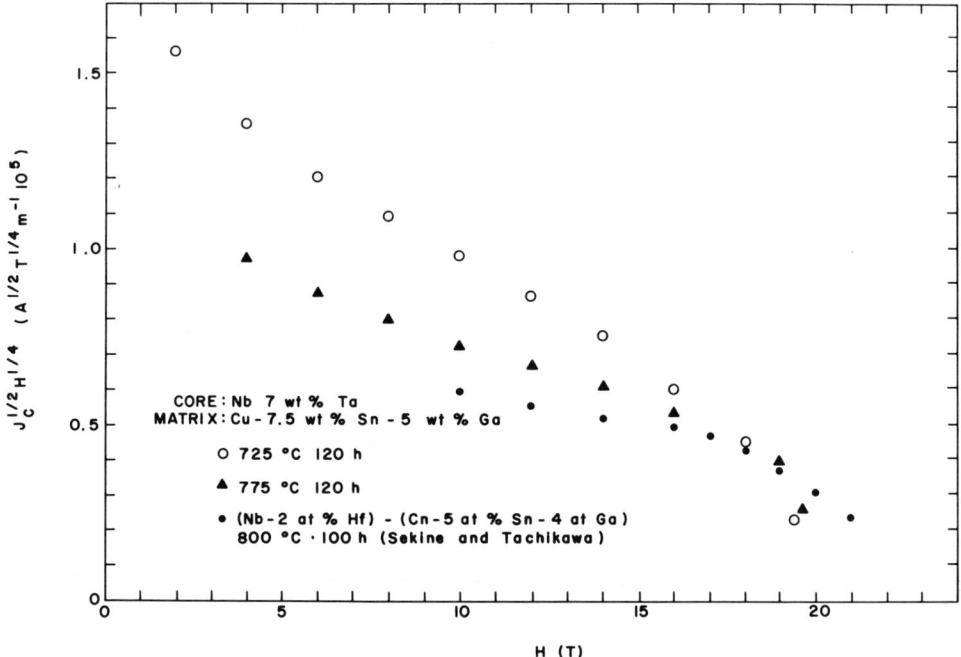

Fig. 4. $J_c^{1/2}H^{1/4}$ vs H for Ga-containing Nb₃Sn wires. These data do not appear to obey Eq. (2).

in J_c of bronze-processed V₃Ga tapes which are measured with applied fields parallel ($J_{c\,||}$) and perpendicular ($J_{c\perp}$) to the surface of the tape. These differences are a factor of two or more, depending on the heat treatments and the value of H,[12] and they are attributed to the columnar structure of V₃Ga grains which grow perpendicular to the substrate. Thus, the effective grain size with H perpendicular to the tape is smaller than that with H parallel. The difference in J_c was accounted for by taking into account the variation in the effective grain size and assuming that J_c is inversely proportional to the grain size. The Kramer scaling law cannot be applied to these results because of paramagnetic limitation in V₃Ga. However such an experiment for Nb₃Sn is revealing.

We have performed a similar experiment for Nb₃Sn using a flattened wire which had a matrix-to-core ratio of ∿15. Pieces of the wire were heated for 32 h at 725°C and 4 h at 775°C. In both specimens, the $J_c^{1/2}H^{1/4}$ vs H plots are straight in the field range of ∿5 to ∿10 tesla regardless of the orientation of H to the tape surface, as shown in Fig. 5. As found by Tanaka et al.,

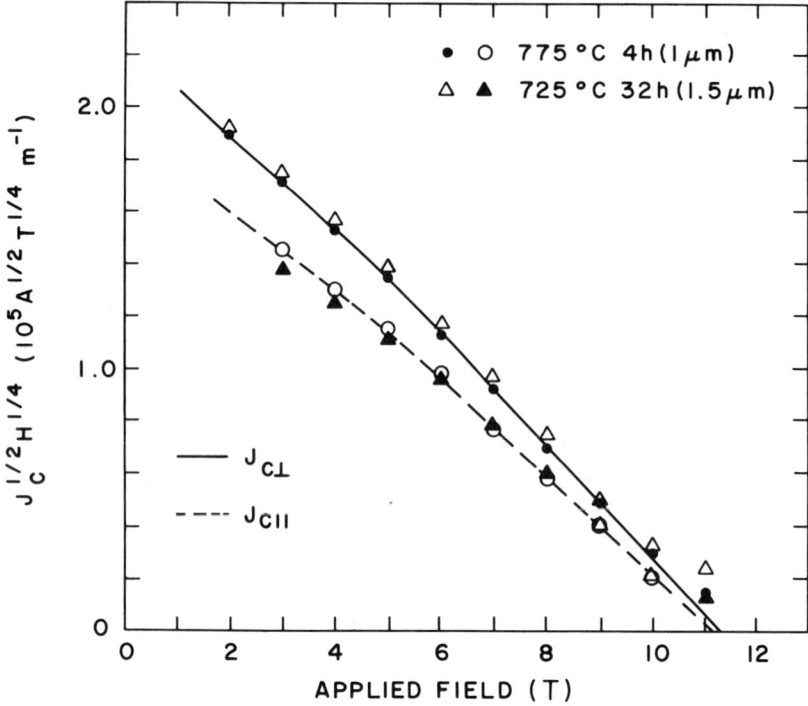

Fig. 5.   Plots for $J_c^{1/2}H^{1/4}$ vs H for H parallel ($J_{c\,||}$) and per-
          pendicular ($J_{c\perp}$) to the surface of Nb$_3$Sn tapes.

$J_{c\perp}$ is larger than $J_{c\,||}$, but only by 30 to 40%. These smaller
differences in J_c are due to nearly equiaxed growth of Nb$_3$Sn
grains in these specimens, as observed with a scanning electron
microscope. However, these results clearly reveal a difficulty
with Eq. (2) as a description of the grain-size dependence of J_c.
Variations of J_c by as much as 30–40% caused by anisotropy in the
grain size imply, in Eq. (2), that the $a_o\sqrt{\rho}$ term must be large
enough that a plot of $J_c^{1/2}H^{1/4}$ vs H would exhibit an easily observable concave-up curvature, as in Figs. 2 and 3. Yet, as seen
in Fig. 5, the results show no such curvature and imply that
$a_o\sqrt{\rho}$ << 1, in which case variations in $a_o\sqrt{\rho}$ would cause a negligible variation in J_c. Clearly there is a contradiction here,
and its resolution is not obvious at present. Anistropy in H_{c2}
(and thus κ_1) with crystallographic orientation is not likely to

be the origin of anisotropy in J_c, since, as seen in Fig. 5, the extrapolated H_{c2} from $J_{c\perp}$ and $J_{c\|}$ is essentially the same in both cases. Unfortunately, the values of H_{c2} for these tapes are significantly lower than those for Nb_3Sn wires with the same matrix-to-core ratio. The cause for the reduction of H_{c2} is not understood at this time. However, we believe that the observed differences in $J_{c\perp}$ and $J_{c\|}$ and our conclusion about the scaling law drawn from the present data are valid in spite of the unexplained reduction in H_{c2} of these specimens.

As a further test of the scaling law, the behavior of the values of κ_1, as determined by Eq. (2) from the slope of $J_c^{1/2}H^{1/4}$ versus H plot, were studied for a set of monofilamentary wires which are heat treated for 6, 16, and 64 h at 725°C and 96 h at 675°C. The ratio of bronze-to-core for these wires was ∿7.6. The data are plotted in Fig. 6 as $J_c^{1/2}H^{1/4}$ vs H, and the lack of concave-up curvature over the field range 8-15 tesla implies that $a_o\sqrt{\rho} \ll 1$. Values of κ_1 obtained from the slopes of these plots, as suggested by Eq. (2), are listed in Table 1. (The data from this set of specimens was chosen as an illustration, but essentially identical behavior was observed for wires with other bronze-to-core ratios.) The values of κ_1 so obtained are smaller than that expected for Nb_3Sn, for which the smallest expected value is ∿20 for very clean specimens.[13] A quantitative discrepancy of this order is perhaps not unreasonable considering the uncertainty in the value of the shear constant, C_{66}, for the flux line lattice which was used in the derivation of Eq. (1). What is a more serious difficulty is the fact that the observed value of the κ_1 increases with heat treatment time while the observed value of (H_{c2}/T_c) is unchanged (see Table 1). This difficulty is best illustrated by examining the relationship between (H_{c2}/T_c) and κ_1[3,13,14]:

$$\frac{H_{c2}}{T_c} \simeq c[N(0)(1+\lambda)]^{1/2} (1-t^2)\kappa_1 = c'\gamma^{1/2}(1-t^2)\kappa_1 \qquad (3)$$

where c and c' are numerical constants, N(0) is the density of states at the Fermi level, λ is the electron-phonon coupling constant, γ is the electronic specific heat coefficient and t = $= T/T_c$. Recent experimental studies[14-17] of the effect of disorder and nonstoichiometry on the superconducting properties of a variety of A15 compounds show N(0), λ, and γ to be monotonically increasing functions of T_c. Our data, Table 1, clearly show that T_c increases by ∿4% on increasing the reaction time from 6 h to 64 h at 725°C. Thus one would expect, on the basis of the experimental results cited above, that N(0), λ, etc. would increase by roughly the same amount. Since our data show H_{c2}/T_c to increase by about 3%, one would then expect κ_1 to remain more-or-less constant, yet, as seen from Table 1, the value of κ_1 deduced with

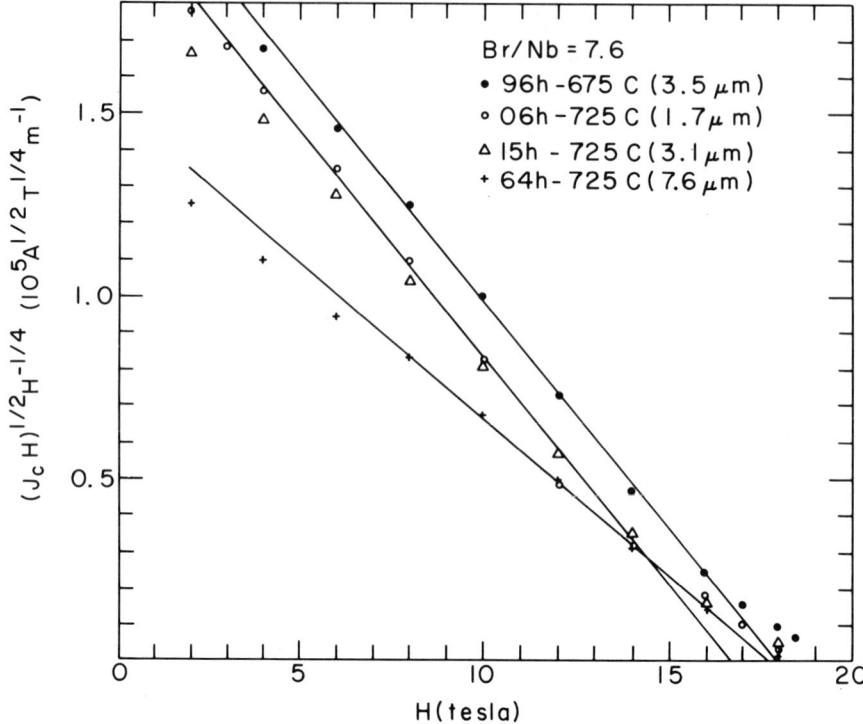

Fig. 6.   Plots for $J_c^{1/2}H^{1/4}$ vs H for a series of Nb$_3$Sn wires which
were heat treated for 6, 16, and 64 h at 725°C and for
96 h at 675°C.

the scaling law increases by ∿50%! This discrepancy suggests
that even though the <u>form</u> of the scaling law for flux pinning at
high fields derived by Kramer seems to be obeyed for Nb$_3$Sn, the
interpretation of parameters such as "κ_1" which appears in it may
not be the same as assumed in Kramer's derivation.

CONCLUSION

The scaling law derived by Kramer for magnetic flux pinning
in high magnetic fields was examined for its applicability to the
magnetic field dependence of critical-current densities in the
bronze processed monofilamentary Nb$_3$Sn wires. From this we

Table 1.   Superconducting Properties of the Nb_3Sn Wires

	6h/725°C	16h/725°C	64h/725°C	96h/675°C
κ_1	9.2	9.9	13.8	8.66
$T_c^{(a)}$	16.15	16.45	16.85	16.80
$H_{c2}^{(b)}$	16.6	16.9	17.5	17.6

[a] The midpoint T_c measured inductively with the bronze matrix still present.

[b] Determined by the linear extrapolation according to Eq. (2).

conclude about the scaling law that: 1) its prediction for the form of the dependence of critical current on magnetic field and grain size $\{|\vec{J}_c \times \vec{H}| \sim h^{1/2}(1-h)^2(1-a_o\sqrt{\rho})^{-2}\}$ was found to be very good in most cases including wires with very small Nb_3Sn grains ($\sim$400 Å).   It was found very useful in comparison of J_c for different wires and in extrapolating to obtain H_{c2} for these wires. 2) However, it could not account consistently for the anisotropy in critical current of a tape which was measured with H applied perpendicular and parallel to the tape face.   3) The values of κ_1 which were determined with the scaling law were too small by a factor of 2 to 3, and the trend in the variation with heat-treating time was opposite to that which is reasonably to be expected.   That the behavior of κ_1 is thus seriously in contradiction with the expected behavior for Nb_3Sn suggests basic faults in the derivation of the scaling equation for critical currents at high magnetic fields.

ACKNOWLEDGMENTS

    The authors appreciate technical assistance by A. Cendrowski, D. Horne, R. Mitchell, F. Perez, and R. Sabatini and the use of the high magnetic field facilities at the Francis Bitter National Magnet Laboratory.

REFERENCES

1. For a review see M. Suenaga, W. B. Sampson, and C. J. Klamut, IEEE Trans. on Magnetics MAG-11:231 (1975).
2. For a review see, J. D. Livingston, Kristal und Technik 13: 1379 (1978).
3. For a review see, D. O. Welch, to be published in Adv. Cryo. Engin. 25 (1980).
4. E. J. Kramer, J. Appl. Phys. 44:1360 (1973).
5. For example, D. U. Gubser, T. L. Francavilla, D. G. Howe, R. A. Muessner, and F. T. Ormand, IEEE Trans. on Magnetics MAG-15:385 (1979); T. S. Luhman, C. S. Pande, and D. Dew-Hughes, J. Appl. Phys. 47:1459 (1976).
6. G. Rupp, E. Springer, and S. Roth, Cryogenics, 17:141 (1977).
7. M. Suenaga, T. Onishi, D. O. Welch, and T. S. Luhman, Bull. Am. Phys. Soc. 23:229 (1978), and unpublished data.
8. C. L. Snead, Jr. and M. Suenaga, Appl. Phys. Lett. 36:474 (1980).
9. D. K. Finnemore and J. D. Verhoeven, to be published in Prog. in Cryo. Engin. 25 (1980).
10. C. L. Snead, Jr. and M. Suenaga, IEEE Trans. Magnetics MAG-15:625 (1979).
11. H. Sekine and K. Tachikawa, Appl. Phys. Lett. 35:472 (1979).
12. Y. Tanaka, K. Itoh, and K. Tachikawa, J. Japan Inst. of Metals 40:515 (1977).
13. T. P. Orlando, E. J. McNiff, Jr., S. Foner, and M. R. Beasley, Phys. Rev. B 19:4545 (1979).
14. H. Wiesmann, M. Gurvitch, A. K. Ghosh, H. Lutz, O. F. Kammerer, and M. Strongin, Phys. Rev. B 17:122 (1978).
15. A. K. Ghosh and M. Strongin, Proc. of the 1979 Conf. on Superconductivity in d- and f-Band Metals, LaJolla, CA (in press).
16. F. Y. Fradin and J. D. Williamson, Phys. Rev. B 10:2803 (1971).
17. R. Viswanathan and R. Caton, Phys. Rev. B 18:15 (1978).

AN AUGER ELECTRON SPECTROSCOPY STUDY OF BRONZE ROUTE NIOBIUM-TIN

DIFFUSION LAYERS

D. B. Smathers and D. C. Larbalestier

Materials Science Center, University of Wisconsin
1500 Johnson Drive
Madison, Wisconsin 53706

INTRODUCTION

Nb_3Sn made by solid state diffusion by the bronze route process
(or one of its variants) is now commercially available from many
sources and filamentary (FM) Nb_3Sn conductors are being delivered
for a wide range of both small and large scale devices. However,
there is much that still remains unclear in the properties of FM
Nb_3Sn composites. The critical current density (J_c) is known to be
sensitive to the state of stress existing in the Nb_3Sn[1,2] and to the
degree of stoichiometry[3] in the layer. It is also to be expected
that the concentration of impurities in the Nb_3Sn layer will be of
importance but as yet little has been reported on this subject. As
a result of this threefold dependence of J_c, most master curves for
J_c for FM Nb_3Sn are rather conservative and the specification of J_c
for a particular composite is still generally a subject of empirical
investigation, after the composite has been made.

Of the three variables noted above, the effect of strain is
now relatively well understood[1,2] but little definite is known
about the chemical compositions, departures from the stoichiometric
Nb_3Sn composition and impurity concentrations in bronze route Nb_3Sn
layers. These questions are both of intrinsic fundamental interest
and related to such practical problems as the specification of air
or vacuum melt bronze and the purity of Sn and Nb which needs to
be chosen for production composites.

Experimental investigation of these points is made difficult
by the extreme thinness of Nb_3Sn layers ($\sim$3 μm down to as little
as 10 nm in some in-situ composites[4-6]). The electron probe microanalyzer (EPMA) does not resolve below 1-2 μm and we have

143

accordingly utilized an Auger electron spectrometer (AES) in both
Auger and ESCA mode to investigate thin layers of Nb_3Sn. Our initial investigations have shown this to be a useful tool for the
investigation of Nb_3Sn layers[7] and it has also yielded useful information on Nb_3Ge layers.[8]

In the present paper we report on the development of composition profiles (Sn,Nb,O,C) as a function of reaction time and temperature and on the dependence of the transition temperature T_C on
composition and impurity level. The work has been performed on a
commercial prototype Brookhaven transmission line tape, in order to
permit us to use a relatively broad beam Auger spectrometer (thus
giving us a larger S/N ratio) and to calibrate our AES results
against those obtainable on thick layers with the EPMA. Since the
quantitative application of fine beam Auger spectroscopy is not yet
a mature science, we believe that this approach is preferable to an
immediate investigation of fine filament Nb_3Sn composites.

EXPERIMENTAL DETAILS

The 60 mm wide tape used for this study was produced by magnetron sputtering 13 wt% Sn bronze onto both sides of an electron
beam melted Nb foil. The thickness of the bronze layers is $\sim$35 µm
and the Nb foil $\sim$23 µm. Samples 5x10 mm were cut from the tape and
sealed into quartz tubes under 50 mTorr of argon. To avoid oxygen
contamination from residual impurities in the argon, etc., Ti rods
were also sealed into the quartz tubes and flamed to red heat before
reacting to form the Nb_3Sn. After reaction the bronze was removed
when desired with a 50% nitric acid solution. The AES sputter profiles were taken using a Physical Electronics Industries Auger spectrometer (Model 548), using a 3 keV primary electron source and a
40 µA beam current. In order to examine the small Auger peaks, the
derivative of the Auger energy spectrum was obtained by modulating
the cylindrical mirror analyzer by 3 eV. The Nb_{167}, C_{270}, Sn_{246},
and O_{510} eV Auger peaks were monitored with a multiplexing unit
while the sample was being ion milled. The ion milling agent was a
beam of Ar^+ ions at a potential of 2 kV and an emission current of
30 mA. Our earlier investigation has shown this to be a satisfactory procedure.[7]

Critical temperatures were measured inductively at a frequency
of 100 Hz and an applied field of 0.3 mT. Critical temperatures are
quoted as the upper limit to the transition, defined as the inter-
section of extrapolations of the normal state and superconducting
state susceptibilities. All T_C values reported here are for samples with the bronze chemically removed, in order to avoid depression of the T_c by the surrounding bronze.[9]

RESULTS

An example of the composition profile obtained during an experimental run is shown in Fig. 1 for a layer reacted 10 h at 650°C. The data is plotted as it comes from the spectrometer, as peak-to-peak Auger signal height versus sputter time. The conversion of the experimental data to composition/depth profiles is discussed later. Time zero on the figure corresponds to the bronze/Nb₃Sn interface exposed by etching the bronze off with nitric acid.

The main features of the profiles are as follows. Within a few atomic layers of the surface are oxides of Sn and Nb, as evidenced by chemical shifts in the position of the Auger peaks. Beyond this there is a region which appears anomalously high in Sn (and deficient in Nb) while, for the layer in Fig. 1, after about 40 min sputtering the composition profile takes on the appearance of a gently sloping "plateau". After about 130 min, the Sn content begins to decline more steeply as the Nb₃Sn/Nb interface is reached. The sputter time-depth calibration is 1.5 nm/min and the position of the Nb₃Sn/Nb interface is defined as midway between points

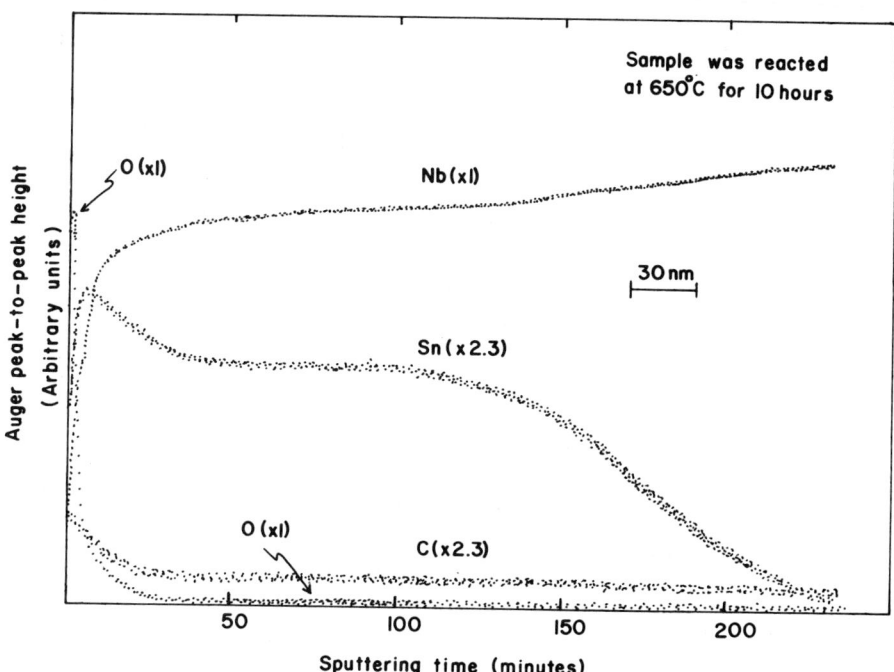

Fig. 1.   The Auger depth profile for a Nb₃Sn layer reacted for
          10 h at 650°C.

corresponding to 10% and 90% of the "plateau" value of the Sn signal. The position of this interface is uncertain to, typically,
100-200 nm. Scanning electron metallography shows that the bronze/
Nb_3Sn interface is uneven to 1 or 2 grain diameters and we may expect a similar roughness at the Nb_3Sn/Nb boundary. Our broad beam
Auger averages over a 100 μm diameter region and thus averages over
many thousand grains. According to West and Rawlings,[10] the grain
size of our samples should be ∿50-200 nm, so that this "interface"
width is not unreasonable.

The interpretation of the Auger signal/sputter time profiles
requires further discussion. In anticipation of this discussion,
we may say that we believe that the initial Sn peak seen in Fig. 1
and in the profiles of Fig. 2 results from the fact that the sputtering process preferentially removes Sn atoms from the surface
and that in the early stages of sputtering the equilibrium Sn to
Nb ratio has not yet been reached. Excepting the very thin oxide
layer right on the as-etched surface, the layer consists only of
Nb_3Sn (of variable composition) and Nb (possibly with some dissolved Sn).

The growth of the Nb_3Sn layer and the flattening of the concentration gradient with increasing reaction time are shown in
Fig. 2, for a sequence of samples reacted 2 to 30 h at 650°C. The
tendency of the Sn concentration gradient to flatten as the reaction proceeds and to assume a "plateau" character is quite evident.
Figure 3 shows a plot of the transition temperatures of samples
reacted at 650, 700, 800, and 900°C versus composition. The composition is plotted both as the ratio of Nb/Sn Auger p-p signal
height and the derived Sn content. The concentration values cor-
respond to a point midway through the Nb_3Sn layer. It can be seen,
as expected, that the T_c peaks at 18.1 K for stoichiometric Nb_3Sn,
declining as the Sn content diminishes.

Figure 4 includes some data on the O and C content of the
layers. It can be seen that the O and C contents (at the mid-
points, not the surface of the Nb_3Sn layers) are rather high (up
to 1.8% O) for short reaction times, declining as the reaction
time increases. The high initial O and C contents at short reaction times appear to correlate with low Sn contents. At longer
reaction times the Sn content increases to 25 at.% Sn while the
O content declines to 0.4%.

DISCUSSION

Concentration/Depth Calibrations

Quantitative analysis with Auger electron spectroscopy is
still at a relatively early stage and in this and our previous

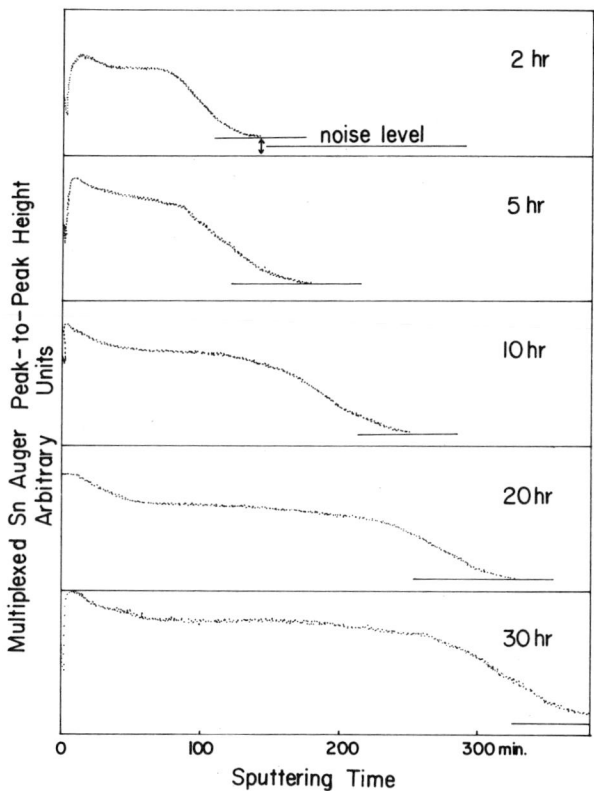

Fig. 2.   Tin Auger signal profiles for samples reacted at 650°C.

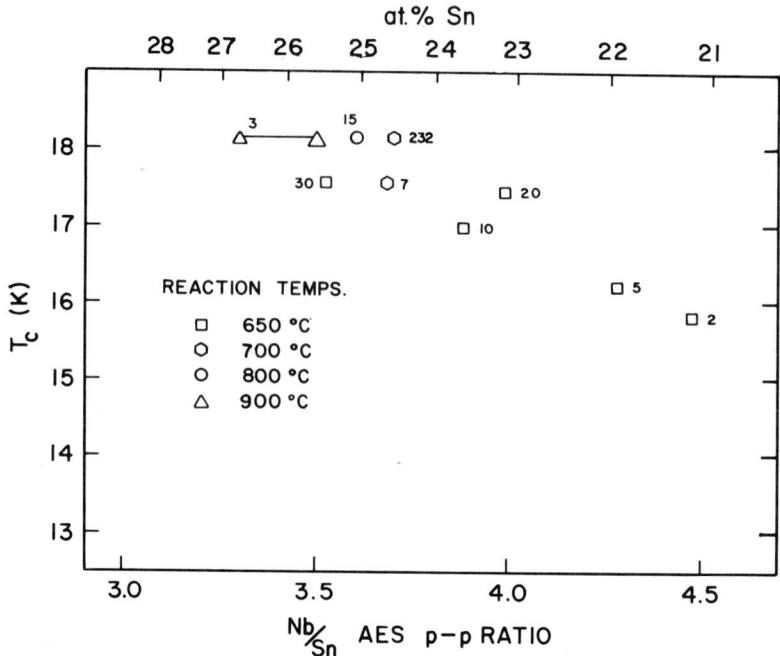

Fig. 3.   Transition temperature versus the Nb/Sn Auger p-p ratio
          and the derived chemical composition. Compositions are
          measured in the middle of the Nb_3Sn layer. Reaction
          times in hours are given by the numbers adjacent to the
          data points.

study[7] we have attempted to cross-check our results against those
provided by other techniques. In our earlier study we showed that
a combination of light microscopy, Rutherford Back Scattering (RBS)
and Auger data yielded a value of 40 nm/min for the sputter rate.[7]
In the present study, the sputter conditions were different from
those previously reported. In this case, we established the sputter rate by comparing the time needed to sputter through an iden-
tically reacted layer under the two sets of conditions. The sputter rate found for this study is 1.5 nm/min.

    We have taken two approaches to converting the Auger p-p sig-
nal heights to chemical composition. The simplest approach is to

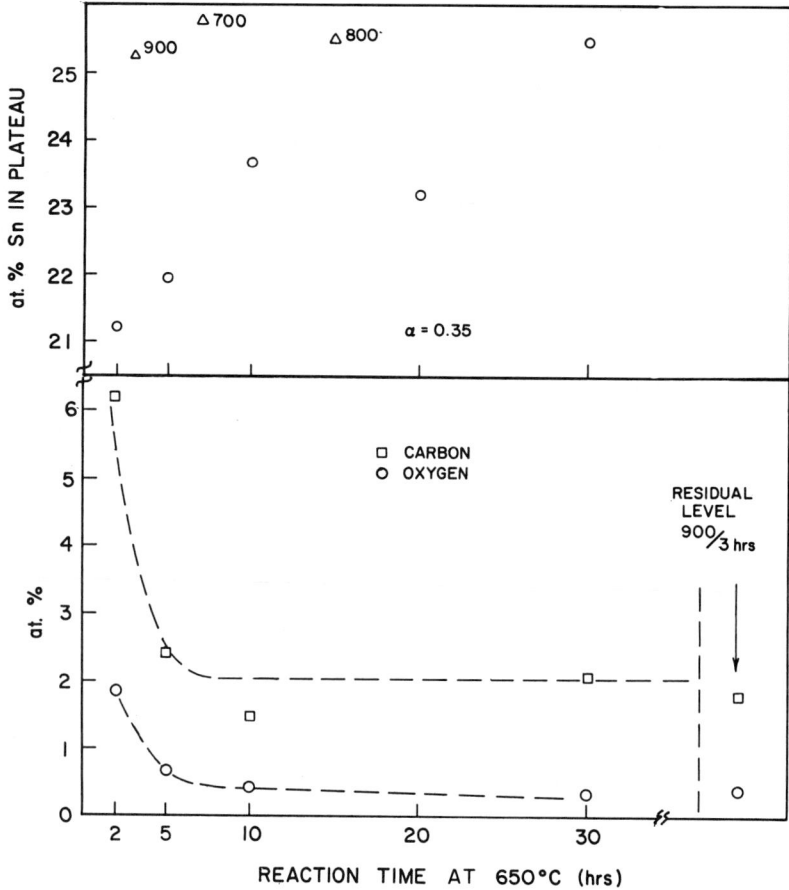

Fig. 4.   Derived chemical compositions in the center of the Nb_3Sn
          layer versus reaction conditions.   Reaction temperatures
          other than 650°C are indicated by the data points.

compare the pure element signal heights quoted in the Physical
Electronics Industries Handbook.[11]   However, this neglects any
synergistic effects due to the combination of the elements into a
compound and also neglects preferential sputtering effects.   Preferential sputtering certainly occurs, as may be seen from the Auger
traces in Fig. 5.   Two composites reacted for 3 h at 900°C were examined.   This high temperature treatment grows large grains, so
that individual grains can be examined with a fine beam Scanning
Auger Microprobe (SAM).   The sample examined in the broad beam AES
had its Nb_3Sn layer sputtered through in the usual way, while the
sample examined in the SAM was cleaved in vacuo while cooled with
liquid nitrogen.   The cleavage plane was normal to the Nb_3Sn layer.

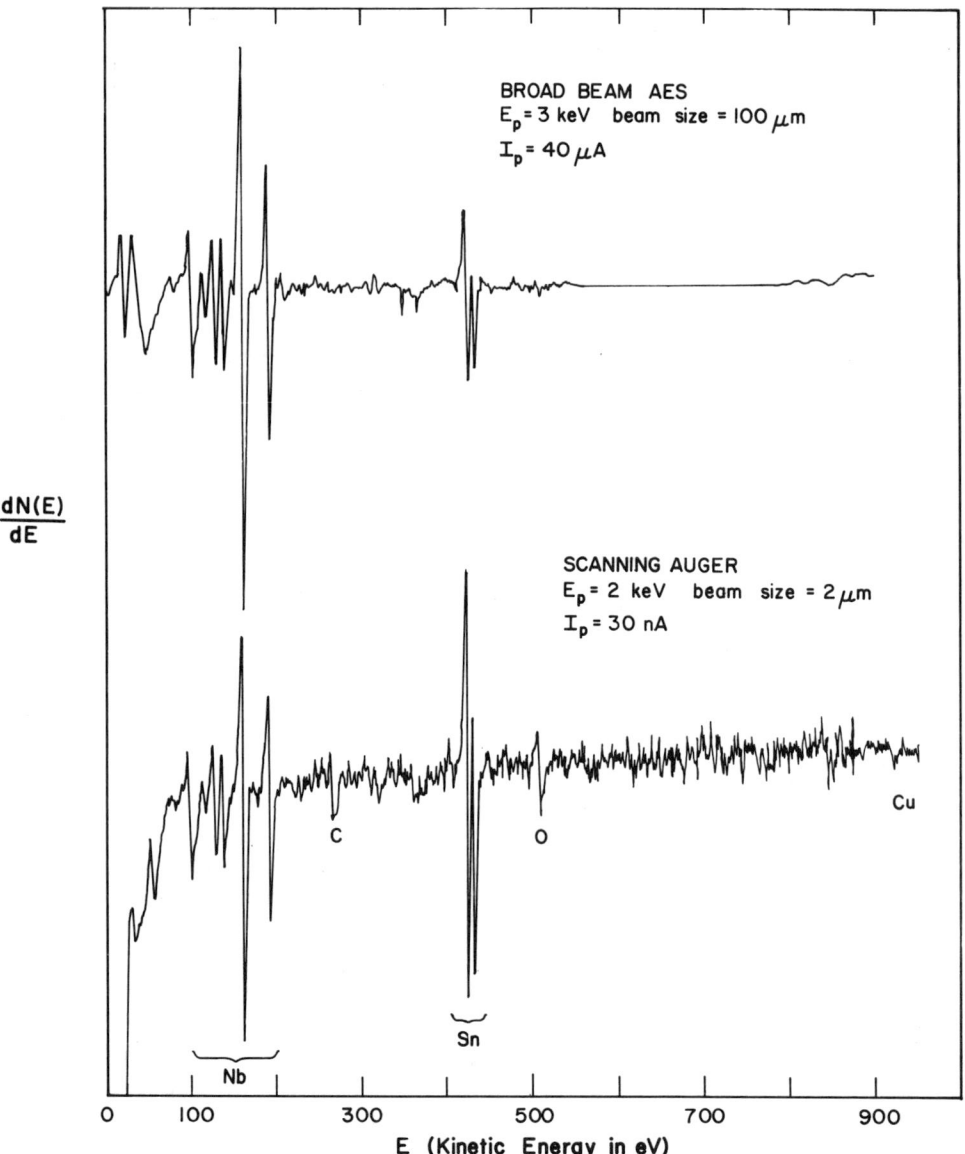

Fig. 5.   dN(E)/dE spectra for samples reacted at 900°C for 3 h.
          The top trace shows the spectra after sputtering with
          2 keV argon ions for several hours.   The bottom trace
          shows the spectra after fracture in vacuo and observa-
          tion of a single grain of Nb_3Sn using a scanning Auger
          microprobe.

The traces in Fig. 5 therefore represent the difference between a sputtered and an unsputtered surface at the same depth in the layer. The Nb:Sn peak-to-peak ratios are seen to be very different in the two cases; in the sputtered case 3.29:1 and in the cleaved case 0.95:1. Perhaps fortuitously, the Handbook values[11] for the pure element signal heights in the cleaved case yield a layer composition of Nb 25.6 at.% Sn, in agreement with that expected from a 18 K T_c and that derived from the electron probe calibration of our sputtered sample. Figure 5 is clear evidence for the existence of substantial sputtering effects and confirms the need for an independent confirmation of the composition of sputtered layers. In the present study, we have again used the electron probe microanalyzer to calibrate the Nb:Sn p-p Auger signal heights obtained from thick layers of substantially constant composition ($\sim$5 μm[7]). Compositions are then derived from peak-to-peak signal heights using the equation:

$$C_i = [A(iNb) \cdot \frac{I_{Nb}}{I_i} \alpha_i Nb + A(iC) \cdot \frac{I_c}{I_i} \alpha_i C + A(iSn) \cdot \frac{I_{Sn}}{I_i} \cdot \alpha_i Sn + A(io) \cdot \frac{I_o}{I_i} \cdot \alpha_{io}]^{-1}$$

C_i is the atom fraction of ith element, A(ij), etc. is the ratio of the elemental sensitivities of the two named elements[11] and I_j/I_i, etc. is the ratio of the p-p Auger signal heights of the two named elements. α_{ij} is the sputtering adjustment parameter required to adjust the observed p-p AES signals to those indicated by the EPMA. This parameter was experimentally determined to have the value 0.35 for Sn and Nb and was taken as unity in other cases. The Sn content was determined with respect to the Nb content from the equation

$$C_{Sn} = [1 + A(Sn/Nb) \cdot I_{Nb}/I_{Sn} \ 0.35]^{-1} \ .$$

Layer Compositions

The increase in T_c with approach to stoichiometry seen in Fig. 1 is a result that is well established on bulk samples[12,13] and it is encouraging to note that it can be reproduced on thin layers, using AES as the analytical technique. AES also gives information about the spatial distribution of light element impurities, for example the significant levels of C and 0 seen in the layers grown at 650°C and reported in Fig. 4.

A curious feature of many investigations of the T_c in bronze route Nb_3Sn is the observation that the T_c does not immediately reach its maximum value but increases steadily with reaction time.[14] Stress effects and the changing bronze:Nb_3Sn ratio clearly play an important role in composites[9] but this cannot be the explanation in the present case since the results in Fig. 3 were obtained after the bronze layer was removed. Although there is a small mismatch between the contraction coefficients of Nb and Nb_3Sn,[15] the total strain on cooling from the reaction temperature to 4 K will not exceed 0.12%, leading to a T_c depression of less than 0.1 K.[9] It is interesting to note, moreover, that layers grown at 700°C with the same thickness as those grown at 650°C have a higher T_c, a result that again suggests that strain cannot be playing a significant role in the depressed T_c values observed for short reaction times.

An alternative explanation for depressed T_c values in thin layers is the proximity effect between the thin Nb_3Sn layer and the Nb core. This effect has been studied by Dickey, et al.[16] It appears that no realistic values of coherence length or mean-free path can produce a T_c depression of the magnitude observed in our layers and we therefore exclude the proximity effect as a cause of our depressed T_c values.

The present results (Figs. 3 and 4) suggest an alternative explanation for this common effect. On thermodynamic grounds one would expect that all Nb_3Sn/bronze compositions in thermodynamic equilibrium at a given temperature should exist in a layer grown at that temperature while the diffusion reaction is still capable of proceeding. Thus stoichiometric Nb_3Sn with an 18 K T_c should form at or close to the bronze/Nb_3Sn interface, even in layers grown for only short times. Figure 4 shows however that it is precisely these short term layers which are high in O and C and which are low in Sn. As the reaction proceeds the Sn composition builds up to stoichiometry and the T_c rises. The O and C content falls off as the reaction proceeds, although at a somewhat more rapid rate than the Sn builds up. A possible interpretation of the effect is that interstitial impurities are gettered to the Nb_3Sn layer and exist in significant concentrations when the layer is thin. At this stage their effect is to cause the Nb_3Sn to exist off-stoichiometry but, as the layer grows, the interstitial can distribute itself through the layer or into the Nb and stoichiometry can be established. These observations may be of relevance if the current switch from vacuum-melted to air-melted bronze is seen to produce any significant effects on the superconducting properties of composites made from such ingots.

A final point is the concern that the O and C levels measured here may be artifacts due to the sputtering process or the electron beam. We do indeed see a residual O and C level in all our samples (see the value for the sample reacted 3 h at 900°C in Fig. 4) but the high values seen at short reaction times are substantially above this level. Dissolved interstitials in the A15 compounds are not unusual[8] and in view of the considerable precautions taken to minimize them in the starting Nb and bronze, their influence deserves further study.

CONCLUSIONS

We have continued our application of Auger Electron Spectroscopy to the study of thin bronze route Nb_3Sn layers and have obtained compositional and T_c information on layers of 0.1 μm thickness and larger. We have shown that significant preferential sputtering of Sn can occur when the layers are depth profiled with Ar^+ ions but that calibrations of thick Nb_3Sn layers with the electron probe enable this effect to be corrected for. An experiment with a fine beam Auger instrument suggests that, in the absence of sputtering, the elemental sensitivity factors are adequate for analysis of Nb_3Sn layers. The Sn content and T_c of thin layers are found to be low and the O and C contents are high. As reaction proceeds, the Sn content increases towards stoichiometry, T_c rises, and the O and C content declines. The effective depth resolution in a broad beam Auger instrument appears to be limited by the grain size and interface roughness and is typically 0.1-0.2 μm, at least an order of magnitude better than an electron probe.

ACKNOWLEDGMENTS

This work has been financially supported by the Department of Energy and the University of Wisconsin Graduate School. We are grateful to Dr. Y-W. Chung of Northwestern University for the use of the SAM, to Dr. Eric Gregory of Airco Superconductors for the Nb_3Sn tape and to Dr. Max Lagally for some helpful discussions.

REFERENCES

1. G. Rupp, IEEE Trans. on Magnetics MAG-15:189 (1979).
2. J. W. Ekin, IEEE Trans. on Magnetics MAG-15:197 (1979).
3. D. C. Larbalestier, Proc. of 6th Intl. Conf. on Magnet Technology MT-6:1080, 1978, ALFA Publishing Co., Bratislava, Czechoslavakia.
4. J. Bevk, J. P. Harbison, and J. L. Bell, J. Appl. Phys. 49: 6031 (1978).
5. S. Foner, E. J. McNiff, B. B. Schwartz, R. Roberge, and J. L. Fihey, Appl. Phys. Letters 31:853 (1977).
6. J. D. Verhoeven, D. K. Finnemore, E. D. Gibson, J. E. Ostenson, and L. F. Goodrich, Appl. Phys. Letters 33:101 (1978).
7. D. B. Smathers and D. C. Larbalestier, to appear in Adv. in Cryogenic Eng. 26 (1980).
8. R. H. Buitrago, L. E. Toth, and A. Goldman, J. Appl. Phys. 50:983 (1979).
9. T. Luhman, M. Suenaga, and C. J. Klamut, Adv. in Cryogenic Eng. 24:325 (1978).
10. A. W. West and R. Rawlings, J. Mat. Sci. 12:1862 (1974).

11. L. E. Davis, N. C. MacDonald, P. N. Palmberg, G. E. Riach,
     and R. E. Weber, "Handbook of Auger Electron Spectroscopy",
     2nd Edition, Physical Electronic Industries, Eden, Prairie,
     MN (1976).

12. J. J. Hanak, K. Strater, and G. W. Cullen, RCA Review 25:342
     (1964).

13. T. H. Courtney, G. W. Pearsall, and J. Wulff, Trans. Met.
     Soc. AIME 233:212 (1965).

14. D. C. Larbalestier, P. E. Madsen, J. A. Lee, M. N. Wilson,
     and J. P. Charlesworth, IEEE Trans. on Magnetics MAG-11:
     247 (1975).

15. H. W. Schadler, L. M. Osika, G. P. Salvo, and V. J. DeCarlo,
     Trans. Met. Soc. AIME 230:1074 (1964).

16. J. M. Dickey, M. Strongin, and O. F. Kammerer, J. Appl. Phys.
     12:5808 (1971).

THE IMPORTANCE OF BEING PRESTRESSED

G. Rupp[†]

Siemens AG, Research Laboratories, D-8520
Erlangen, F.R. Germany

INTRODUCTION

Nb_3Sn is known as a brittle material. Some measures have to be taken to fabricate a commercial superconductor which can be handled on an engineering scale. One way of solving the problem is to use an additional supportive material besides the superconductor which takes care of the forces acting on the conductor; the disadvantage is that the overall current density is reduced by this method. Fortunately there is a built-in mechanism in filamentary Nb3Sn conductors which considerably increases the flexibility of these conductors. Nb_3Sn is under a compressive strain resulting from the larger thermal contraction of the surrounding bronze.

The effect of a compressive strain on the critical current of Nb_3Sn wires was first detected in 1965 by E. Buehler and H.J. Levingstein[1] who measured a slight maximum in critical current under tensile stress. The first results on filamentary Nb3Sn conductors under tensile stress were published ten years later.[2-8] They showed a decrease in critical current with stress (or only a slight increase before decrease) which was discouraging for future applications of these conductors. Fortunately later on considerable current increase under strain was reported.[9-13] These and further investigations[14-19] showed that the compressive strain of Nb_3Sn in a composite conductor acts as a mechanical reserve which

[†]Visiting scientist at Francis Bitter National Magnet Laboratory, M. I. T., Cambridge, Massachusetts 02139

allows the conductor to be strained far beyond the strain where
bare Nb_3Sn fractures.

It is well known that under both compressive and tensile forces
the transition temperature,[20-22] the critical field[23] and the
critical current[22-24] of Nb_3Sn are reduced. Therefore the compressive strain of Nb_3Sn in the composite results in a degradation
of these parameters. An externally applied tensile stress reduces
the compressive strain thereby increasing these parameters. When
the compressive stress is compensated by the external tensile
stress (at least in axial direction) the parameters reach their
maximum values and decrease again when upon further increase of the
external tensile stress the Nb_3Sn itself is under tensile stress.

The following paper reports about the measured dependence of
the critical current on the strain of multifilamentary Nb_3Sn conductors. It will be shown that the main parameter determining
the strain dependence of the critical current is the critical
field and its strain dependence. The questions will be investigated how the prestress occurs in the composite and which parameters affect the amount of compressive strain. It turns out that
the mechanical properties of the bronze play the major part.
These investigations provide the knowledge of how to design a
conductor when specified values of critical current and strain
tolerance have to be reached.

CRITICAL CURRENT UNDER STRAIN

The compressive strain of Nb_3Sn in a conductor consisting of
Nb_3Sn filaments embedded in a bronze matrix results from different
thermal expansion coefficients of Nb_3Sn and bronze. The heat
treatment of the conductor which is necessary to form the Nb_3Sn
usually takes place at about 700° C. At this temperature both
materials in the composite are assumed not to be exposed to tensile
or compressive forces. When after heat treatment the conductor
begins to cool, the bronze tries to contract stronger than the
Nb_3Sn. In Fig. 1 the upper and lower curve show the thermal
contraction of Nb Sn and bronze, respectively, when they separately
cool down.[25] The difference in length Δ between bronze and Nb_3Sn
increases with falling temperature and amounts to 1.05% at 4 K.
In a composite with a good bond between filaments and bronze the
contraction of the composite is somewhere in between depending on
the volume fraction of the constituents and their mechanical properties as will be shown later. The curve in Fig. 1 for a composite
with a bronze to Nb ratio α of 3.5 has been calculated assuming
elastic behavior of Nb_3Sn and bronze. As Fig. 1 shows the Nb_3Sn
in this conductor is under a compressive strain of 0.7% while the
bronze is under a tensile strain of 0.3% at 4 K. Plastic deformation of the bronze under this tensile strain can reduce the
compressive strain of the Nb3Sn as indicated by the dashed curve
(schematic) and discussed later.

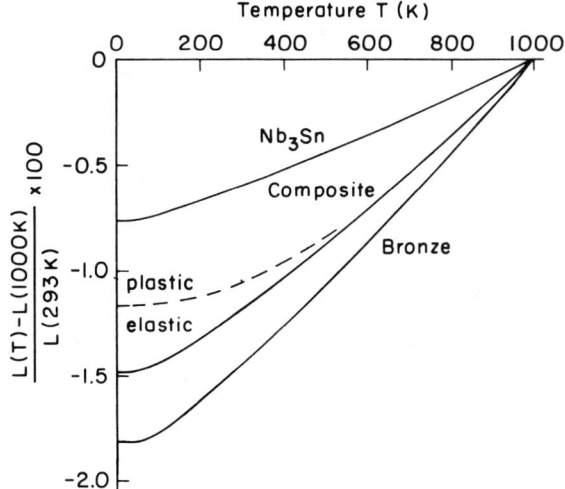

Fig. 1   Relative thermal contraction of Nb_3Sn, bronze and a
composite ($\alpha=3.5$) beginning at 1000 K.

The conductors investigated were manufactured by Vacuumschmelze
GmbH, Hanau. Hillmann et al. reports about the manufacturing and
general properties of these conductors in a separate contribution
to this conference. They are of the bronze process type consisting
of up to 10000 filaments embedded in a bronze matrix with an
initial Sn content of 13.5 wt % Sn. The conductors are twisted
with a twist pitch of 20 mm. The filament diameter is in the
range of 1 to 5 μm depending on the number of filaments and the
reduction ratio. None of the investigated conductors contained
any material for thermal stabilization like Cu protected by Ta.

The apparatus is described in detail elsewhere.[14] It allows
simultaneous measurements of critical current and tensile stress
as a function of strain on a straight piece of conductor about
45 mm long. The measurements were carried out in a superconducting
solenoid capable of producing 14 T. Additional 2 T were applied
by using Ho-cores as flux concentrators. All measurements were
performed at 4.2 K. The critical current is defined by a potential
drop of 2 μV/cm. The strain was measured by strain gages attached
to a small bronze plate positioned next to the conductor; this
plate is bent when the sample is strained.

The critical current I_c of a conductor was measured under
constant externally applied tensile stress σ and, hence, constant
strain ε. After each measurement of I_c, ε and σ the load was
increased by a discrete amount. In Fig. 2a, I_c at 16 T is shown

as a function of ε for several Nb_3Sn multifilamentary conductors, each having a different thickness of Nb_3Sn layer resulting from different heat treatments. I_c has been normalized to the initial value I_{c0} with zero applied tensile stress. For all samples I_c increases with increasing strain and passes through a maximum. Both the amplitude of the maximum and the strain ε_m at which the maximum occurs depend upon the diffusion conditions. At 16 T a factor of about two in critical current increase at ε_m between 0.4% and 0.5% is reached for conductors of technical interest which have nearly fully reacted filaments. The reduction of I_c beyond the maximum is not due to damages occuring in the conductor; after removal of the load the critical current changes almost reversibly. It is even possible to reversibly drive the sample into the normal conducting state as can be seen by the zero current data in Fig. 2a and b.

The characteristic dependence of I_c on ε can be explained by the internal compressive strain of the Nb_3Sn and the reduction of this compression by external tensile stress. At the I_c maximum, a quasi stress-free state is supposed to exist for the Nb_3Sn such that the internal compression and external tension just compensate each other in axial direction. In this state, the superconducting parameters of Nb_3Sn reach its highest value. For strains larger than ε_m, the Nb_3Sn itself is under tensile stress.

Although the conductors of Fig. 2a have the same ratio of bronze to Nb, ε_m changes considerably. As will be shown later, it is primarily not the varying thickness of Nb_3Sn which alters ε_m, but the changing mechanical properties of the bronze when more Sn is depleted from the bronze.

The Critical Field. The increase of the critical current with strain depends strongly on the applied magnetic field. Fig. 2b shows for one of the conductors of Fig. 2a the critical current as a function of strain for different flux densities between 10.5 T and 16 T. The higher the flux density the more strain sensitive is the critical current. The reason for this is that the critical field is affected by the strain.

Using the flux pinning theory of E.J. Kramer,[26] a critical field B_{c2}^* can be extrapolated from the I_c dependence on magnetic field. If this extrapolation is carried out at fixed strain, one gets B_{c2}^* as a function of strain. Fig. 2c shows the results for the conductors of Fig. 2a. The intrinsic strain ε_o of the Nb_3Sn defined by $\varepsilon_o = \varepsilon - \varepsilon_m$ was chosen as abcissa. It is obvious from this diagram that B_{c2}^* reaches a maximum value when the Nb_3Sn is in its stress-compensated state. This maximum critical field B_{c2m}^* is not constant for all conductors but depends on the diffusion conditions. Presumably, this reflects the influence of improved stoichiometry and long range order: the longer the heat treatment, the

more perfect the Nb_3Sn, the higher B^*_{c2m}. It should be pointed out
that B^*_{c2} is not the upper critical field B_{c2} where all superconductivity disappears, as measurements in higher fields show,[27] but
it serves as a useful parameter characterizing a conductor.

The question arises as to whether B^*_{c2} is the only parameter
whose change with strain has to be taken into account to calculate
the change of I_c with strain.

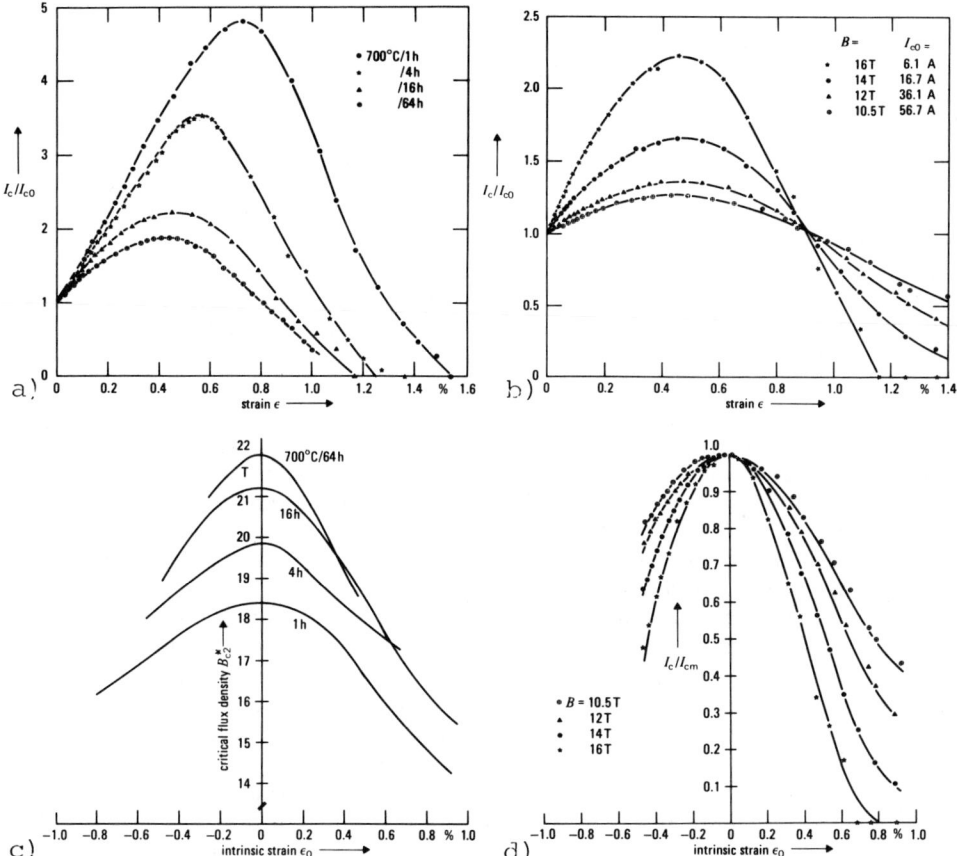

Fig. 2   Results of critical current measurements under stress on a
         Nb_3Sn conductor with 1615 filaments, $\alpha=4$ and $0.65 \times 0.25mm^2$
         cross-section: critical current I_c, normalized to the
         initial value, I_{c0}, as a function of strain of differently
         heat-treated conductors (a) and at different magnetic
         fields (b); extrapolated critical field B^*_{c2} (c) and I_c,
         normalized to its maximum value I_{cm}, together with calcu-
         lated curves at different fields (d) as a function of
         the intrinsic strain ε_o of the Nb_3Sn.

Calculation of the Strain Dependence of I_c. In Fig. 2d the critical current I_c normalized by its maximum value I_{cm} is plotted as a function of intrinsic strain ε_o of the Nb_3Sn for the conductor of Fig. 2b. The points indicate the measured data. The curves are calculated from the formula

$$\frac{I_c(\varepsilon_o)}{I_{cm}} = \left(\frac{1 - B/B^*_{c2}(\varepsilon_o)}{1 - B/B^*_{c2m}} \right)^2 \qquad (1)$$

using the strain dependence of B^*_{c2} as plotted in Fig. 2c. It is obvious that the measurement data are very well described by this formula. Equation (1) is derived from Kramer's theory assuming no change in the pinning behaviour and thermodynamic critical field B_c with strain.[14] It should be pointed out that equation (1) can not be applied in fields higher than about 16 T. Close to the upper critical field B_{c2}, the field dependence of I_c of these conductors deviates from that calculated from Kramer's theory[27]; therefore, B^*_{c2} is not a useful parameter in this field region. The recently published difference between the strain of the I_c maximum and that of the B_{c2} maximum, and the change in the pinning behaviour with strain in monofilamentary wires at temperatures near T_c could not be observed in the investigated conductors at 4 K.[28]

Therefore, in the temperature and field regime of practical interest, it seems to be sufficient to know the strain dependence of the extrapolated critical field B^*_{c2} in order to be able to calculate the strain and field dependence of I_c with equation (1). When dealing with fully reacted filaments where stoichiometry and long-range order seem to be optimal, the critical field B^*_{c2} normalized by its maximum value B^*_{c2m} is nearly independent of the reaction conditions and depends only on the intrinsic strain ε_o of the Nb_3Sn. Fig. 3 shows the correlation between B^*_{c2}/B^*_{c2m} and ε_o which is an average over many investigated conductors. This result is similar to that obtained by Ekin[15] from extrapolation of low field critical currents. If, therefore, I_c and B^*_{c2} are measured on a fully reacted conductor with no external stress applied, the strain and field dependence of I_c can roughly be calculated with equation (1) using the correlation of Fig. 3 and setting B^*_{c2m} equal to 23 T.[23]

MECHANICAL PROPERTIES

The Force Balance. The compressive strain ε_m results from the force balance between the stronger contracting bronze and the filaments during cooling from the heat-treatment temperature to 4 K. Assuming a totally elastic behaviour of bronze and filaments in the composite during cooling, the force balance at every temperature can be calculated from the Young's moduli and the difference in

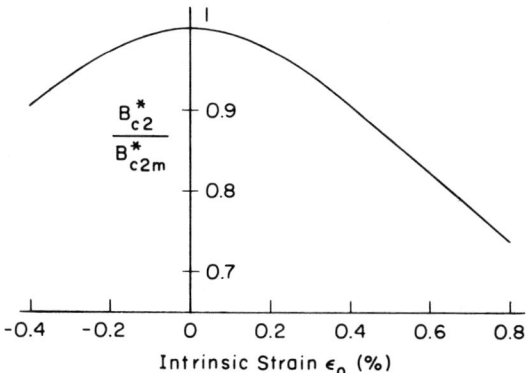

Fig. 3   Extrapolated critical field B^*_{c2}, normalized to its maximum
value B^*_{c2m}, as a function of intrinsic strain ε_o of the
Nb_3Sn in fully reacted multifilamentary conductors at 4.2K.

length Δ between separately cooled bronze and Nb_3Sn at that temperature.   Fig. 4 shows schematically how the force balance builds
up.   The bronze is stretched and the Nb_3Sn is compressed according
to their stress-strain diagrams so that tensile force and compressive force are equal.   The compressive strain which the Nb_3Sn
suffers at 4 K is assumed to be equal to ε_m, the strain to which
the Nb_3Sn has to be elongated to reach the maximum critical
current.   (If forces other than in axial direction have to be
taken into account, the problem may be more difficult.)

From Fig. 4 it is obvious that ε_m is determined by the
Young's moduli E_f and E_b of the Nb_3Sn filaments and the bronze,
respectively, and by the ratio $\alpha = a_b/a_f$ of the cross-sectional
areas.   From the triangles involved in Fig. 4, the following
equation can be deduced

$$\varepsilon_m = \left(1 + E_f/\alpha E_b\right)^{-1} \Delta \; . \tag{2}$$

The Young's moduli can be measured on the real conductor and α is
given by the design of the conductor.   Therefore, it should be
possible to compare the measured values of ε_m with that calculated
from the elastic model.

The Stress-Strain Diagram.   The upper curve of Fig. 5 shows
a stress-strain curve measured at 4 K on a conductor with 10000
fully reacted filaments and $\alpha = 3.5$.   The shape of the curve is
typical for this kind of conductor.   It consists of two linear
parts.   One can be seen at low strain values; it is sometimes very
short or even not existent.   The other occurs in the vicinity of
1% strain; it is attributed to the elastic behavior of the Nb_3Sn

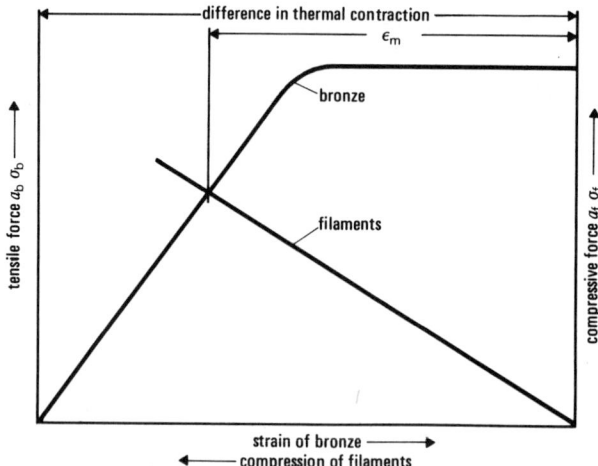

Fig. 4   Schematic presentation of the force balance in a composite
          between bronze under tension and filaments under compres-
          sion, resulting in a compressive strain ε_m of the filaments.

filaments.   The bronze shows plastic behaviour for this strain
level with nearly no work hardening.   The linear part at small
strain results from the elastic behaviour of both bronze and
filaments.

        In the case of a composite, the stress acting on a sample is
the sum of the individual stresses σ_i acting on the i-th component
weighted by the relative cross-sectional area a_i/a:

$$\sigma \; = \; \sum_i \; (a_i/a) \; \sigma_i \; .$$

With two components in the composite, one of which supposedly
behaves elastically over the entire strain range, the separation
of the individual σ–ε diagrams is possible.

        In Fig. 5 the two lower curves show the stress-strain
diagrams of the filaments and the bronze weighted by the relative
cross-sectional area.   They have been achieved from the upper
curve by supposing the stress-strain diagrams of the filaments to
be a straight line with the same slope as the linear part of the
composite diagram.   The bronze diagram gained by this procedure
shows the usual characteristics: elastic behaviour at low strain
values and plastic deformation at high elongation.   However, one
should bear in mind that zero strain in Fig. 5 does not mean zero
intrinsic strain for the components.   The bronze is under tension;
therefore, the σ–ε diagram of the bronze shown in Fig. 5 is only
the part of the total σ–ε diagram with ε larger than ε_b, the strain

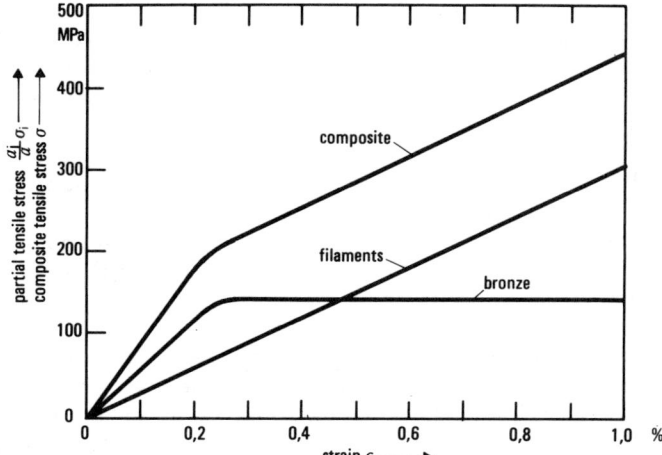

Fig. 5   Measured stress–strain curve of a Nb_3Sn conductor (10000
         filaments, α = 3.5, diameter 0.4 mm) and separated partial
         stress–strain curves of the filaments and the bronze of
         this conductor at 4.2 K.

of the bronze under the force balance.

        This analysis of σ–ε diagrams measured at 4 K provides useful
mechanical data. The Young's modulus E_f of the filaments turns
out to be 130 ± 10 GPa regardless of filament diameter and heat-
treatment.[*]   The Young's modulus E_b of the bronze is not a constant.
It varies from about 80 GPa for only partly depleted bronze to
about 55 GPa for fully reacted conductors; these values are lower
than the expected value of 103 GPa[31] but in accordance with values
measured on bronze wires.[32]   Another important parameter is the
yield strength $\sigma_{.02}$ of the bronze. It is defined as the tensile
stress where the plastic deformation of the bronze amounts to
0.02%. The yield strength can not be revealed from the composite
diagram because the zero intrinsic strain of the bronze is not
known. Together with the simultaneously measured ε_m, the zero
intrinsic strain can be determined approximately.[33]   The yield
strength, then calculated, increases with the Sn content of the
bronze and lies within the 70 to 240 MPa margin measured on bronze
wires with up to 8% Sn by Cogan.[32]

        The Validity of the Elastic Model. As mentioned before,
within the elastic model the compressive strain ε_m at each temperature can be calculated with equation (1) by using E_f, E_b and Δ at
that temperature. E_f and E_b can be evaluated from the stress-strain

───────────────

[*]This value lies within the range measured for Nb_3Sn.

curve of the composite and Δ is known from Fig. 1. In Fig. 6
calculated values of ε_m are compared with measured ε_m at 4 K. The
curve represents the calculated values for the compressive strain
and the points show the measured results from different conductors
including those of other authors. (The squares include the conductors of Fig. 2a.) Nearly all measured ε_m lie below the calculated
ones regardless of the kind of conductor and the heat-treatment.
Therefore, the elastic approach is not capable of accurately
calculating the compressive strain ε_m. A considerable amount of
plastic deformation of the bronze during cooling has to be taken
into account.

The Plastic Deformation of the Bronze. The mechanical properties of bronze vary over a large range depending on composition,
homogeneity, grain-size, heat-treatment, amount of coldwork and
temperature. The bronze used in Nb_3Sn conductors generally has a
relatively low yield strength $\sigma_{.02}$ because of the low tin content
after reaction, the high temperature annealing and the existence
of Kirkendall voids. Very little data exist, especially on low
temperature data of bronze. Fig. 7 shows $\sigma_{.02}$ at 4 K for annealed
bronze containing between 0 and 8 wt % Sn.[32] Since no data are
available on the temperature dependence of $\sigma_{.02}$ of these bronzes,
the ones of spring tempered bronze is plotted in Fig. 7 for
comparison.[34]

Whenever the tensile stress which acts on the bronze in the
force balance with Nb_3Sn exceeds the yield strength $\sigma_{.02}$, the
bronze is plastically deformed by 0.02%. At higher stress the
plastic deformation is still larger. Each plastic deformation
reduces Δ, the difference in thermal contraction. In Fig. 7

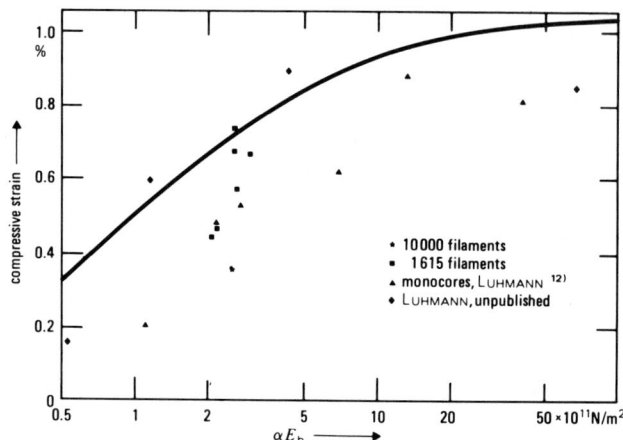

Fig. 6 Observed ε_m as a function of αE_b for different Nb_3Sn conductors and ε_m calculated with an elastic model (curve).

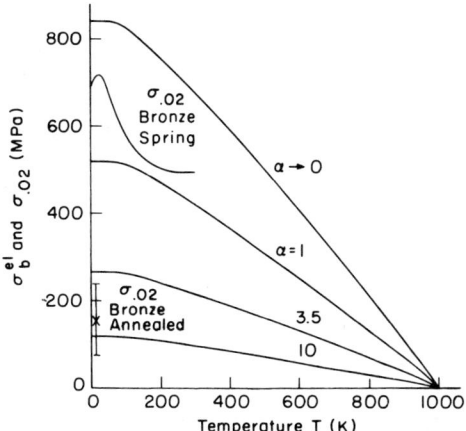

Fig. 7   Tensile strength σ_b^{el} to which the bronze in different
composites is subjected to (according to an elastic model)
as a function of temperature, and yield strength $\sigma_{.02}$ of
different bronzes.

the tensile stress σ_b^{el} acting on bronze in the force balance with
Nb_3Sn calculated from the elastic model is plotted as a function
of temperature for different bronze to Nb ratios α. As can be seen
from this diagram, σ_b^{el} exceeds $\sigma_{.02}$ of annealed bronze except for
composites with very high values of α ($\alpha > 10$). Therefore, a
plastic deformation of the bronze during cooling should be a
general feature of multifilamentary Nb_3Sn conductors. Assuming a
temperature independent $\sigma_{.02}$ of 106 MPa for a bronze with 3% Sn,
the bronze is plastically deformed by 0.02% when it reaches about
520 K during cooling down. This temperature is even higher if a
decrease of $\sigma_{.02}$ with increasing temperature is assumed as indicated by the spring bronze. In this case, the bronze can behave
elastically in the force balance with Nb_3Sn at 4 K (Fig. 4) although there was plastic deformation at high temperatures. It is
not possible to accurately calculate the amount of plastic deformation as long as the temperature dependence of $\sigma_{.02}$ is not known.
This is confirmed by the attempts to calculate the compressive
strain from room temperature and 4 K mechanical data by Ziegler[35]
and Easton et al.,[36] respectively. The approximate amount of
plastic deformation can be seen from the results shown in Fig. 6;
it is equal to the difference between the measured compressive
strain and the one calculated with the elastic model; it amounts up
to 0.4%. It decreases when $\sigma_{.02}$ is increased as can be seen from
the conductors of Fig. 2a (squares): the measured ε_m is closer to
the calculated one the higher $\sigma_{.02}$ because of less depleted bronze
resulting from thinner Nb_3Sn layer thickness. The plastic deformation may be smaller, or may even not exist in composites with
still higher values of $\sigma_{.02}$, e.g. conductors with Be bronze[37] and

in-situ conductors.[38]

Therefore, the main parameters affecting the compressive
strain are the bronze-to Nb ratio and the yield strength of the
bronze. The ratio α is determined by the design of the conductor,
while $\sigma_{.02}$ can be considerably changed by well known metallurgical
methods.

THE IMPORTANCE OF BEING PRESTRESSED

Although it is commonly used, the term "prestressed" has not
been used in this paper, except in the title. The term "precompressed" is more precise and hits the essential point.

Strain to Irreversible I_c Degradation. Why is it important
to have the Nb_3Sn precompressed? It is known that bare Nb_3Sn
breaks at a strain of approximately 0.2%. The compressive strain
of the Nb_3Sn in a composite acts as a mechanical reserve during
elongation of the conductor. The strain to failure of the Nb_3Sn
is consequently increased so that the conductor can tolerate,
without damage, a strain in excess of the breaking strain of bulk
Nb_3Sn. Values of approximately 0.7% have been reported prior to
the concept of precompression.[39,40] In Fig. 8 several strain
parameters are plotted which illustrate the improvement. These
measurements were performed on a conductor with varying layer
thicknesses produced by different heat-treatment; therefore, the
Nb_3Sn-to-bronze ratio was chosen as abcissa. ε_m is slightly
decreasing from 0.7% to 0.4% with increasing layer thickness (see
Fig. 2a) because of decreasing yield strength and Young's modulus
of the bronze and increasing ratio of Nb_3Sn to bronze as explained
above. The strain ε_d where the critical current is irreversibly
degraded is indicated by bars. At this strain the critical
current does not reversibly recover after removal of the load due
to formation of cracks in the Nb_3Sn layer. ε_d is lower or equal
to the breaking strain δ of the composite and varies from sample
to sample. The dashed curve represents the lower limit curve for
ε_d. The difference between the values on this curve and ε_m is due
to the ductility of Nb_3Sn between the force-compensated state and
the strain to failure. Values for this ductility between about
0.5% for partially reacted filaments and 0.3% for fully reacted
conductors can be derived from Fig. 8. Therefore, the total strain
the fully reacted conductor can be subjected to amounts to about
0.8% and more, which is a considerable improvement compared to 0.2%
for bulk Nb_3Sn. Without this increase Nb_3Sn conductors would not
have reached the level of technical superconductors.

Recovery of the Compressive Strain. When a multifilamentary
Nb_3Sn conductor is elongated to a relatively high strain and the
load is subsequently removed, a residual strain remains because
of the plastic deformation of the bronze during the elongation.

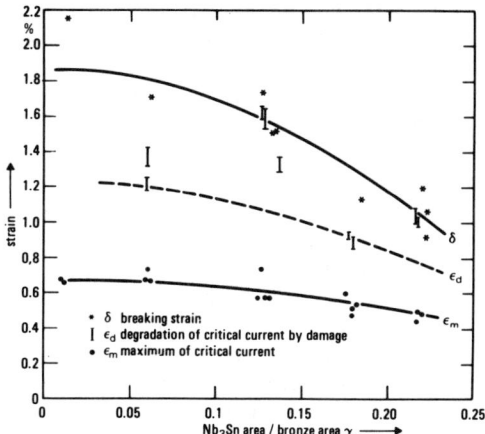

Fig. 8   Strain parameters ε_m, ε_d, δ versus cross-sectional area
ratio of Nb_3Sn and bronze of differently heat-treated
Nb_3Sn conductors with 1615 filaments.

The critical current is then determined according to the intrinsic
strain of the Nb_3Sn.   The residual intrinsic strain can be a
tensile strain depending on the amount of preceding tensile stress.
If this is the case, the mechanical reserve has been lost and the
conductor is now very sensitive to further stress application.
Experiments, however, show that part of the compressive strain of
the Nb_3Sn can be regained by warming the conductors up to room
temperature.   The reason for this is the fact that the bronze gets
under compression when the conductor is heated to room temperature
because of the mismatch in thermal expansion.   This compression
causes a plastic deformation of the bronze making the temperature
cycling an irreversible process for the intrinsic strain of the
bronze.

This self-healing of an overloaded conductor may provide an
explanation for an effect which sometimes can be observed in superconducting solenoids wound with multifilamentary Nb_3Sn conductors[41]:
After several months of permanent use of the magnet, the maximum
field starts to decrease but it can be recovered by warming the
whole magnet to room temperature.

SUMMARY

The larger thermal contraction of bronze as compared to Nb_3Sn
has two important consequences for a Nb_3Sn composite conductor: the
critical current is decreased because of a compressive strain on
the Nb_3Sn and the mechanical properties are improved because of the
same compressive strain.

The decrease in critical current is mainly caused by a decrease in the critical field. A formula has been derived which allows one to approximately calculate the strain and field dependence at 4 K from the knowledge of the critical current and the critical field at zero applied stress.

The compressive strain resulting from a force balance between the stronger contracting bronze and the Nb_3Sn filaments acts as a mechanical reserve for the brittle Nb_3Sn, thus considerably increasing the strain where damaging of the Nb_3Sn degrades the critical current. The compressive strain cannot be calculated from an elastic model. A considerable amount of plastic deformation of the bronze starting at high temperatures reduces the compressive strain calculated from the elastic model. Methods for increasing the yield strength of the bronze increase the measured compressive strain. An overloaded conductor which has lost its compressive strain but is still undamaged can regain part of this compressive strain by being heated to room temperature.

ACKNOWLEDGMENTS

I would like to thank Drs. H. Pfister and K. Wohlleben for many stimulating discussions, and Drs. E. Springer (Vacuumschmelze) and M. Wilhelm for manufacturing and preparation of the conductors. Thanks are further due to Mrs. F. Wellhoefer and to G. Friedrich who carried out the measurements with particular care. A special note of thanks is given to S. Foner and E. J. McNiff, Jr. for valuable and helpful discussions.

Part of this work has been supported by the technological program of the Federal Department of Research and Technology of the Federal Republic of Germany.

REFERENCES

1. E. Buehler and H. J. Levinstein, J. Appl. Phys. 36:3856 (1965).
2. I. L. McDougall, IEEE Trans. Magn. MAG-11:1467 (1975).
3. A. F. Clark, Cryogenics 16:632 (1976).
4. D. C. Larbalestier, J. E. Magraw, and M. N. Wilson, Proc. 6th Int. Cryogenic Eng. Conf. ICEC-6:387 (1976).
5. I. L. McDougall, Proc. 6th Int. Cryogenic Eng. Conf. ICEC-6: 396 (1976).
6. J. W. Ekin, Appl. Phys. Letters 29:216 (1976) and IEEE Trans. Magn. MAG-13:127 (1977).
7. D. C. Larbalestier, J. E. Magraw, and M. N. Wilson, IEEE Trans. Magn. MAG-13:462 (1977).
8. D. S. Easton and R. E. Schwall, Appl. Phys. Letters 29:319 (1976).

9. H. Hillmann, H. Kuckuck, H. Pfister, G. Rupp, E. Springer, M. Wilhelm, K. Wohlleben, and G. Ziegler, IEEE Trans. Magn. MAG-13:792 (1977).

10. G. Rupp, IEEE Trans. Magn. MAG-13:1565 (1977) and J. Appl. Phys. 48:3858 (1977).

11. J. W. Ekin, Adv. Cryogenic Eng. 24:306 (1978).

12. T. Luhman, M. Suenaga, and C. J. Klamut, Adv. Cryogenic Eng. 24:325 (1978).

13. D. W. Deis, D. N. Cornish, A. R. Rosdahl, and D. G. Hirzel, Proc. 6th Int. Conf. Magnet Technology MT-6:1028 (1977).

14. G. Rupp, IEEE Trans. Magn. MAG-15:189 (1979).

15. J. W. Ekin, IEEE Trans. Magn. MAG-15:197 (1979).

16. D. S. Easton and D. M. Kroeger, IEEE Trans. Magn. MAG-15:178 (1979).

17. T. Luhman, M. Suenaga, D. O. Welch, and K. Kaiho, IEEE Trans. Magn. MAG-15:699 (1979).

18. R. Roberge, S. Foner, E. J. McNiff, Jr., B. B. Schwartz, and J. L. Fihey, IEEE Trans. Magn. MAG-15:687 (1979).

19. R. Flukiger, R. Akihama, S. Foner, E. J. McNiff, Jr., and B. B. Schwartz, Appl. Phys. Letters 35:810 (1979).

20. T. F. Smith, J. Low Temp. Phys. 6:171 (1972).

21. W. A. Pupp, W. W. Sattler, and E. J. Saur, J. Low Temp. Phys. 14:1 (1974).

22. C. B. Muller and E. J. Saur, Rev. Mod. Phys. 36:103 (1964).

23. M. Pulver, Z. Phys. 257:261 (1972).

24. C. B. Muller, E. J. Saur, Adv. Cryogenics Eng. 9:338 (1964).

25. Y. S. Touloukian, R. K. Kirby, R. E. Taylor, and P. D. Desai, "Thermophysical Properties of Matter," IFI/Plenum, New York (1975).

26. E. J. Kramer, J. Appl. Phys. 44:1360 (1973).

27. G. Rupp, E. J. McNiff, Jr., and S. Foner, Proc. Appl. Sc. Conf., to be published.

28. D. M. Kroeger, D. S. Easton, A. DasGupta, C. C. Koch, and J. O. Scarbrough, J. Appl. Phys. 51:2184 (1980).

29. G. Rupp, Adv. Cryogenic Eng. 26, in press.

30. D. C. Larbalestier, V. W. Edwards, J. A. Lee, C. S. Scott, and M. N. Wilson, IEEE Trans. Magn. MAG-11:555 (1975).

31. R. P. Reed and R. P. Mikesell, "Low Temperature Mechanical Properties of Copper and Selected Copper Alloys," NBS Monograph 101:53 (1967).

32. S. F. Cogan, Thesis, MIT (1979) and S. F. Cogan and R. M. Rose, submitted to Cryogenics.

33. G. Rupp, Cryogenics 18:663 (1978).

34. F. R. Schwartz, "Cryogenic Materials Data Handbook," Techn. Doc. Rep. AFML-TDR-64-280, Vol. 2:271 (1970).

35. G. Ziegler, J. Appl. Phys. 49:4141 (1978).

36. D. S. Easton, D. M. Kroeger, W. Specking, and C. C. Koch, submitted to J. Appl. Phys.

37. T. Luhman, K. Kaiho, and M. Suenaga, Adv. Cryogenic Eng. 26, in press.

38. R. Roberge, Adv. Cryogenic Eng. 26, in press.
39. C. F. Old and J. P. Charlesworth, Cryogenics 16:469 (1976).
40. D. C. Larbalestier, J. E. Magraw, and M. N. Wilson, IEEE
    Trans. Magn. MAG-13:462 (1977).
41. H. Hillmann, H. Kuckuck, E. Springer, H.-J. Weisse, M. Wilhelm,
    and K. Wohlleben, IEEE Trans. Magn. MAG-15:205 (1979).

STUDIES OF THE STRAIN-DEPENDENT PROPERTIES OF A15 FILAMENTARY

CONDUCTORS AT BROOKHAVEN NATIONAL LABORATORY*

Thomas Luhman and David O. Welch

Brookhaven National Laboratory
Division of Metallurgy and Materials Science
Upton, New York 11973

INTRODUCTION

Strain-dependent properties of A15 filamentary conductors can be divided into two broad categories of effects: elastic or "intrinsic" strain effects and irreversible behavior, usually associated with the formation of cracks in the A15 compound. The term "strain tolerance" of conductors is often used without properly distinguishing between the two types of effects with resulting confusion and conflicting claims from proponents of various types of conductors. This may be illustrated by a comparison between the behavior of Nb_3Sn and V_3Ga conductors. The intrinsic response of the "thermodynamic" superconducting properties, such as critical temperature and magnetic field, to elastic strain-produced changes in size and shape of the crystal lattice differs between the two compounds by nearly an order of magnitude, with Nb_3Sn having the larger response,[1] yet the degree to which the crystal lattice may be distorted from its equilibrium state before brittle fracture occurs is essentially the same for the two compounds. Therefore, when bronze-processed composite conductors are intentionally strained, the superconducting critical parameters of a V_3Ga conductor will be relatively insensitive to applied strain compared to a Nb_3Sn conductor. However, given equivalent initial strain states in the composite conductors, the applied strain at which irreversible behavior is observed will be essentially the same for V_3Ga and Nb_3Sn conductors, since the fracture strains of the two

*Work performed under the auspices of the U.S. Department of Energy.

A15 compounds are equal. Furthermore, the actual value of the applied strain at the onset of cracking is determined by the initial
internal strain state of the composite conductor, and this must be
properly accounted for in comparisons of the strain-tolerance of
conductors.

In this paper, we will review briefly recent work at Brookhaven National Laboratory pertaining to the strain response of
filamentary bronze-processed conductors. In this work we have
touched upon all the aspects discussed above: the intrinsic strain
dependence of the critical properties of A15 structure compounds,
the nature of the initial internal strain state of composite conductors, and the interplay between these residual strains and applied strains which governs the response of the conductor to external strain. We will also discuss briefly some factors which
can enhance the strain tolerances of filamentary conductors.

INTRINSIC STRAIN DEPENDENCE OF A15 COMPOUNDS

Filamentary bronze-processed conductors provide a unique op-
portunity to study the strain dependence of the superconducting
properties of A15 compounds, since the composite nature of these
conductors allows much larger strains to be obtained before fracture than is possible with free-standing material. Historically
the first experimental evidence of residual strains being associated with the composites took the form of measurements of the
transition temperature, T_c, on monofilamentary conductors.[2,3] The
superconducting transition temperature showed significant changes
when the bronze matrix was removed by etching. The data of Fig. 1
illustrate the changes in T_c observed in this type of experiment.[4]
The source of these changes in T_c is a compressive strain on the
Nb$_3$Sn compound induced by the relatively larger thermal contraction of the outer bronze matrix, and increasing the matrix-to-core
ratio increases the strain in the compound and consequently the
depression of T_c, as seen in Fig. 1. A quantitative analysis of
such experiments requires knowledge of the functional form of the
dependence of T_c on strain and of the nature of the triaxial
strain state in the composite.

The strain dependence of the critical temperature of a single
crystal can be shown on rather general grounds to be expandable
in powers of the strain components, and the number and form of
nonzero terms of the series are determined by crystal symmetry.[5,6]
For a non-textured polycrystal in a cylindrically symmetric strain
field this series, when truncated past quadratic terms and appropriately averaged with respect to orientation, reduces to[1]:

$$T_c - T_c(o) = \Gamma_1 \varepsilon (1+2\xi) + \varepsilon^2 \{\tfrac{1}{2}\Delta_v (1+2\xi)^2 + \tfrac{4}{15}\Delta_t (1-\xi)^2\} \qquad (1)$$

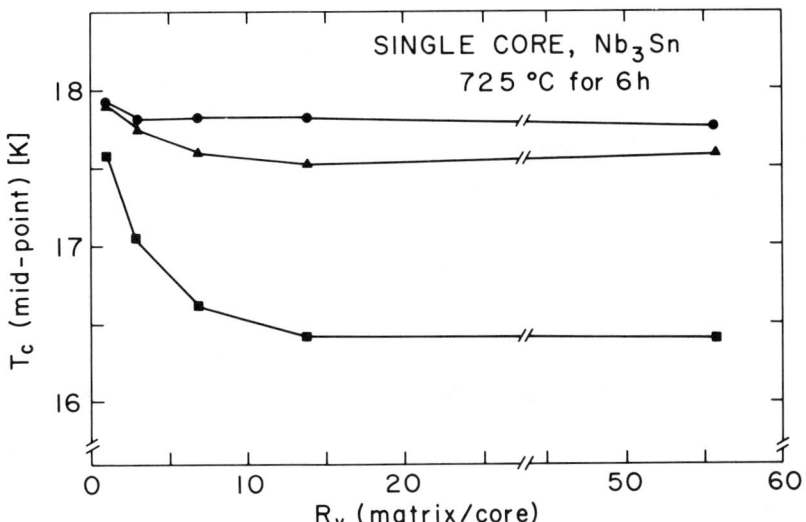

Fig. 1.   The superconducting transition temperature, inductive
          midpoint, as a function of the bronze-to-niobium ratio
          for a series of monofilamentary Nb_3Sn conductors. ■ ,
          represents as heat treated material; ▲, represents con-
          ductors whose bronze matrix has been removed by etching;
          ●, etched material following a short anneal, 1/2 h at
          700°C.

where ε is the axial strain in the compound, ξ is the ratio of the
radial (or tangential) to the axial strain, and Γ_1, Δ_v, and Δ_t are
material constants, the first two of which describe the dependence
on hydrostatic strain and the last the dependence on the non-
hydrostatic or deviatoric components of strain and which must be
measured experimentally or calculated from microscopic theory.   A
comparison of the efficacy of hydrostatic pressure and uniaxial
strain in changing T_c shows that at least for Nb_3Sn and V_3Si and
probably for V_3Ge and V_3Ga, the hydrostatic term, Γ_1 and Δ_v can
be neglected in comparison with the uniaxial term Δ_t[1]; perhaps
this behavior is generally true for A15 compounds.   The essentially quadratic nature of the strain dependence is illustrated
by the results of a series of experiments where inductive in situ
T_c measurements were made while samples were strained in tension.[7,8]   The functional dependence of T_c on strain is shown in
the normalized plot of Fig. 2.   A parabolic strain dependence is
evident over a wide range of strain, consistent with Eq. (1) when
modified to account for the presence of both applied and residual
strains.

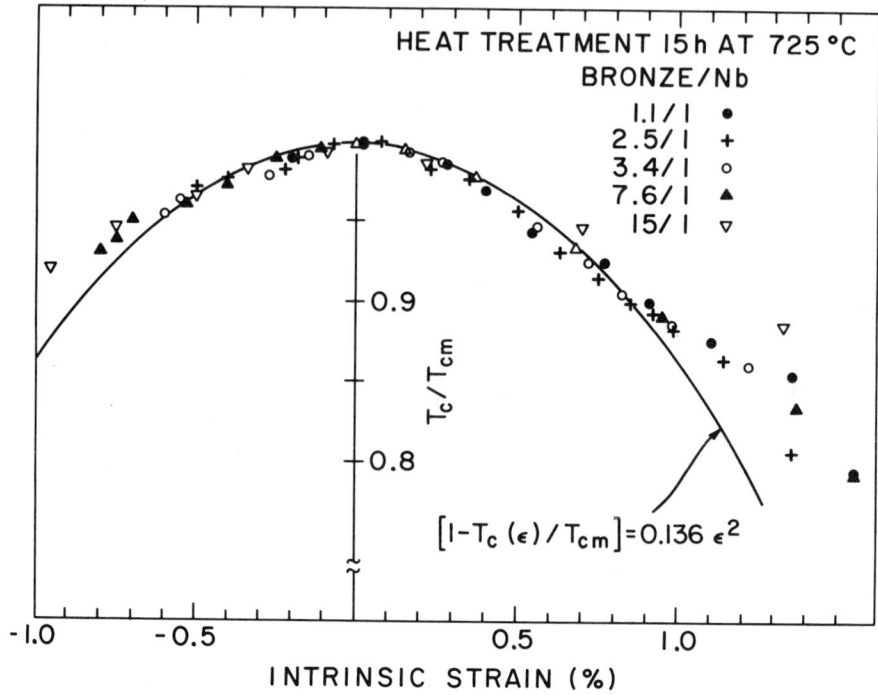

Fig. 2. Normalized superconducting transition temperature data,
         T_c/T_{cm}, where T_{cm} is the value of T_c at its maximum, as
         a function of strain.

        Whereas experiments on composite conductors to determine
intrinsic strain properties are facilitated by the large elastic
strains achievable in such composites, this is vitiated to some
degree by a lack of precise knowledge of the strain state, especially in the non-axial components. (This same lack of knowledge of the strain state is a problem in developing models of the
strain dependence for practical conductors but has usually been
dealt with by ignoring the triaxiality. Recently progress has
been made in the use of a computer to obtain numerical solutions
of the triaxial strain state in conductors of complex geometry,
as discussed by Scanlan et al.[9]) For the geometry of simple
monofilament conductors, detailed calculations of the effects of
triaxiality show that when Δ_t is the only nonzero coefficient
(i.e. only non-hydrostatic strain is important) in Eq. (1), only
minor errors will be made by treating the strain state as uniaxial.[1] This is probably true for multifilamentary conductors as
well; however it need not be the case when $\Delta_v \neq 0$ or for other
geometries, e.g. "internal bronze" conductors.[9]

Bearing these qualifications about triaxiality in mind, the initial axial strain in the superconductor in a bronze-processed wire conductor is given approximately by[1]:

$$\varepsilon_o \simeq -\left|\frac{\Delta\ell}{\ell}\right| \frac{R_v C}{1+R_v C} \qquad (2)$$

where $\Delta\ell/\ell$ is the fractional difference in thermal contraction, R_v is the volume ratio of bronze to core, and C is the ratio of effective "elastic" modulus of the bronze to that of the core. The latter ratio is an effective ratio of the moduli over the temperature range from the heat treatment temperature to the cryogenic testing temperature. This strain can be sufficiently large as to cause some plastic flow in the underlying unreacted niobium. Thus in Fig. 1 even after the bronze is removed T_c remains less than its unstrained value until a final stress relieving heat treatment is applied to remove the plastic set in the niobium.

Combining this estimate (Eq. (2)) for the initial strain in the compound with Eq. (1) (and neglecting the effect of hydrostatic strain and triaxiality) yields an expression for the critical temperature of the composite conductor:

$$\frac{R_v}{\left|T_c-T_c(o)\right|^{1/2}} \simeq \frac{2}{C\left|\frac{\Delta\ell}{\ell}\right|\left|\Delta_t\right|^{1/2}} + \frac{2}{\left|\frac{\Delta\ell}{\ell}\right|\left|\Delta_t\right|^{1/2}} R_v . \qquad (3)$$

Replotting the experimental data of Fig. 1 indeed yields the straight line of Eq. (3), see Fig. 3. This result underscores the appropriateness of the preceding analysis.

The critical magnetic field of A15 compounds is also a function of elastic strain. Considerations based on GLAG theory suggest that in cases where paramagnetic limitation is not a factor the dependence of the thermodynamic critical field H_c on non-hydrostatic strain should be essentially parabolic and should be stronger than the dependence of T_c because of the strain dependence of the renormalized electronic density of states and that the variation with strain of the upper critical field, H_{c2}, will be even stronger than that of H_c if the Ginsburg-Landau parameter κ_1 varies with strain. Such behavior is exhibited by Nb_3Sn, as shown in Fig. 4.

The results described above suggest that for bronze-processed filamentary A15 conductors (excluding those produced by the "internal bronze" technique[9]), and probably for so-called in-situ conductors as well, the strain state can be treated as essentially one-dimensional with little error, and that the thermodynamic critical properties are essentially dependent on the square of the

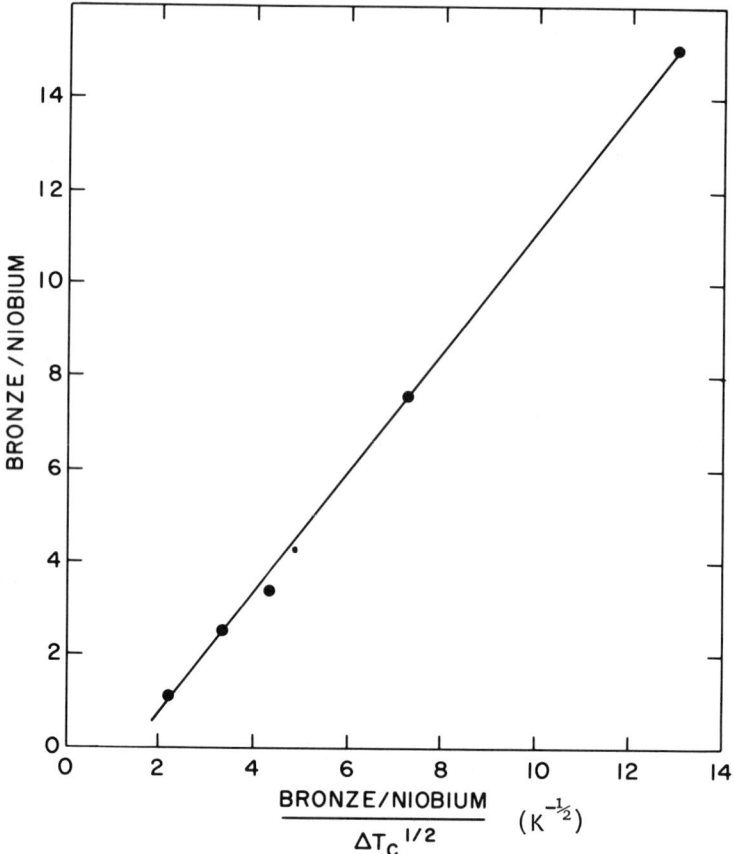

Fig. 3.   A straight line is obtained when $(R_v/\Delta T_c^{1/2})$ is plotted
          as a function of R_v. ΔT_c is the magnitude of the T_c
          depression caused by the residual stress in monofilamen-
          tary Nb_3Sn conductors and was obtained by etching off the
          bronze from a series of samples containing different
          bronze-to-niobium ratios; see Fig. 1.

total axial strain, residual plus applied. Thus applied tensile
strains counterbalance the thermally-induced residual strains in
such a way that the superconducting properties pass through an
extremum, usually but not always a maximum, (essentially characteristic of the strain-free state) as a function of applied
strain.   The data in Fig. 2 for Nb_3Sn conductors are consistent
with such a picture.

     Understanding the intrinsic (reversible) strain dependence of
the critical current density, J_c, requires the addition of yet

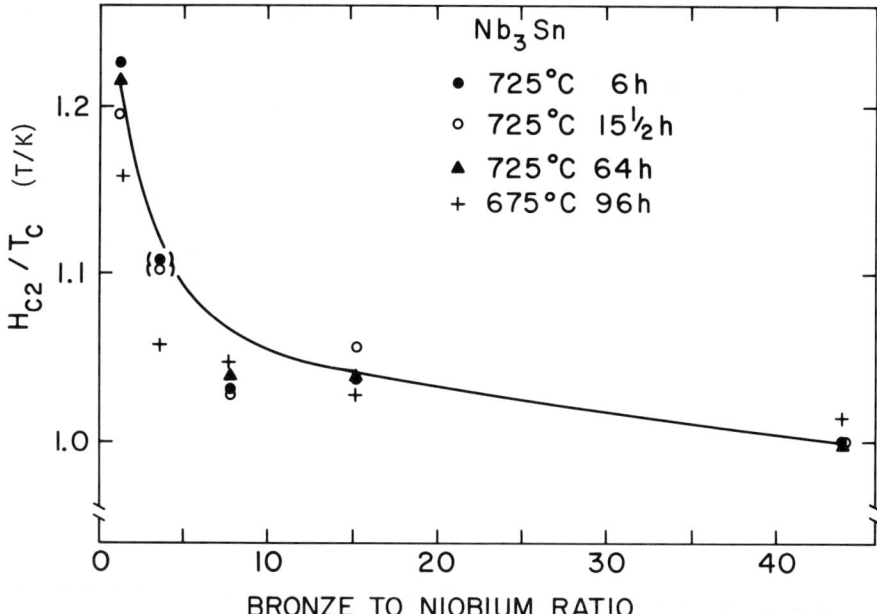

Fig. 4.   The upper critical magnetic field, H_{c2}, of Nb_3Sn normal-
          ized by T_c, as a function of R_v. (H_{c2} was obtained by
          extrapolating high-field critical current data.)   These
          data show that H_{c2} varies more strongly with strain than
          does T_c.

another level of theory: the theory of flux flow and its dependence
on strain.   Because of the influence of metallurgical factors on
flux pinning, this is a more difficult task than understanding the
strain dependence of the thermodynamic properties and much remains
to be done in this area.   An illustration of the differences in behavior of T_c and J_c can be seen by comparing Figs. 2 and 5; it is
clear that the J_c data do not follow a simple quadratic variation
with strain about the maximum.

Scaling laws, especially the one developed by Kramer for the
high-field behavior of hard superconductors,[10] have played an important role in correlating the critical current behavior of filamentary conductors, and several authors[11,12] have used Kramer's
scaling law to link the strain dependence of J_c with that of H_{c2}.
If Kramer's high-field scaling law holds, the critical current in
the strained state, J_c, will be related to that in the strain-free
state, J_{cm}, (the subscript m being used to denote the strain-free
state) by:

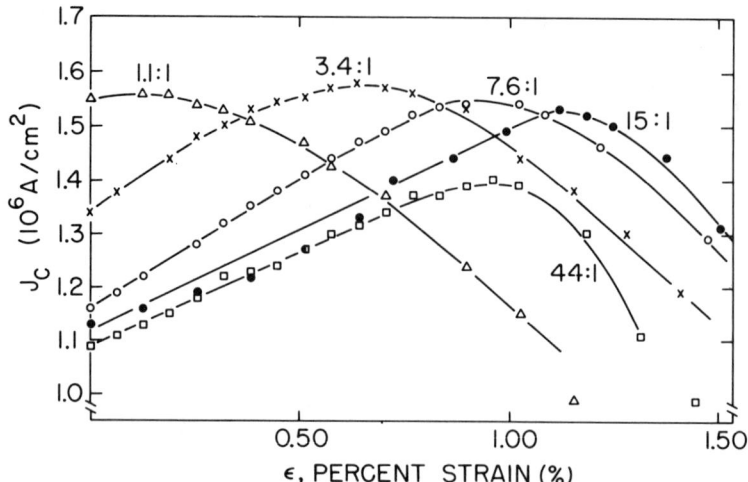

Fig. 5.   Critical current as a function of applied tensile strain.
Measurements made in situ at 4.0 Tesla for a series of
monofilament conductors with varying bronze-to-niobium
ratios.

$$\frac{J_c}{J_{cm}} = \frac{\kappa_{1m}^2}{\kappa_1^2} \frac{(H_{c2}-H)^2}{(H_{c2m}-H)^2} \frac{(H^{1/2}-\sqrt{\phi_o \rho_m})^2}{(H^{1/2}-\sqrt{\phi_o \rho})^2} \tag{4}$$

where $\kappa_1 (\equiv H_{c2}/H_c)$ is the Ginsburg–Landau constant, ϕ_o is the flux
quantum, and ρ is the density of flux-pinning sites. There are
three material parameters in the scaling law: κ_1 (or H_c), H_{c2}, and
$\sqrt{\rho}$. All in principle can vary with strain. Rupp[11] has successfully described the strain dependence of J_c for multifilamentary
Nb_3Sn conductors for $H \gtrsim 8$ T by assuming that

$$(\kappa_{1m}/\kappa_1)^2 (H^{1/2}-\sqrt{\phi_o \rho_m})^2 (H^{1/2}-\sqrt{\phi_o \rho})^{-2} \propto H_{c2m}^2/H_{c2}^2 ,$$

which is equivalent to H_c and $\sqrt{\rho}$ being independent of strain or
depending on strain in a compensating fashion. This leads to the
simple equation:

$$\frac{J_c}{J_{cm}} \simeq \frac{(1-H/H_{c2})^2}{(1-H/H_{c2m})^2} . \tag{5}$$

We find that this does not describe the residual-strain dependence of J_c of monofilament Nb_3Sn conductors for H=4 T, as may be seen in Fig. 6.[13]

Empirically we find a linear relation between J_c/J_{cm} and $(H_{c2m}-H_{c2})$; see Fig. 6. A theoretical framework for discussing such a relation can be obtained from Kramer's high-field scaling law by expanding Eq. (4) to first order in ε^2 and replacing ε^2 itself by the equivalent value of $(H_{c2m}-H_c)$:

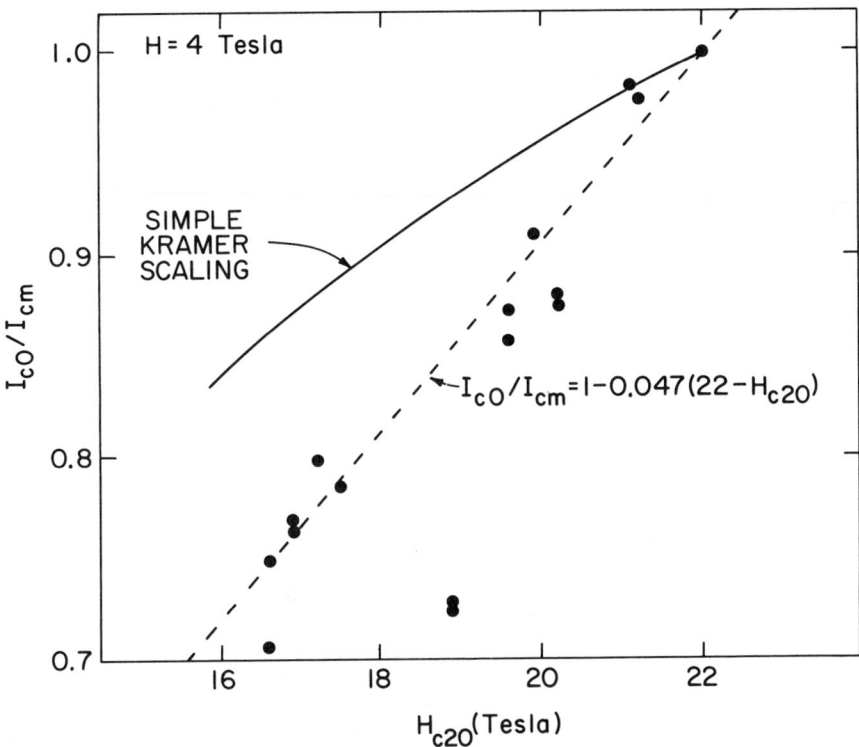

Fig. 6.   The relationship between the residual strain-depressed critical current, I_{co}, and upper critical field, H_{c2o}, of as-processed Nb_3Sn monofilamentary conductors. The subscript m refers to the strain-free state. $H_{c2m} \simeq 22T$. The solid curve is described by Eq. (5) and the dashed line is an empirical fit.

$$\frac{J_c}{J_{cm}} \simeq + \frac{\left\{2\dfrac{H_{c2m}}{H_{c2m}-H}\dfrac{d\ell nH_{c2}}{d\varepsilon^2} - 2\dfrac{d\ell n\kappa_1}{d\varepsilon^2} + \dfrac{\sqrt{\phi_o\rho_m}}{H^{1/2}-\sqrt{\phi_o\rho_m}}\dfrac{d\ell n\rho}{d\varepsilon^2}\right\}}{\dfrac{d\ell nH_{c2}}{d\varepsilon^2}\cdot H_{c2m}} (H_{c2m}-H)$$

$$(6)$$

where the derivatives are taken at zero strain. Studies of Kramer-
scaling in wires with the grain size of the specimens used to
obtain the data of Fig. 6 indicate that the flux-pinning term
(the last term in brackets) is probably negligible,[14] while
$d\ell nH_{c2}/d\varepsilon^2 \simeq -23\%$ per $(\%\varepsilon)^2$, as estimated from the data shown in
Fig. 4. Thus the empirical relation shown in Fig. 6 can be in-
terpreted with Eq. (6) to imply that $d\ell n\kappa_1/d\ell n\varepsilon^2 \simeq -16\%$ per $(\%\varepsilon)^2$,
whereas the assumption of Rupp, $\kappa_{1m}/\kappa \simeq H_{c2m}/H_{c2}$, leads to a value
of -23% per $(\%\varepsilon)^2$ for $d\ell n\kappa_1/d\varepsilon^2$. Whether this discrepancy is real
or a consequence of applying the Kramer scaling law at too low a
field (4 T) for it to be applicable is not known. Furthermore the
general applicability and literal interpretation of Kramer scaling
is subject to some doubt.[14] Clearly much remains to be done toward
understanding the intrinsic strain dependence of J_c.

IRREVERSIBLE EFFECTS

     When used in magnets, bronze-processed conductors must be
limited to strains within the elastic range, an upper limit to
this range is set by compound cracking and the onset of irreversible strain effects.[15] Details of the association between compound cracking and irreversible strain behavior may be obtained
from Ref. 15. One factor determining the cracking strain is the
bronze-to-niobium ratio. Figure 7 illustrates the dependence of
cracking and peak I_c strains on R_v. Both strains increase with
increasing R_v values, saturating when R_v reaches $\sim$ 10. At this
value all of the thermal contraction difference between bronze and
niobium has been taken up in the compressed filament. Under the
action of an applied tensile load, compression in the filament
must first be counterbalanced before the filament can be strained
into tension. As seen from Fig. 7 the tensile domain, that range
of strain intermediate between the peak I_c strain and the cracking strain, remains essentially constant with increasing R_v values.
Thus, by increasing the amount of residual compression larger R_v
values enhance a conductor's overall tolerance to tensile straining.

     In order to be used successfully in magnets these A15 conduc-
tors must also possess reasonable room temperature bend-strain
tolerances. In comparing critical current bend and tensile test

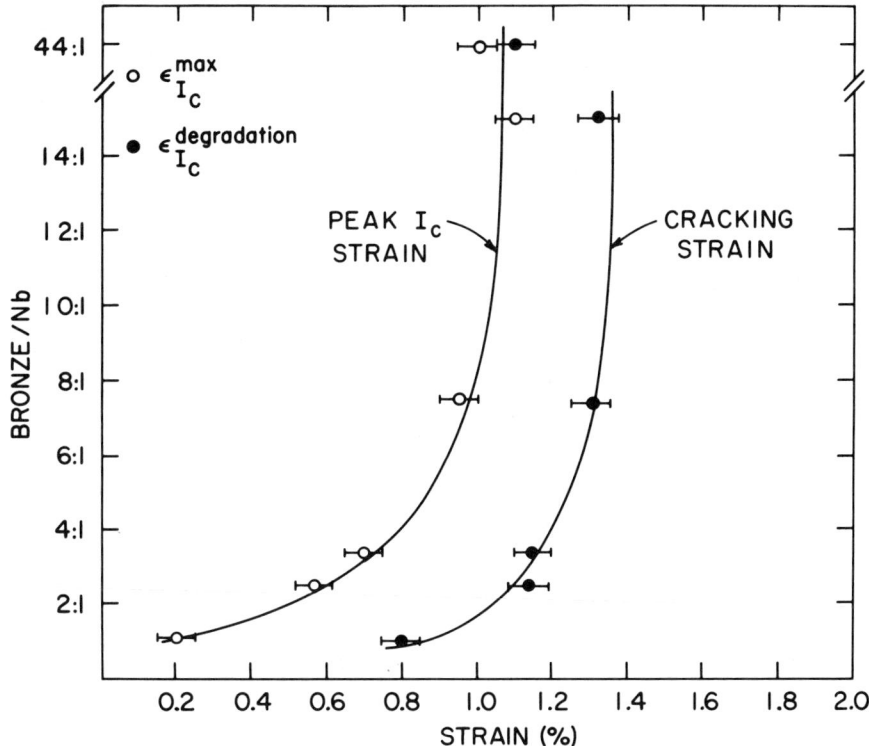

Fig. 7.   Strains associated with peak I_C values and cracking (ir-
          reversible behavior) as a function of the bronze-to-
          niobium ratio for Nb_3Sn monofilamentary conductors heat
          treated 15 h at 725°C, 4.0 Tesla.

data it was found that tolerance to bending strains was somewhat
less than expected on the bases of tensile test data.[16,17] An
example of the different responses of critical current to bending
and tensile strains is presented in Fig. 8. As may be seen irreversible behavior sets in much sooner under bending strain.

     We believe that the primary origin of the difference in strain
tolerances is a shift in the neutral axis of the conductor during
bending. Unless the neutral axis shift is taken into account, the
calculated strain in the compound during bending is less than the
strain actually applied. The shift occurs as a result of interaction between applied and residual strains. Because the matrix is
in tension in the as-heat-treated condition, tensile strains associated with bending result in plastic flow of the tensioned

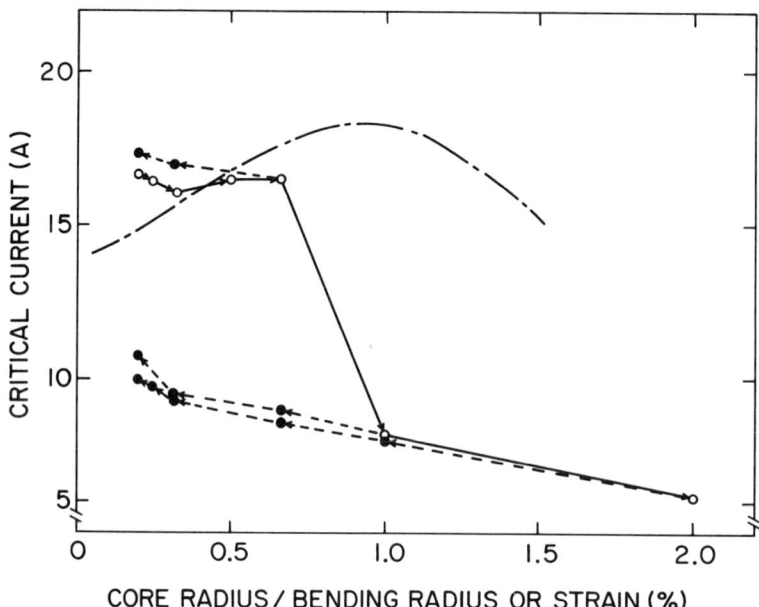

Fig. 8. Comparison of critical currents as a function of bending
and tensile strains for a monofilamentary Nb$_3$Sn conductor
heat treated 15.5 h at 725°C, 4.0 Tesla, R$_v$=13. The dot-
dash line shows the tensile-strain dependence.

bronze before plastic flow occurs in the compression side of the
conductor. The neutral axis is established by a force balance between the compression and tensile zones and thus with plastic flow
in the tension zone, and consequently a lower effective modulus,
the neutral axis will shift toward the center of bend curvature.
This shift of the neutral axis means that the Nb$_3$Sn compound will
be under a larger tensile strain than calculated by geometric considerations alone, and cracking under bending conditions will occur
at seemingly lower applied bending strains than those associated
with tensile tests.

A test of these ideas was made by constructing a model in which
the critical current density in a filament is assumed to be given
by an average of the strain-dependent critical-current density
integrated over the cross section of the filament which is still
uncracked, the fracture strain in the compound is assumed to be

the same as that found in tension, and the plasticity of bronze, i.e. the neutral axis shift, is accounted for. The dependence of critical current on bending strains was calculated and is compared with experimental data[16] in Fig. 9. The agreement is reasonably good and basically confirms these ideas, although somewhat too large values are predicted at the peak in critical current. This is believed to be a result of an assumption in the calculation where the critical current is represented by two linear functions for the tensile and compressive domains rather than one continuous quadratic function.

ON IMPROVING THE STRAIN TOLERANCE AND OTHER MATTERS

As pointed out above, the strain tolerances of A15 conductors may be increased by increasing the bronze-to-niobium ratio.

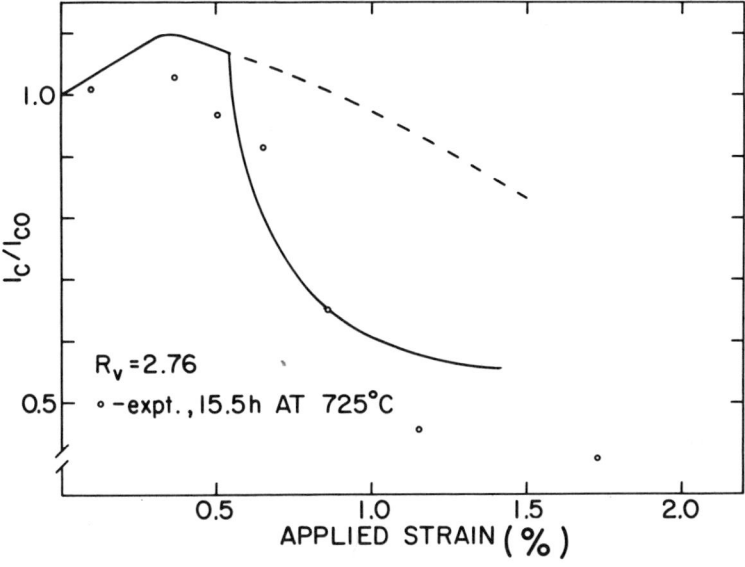

Fig. 9.   Normalized critical current behavior as a function of bending strain. The solid curve represents calculated behavior including a shift in the conductor's neutral axis. The dashed curve is expected behavior should the compound not crack.

Unfortunately adding more bronze dilutes the conductor's overall current density. Some efforts have been made to metallurgically enhance the strain tolerances without diluting the overall current density. The most promising appears to be the alloy addition of Be to the Cu/Sn bronze.[18] The high temperature strength of the bronze is increased through this approach thus eliminating any accommodating plastic flow in the bronze during cool down from the reaction heat treatment temperature. The net effect of plastic flow in the bronze is to minimize the amount of residual compressive strain induced in the A15 compound. Thus the Be addition, by virtue of the fact that plastic flow in the bronze is eliminated, can increase the amount of residual compressive strain without the necessity of increasing the bronze-to-niobium ratio. Figure 10 illustrates the effect of a 0.18 wt% Be addition to a Cu+11.2 wt% Sn bronze matrix. With this alloy addition all of the available difference in thermal contractions, $\sim$ 1%, is found as a compressive strain in the compound.

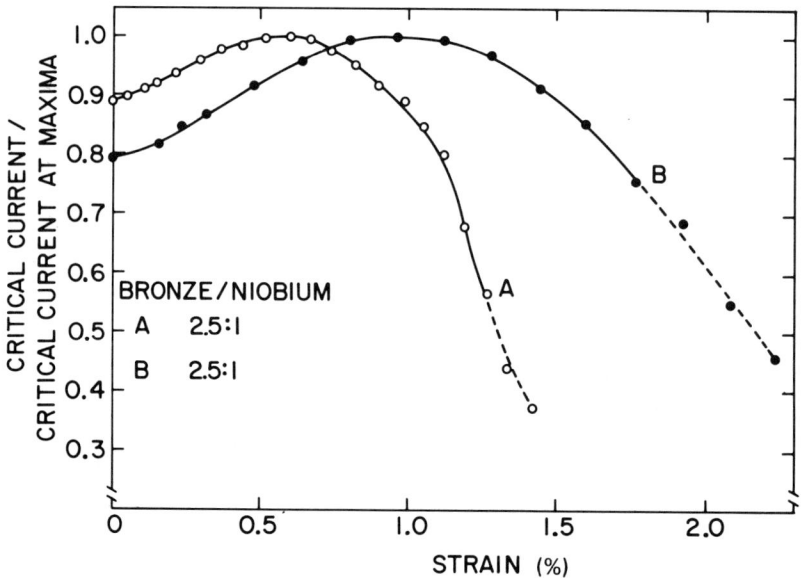

Fig. 10. Superconducting critical current, normalized to the maximum critical current value, as a function of applied tensile strain at 4.2 K and 4.0 T. Sample A represents a Cu+11.2 wt% Sn bronze, sample B a Cu+11.2 wt% Sn+0.18 wt% Be bronze. Samples were heat treated for 15 h at 725°C.

It is evident from the preceding discussions that the composite nature of bronze-processed A15 conductors, in particular their
internal residual strains, play an important role in determining
strain tolerances.   This is also true for mechanical properties.
The existence of internal residual strains prohibits application
of a simple law of mixtures.   For each conductor with a different
R_v value the contribution from the bronze constituent to the overall mechanical strength is different because of the different
plastic-strain history and possibly different work hardening rate
in each case.   This results in anomalous stress-strain relationships.[19]   Figure 11 illustrates how the bronze component varies in
its contribution to the conductor's overall mechanical strength.
The bronze has a different stress-strain relationship depending on
the bronze-to-niobium ratio, that is, the tensile prestrain in the
bronze.   This is an important consideration when developing models

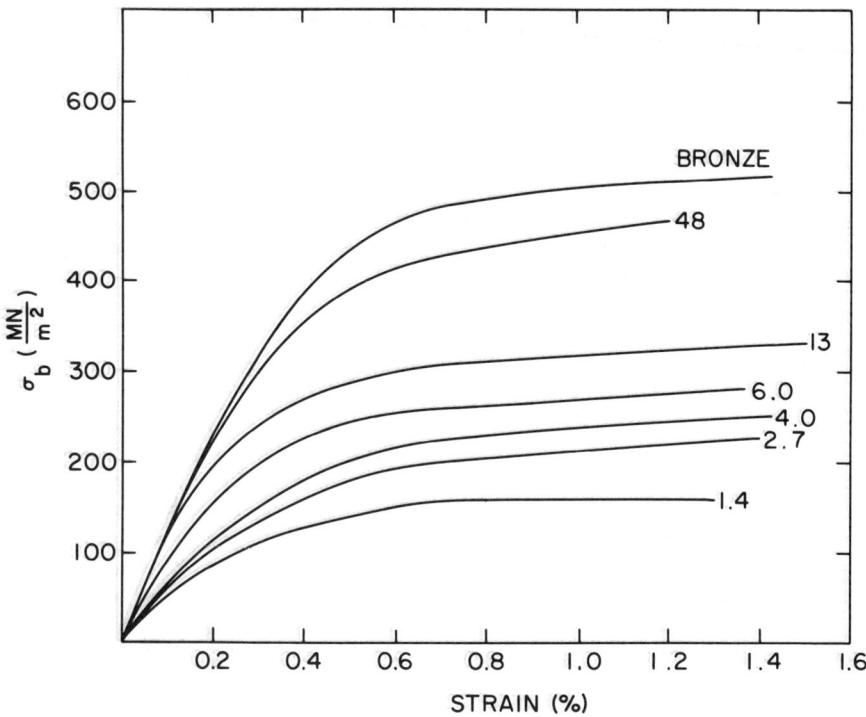

Fig. 11. Derived[20] stress-strain relationships for the bronze con-
        stituents of a series of monofilament conductors.   Mea-
        surements made at 4.2 K.[19]

of the mechanical properties of these composite conductors for use in predicting their behavior under straining conditions.

ACKNOWLEDGMENTS

We wish to express our appreciation to all our colleagues at Brookhaven for their participation in the research referred to in this review.

REFERENCES

1. D. O. Welch, to be published in Adv. in Cryo. Engin. 26 (1980).
2. Thomas Luhman and M. Suenaga, Appl. Phys. Letters 29:61 (1976).
3. Thomas Luhman and M. Suenaga, IEEE Trans. on Magnetics MAG-13: 800 (1977).
4. K. Aihara, M. Suenaga, and Thomas Luhman, to be published in Adv. in Cryo. Engin. 26 (1980).
5. L. R. Testardi, in "Physical Acoustics," W. P. Mason and R. N. Thurston, eds., Academic Press, New York (1973).
6. M. Weger and I. B. Goldberg, "Advances in Solid State Physics," F. Seitz and D. Turnbull, eds., Academic Press, New York (1978).
7. Thomas Luhman, M. Suenaga, and C. J. Klamut, Adv. in Cryo. Engin. 24:306 (1978).
8. M. Suenaga, T. Onishi, D. O. Welch, and Thomas Luhman, Bull. Am. Phys. Soc. 23:229 (1978), unpublished.
9. R. M. Scanlan, R. W. Hoard, D. N. Cornish, and J. B. Zbasnik, Proc. of this Conference (1980).
10. E. J. Kramer, J. Appl. Phys. 44:1360 (1973).
11. G. Rupp, IEEE Trans. on Magnetics MAG-15:189 (1979), and the Proc. of this Conference (1980).
12. J. W. Ekin, IEEE Trans. on Magnetics MAG-15:197 (1979), and the Proc. of this Conference (1980).
13. K. Aihara (1979), unpublished.
14. M. Suenaga and D. O. Welch, Proc. of this Conference (1980).
15. Thomas Luhman, M. Suenaga, D. O. Welch, and K. Kaiho, IEEE Trans. on Magnetics MAG-15:699 (1979).
16. Thomas Luhman and D. O. Welch, Eighth Symposium on Engineering Problems of Fusion Research, The Institute of Electrical and Electronics Engineers, Inc. (1979).
17. K. Kaiho, T. S. Luhman, M. Suenaga, and W. B. Sampson, Appl. Phys. Letters 36:223 (1980).
18. Thomas Luhman, J. Appl. Phys. 50:3766 (1979).
19. T. Luhman, M. Suenaga, and C. J. Klamut, Adv. in Cryo. Engin. 24:325 (1978).
20. G. Rupp, Cryogenics 18:663 (1978).

STRAIN SCALING LAW AND THE PREDICTION OF UNIAXIAL AND BENDING

STRAIN EFFECTS IN MULTIFILAMENTARY SUPERCONDUCTORS

J. W. Ekin

Thermophysical Properties Division
National Bureau of Standards
Boulder, CO

INTRODUCTION

Recently it was shown[1] that at 4.2 K the critical pinning-force density F in a wide range of practical conductors obeys a scaling law of the form:

$$F = [B_{c2}^*(\varepsilon)]^n \, f(b) \qquad (1)$$

where B_{c2}^* is the bulk upper-critical field at 4.2 K and $f(b)$ is a function only of the reduced magnetic field $b \equiv B/B_{c2}^*$. For Nb_3Sn conductors, $n = 1 \pm 0.3$. This relation was found to hold for both compressive and tensile intrinsic strain, and is expected to be limited in validity only by the irreversible strain point[2] where filament breakage becomes significant (typically at intrinsic tensile strains greater than 0.4 – 0.7% for most highly-reacted Nb_3Sn conductors[2]).

In this paper, we demonstrate how this strain scaling law may be used to determine both uniaxial and bending strain degradation of the critical current in practical multifilamentary Nb_3Sn conductors. The results are used to obtain uniaxial-strain limits and bending-strain limits for use in magnet design at fields from 4 to 16 T.

PREDICTION OF UNIAXIAL-STRAIN I_c DEGRADATION

When uniaxial strain is applied to a multifilamentary Nb_3Sn conductor, the critical-current density J_c typically increases to a maximum value J_{cm} at some strain ε_m and then decreases

under further uniaxial tension. This maximum in J_c arises from
compressive prestress[3-5] which the bronze matrix exerts on the
Nb$_3$Sn reaction layer because of the difference in thermal contraction between the two materials on cooldown after the reaction heat
treatment. Presumably the maximum occurs where the Nb$_3$Sn experiences the smallest magnitude of intrinsic strain. It is convenient throughout the remainder of this paper to define an intrinsic strain[4] ε_o to denote the uniaxial-strain state of the
Nb$_3$Sn itself: $\varepsilon_o \equiv \varepsilon - \varepsilon_m$. This parameterization of strain has
been found to remove the discrepancy between the results of a
number of different investigations of J_c vs ε in specimens with
varying amounts of prestrain.[5] Negative values of ε_o indicate
the approximate amount of uniaxial compressive strain the Nb$_3$Sn
experiences, positive values the amount of uniaxial tensile strain.

Ideally, in designing a magnet using multifilamentary conductors, the sum of all expected uniaxial strains from fabrication,
differential-thermal contraction and magnetic forces should place
the conductor in a uniaxial-strain state near ε_m (i.e., $\varepsilon_o = 0$)
where J_c is maximum. This uniaxial-strain criterion for using
multifilamentary conductors was proposed earlier[5]; we now address
the question of how critically it must be applied. That is, how
far from ε_m can the strain be before the J_c strain degradation
becomes significant. As we shall see, the answer depends on the
magnetic field; the margin for error becomes quite small at high
fields.

To determine the magnetic field dependence of the strain
limits, we need to evaluate the uniaxial-strain degradation of
J_c relative to its maximum J_{cm} for a wide range of magnetic
fields. As described in Ref. 1, an expression for J_c/J_{cm} can
be obtained simply by dividing Eq. (1) by itself evaluated at the
strain ε_m where the critical current is a maximum:

$$\frac{J_c}{J_{cm}} = \left[\frac{B_{c2}^*(\varepsilon)}{B_{c2m}^*} \right]^n \frac{f(b)}{f(b_m)} . \qquad (2)$$

Here use has been made of the relation $F = J_c B$. [In Eq. (2) and
throughout this paper, the subscript m is used to indicate that a
quantity has been evaluated at the strain ε_m where the critical
current is maximum.] Pinning theories[6-10] suggest that $f(b)$
is proportional to $b^p(1-b)^q$ where p and q are exponents whose
values vary depending on the particular pinning model. This form
in fact described most of the multifilamentary Nb$_3$Sn conductors
tested quite well, with values of p ranging from 0.5 to 0.7 and
values of q ranging from 2 to 2.4.[1] In the calculation of
J_c/J_{cm}, however, p can be taken to be 0.5 and q to be 2.0 with
little loss of accuracy in the final result. Substituting this

form for f(b) into Eq. (2) results in:

$$\frac{J_c}{J_{cm}} = \left[\frac{B_{c2}^*(\varepsilon)}{B_{c2m}^*}\right]^{n-p} \left[\frac{1 - B/B_{c2}^*(\varepsilon)}{1 - B/B_{c2m}^*}\right]^q \tag{3}$$

where, for Nb$_3$Sn, n $\cong$ 1, p $\cong$ 0.5, and q $\cong$ 2.0.

We thus have in Eq. (3) a simple means of calculating the dependence of J_c/J_{cm} on both field and strain. All that is needed for a given conductor is the strain dependence of the upper critical field B_{c2}^*. The strain dependence of B_{c2}^* varies some-what depending on whether the conductor is highly reacted or only partially reacted, in-situ cast or conventionally processed. For-tunately, however, for highly-reacted conventionally-processed conductors the dependence of B_{c2}^* on intrinsic strain ε_0 is nearly universal.[5,1] This is shown in Fig. 1 for a number of

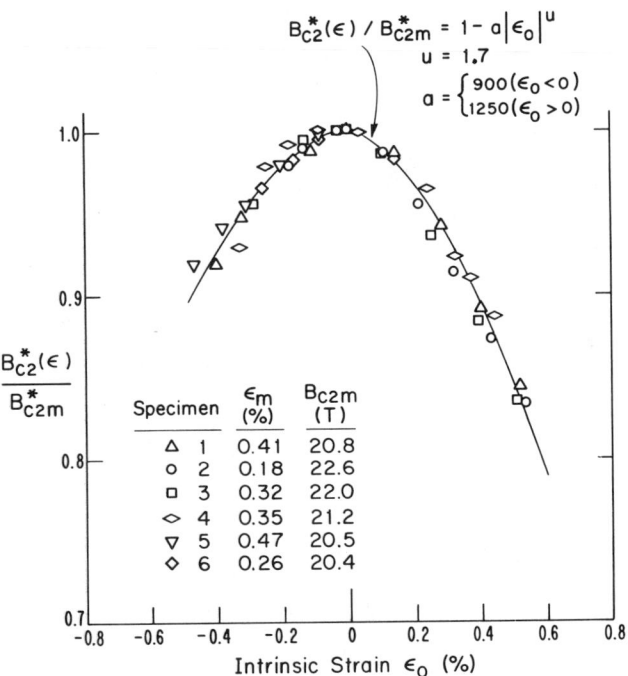

Fig. 1 Normalized upper critical field as a function of intrinsic strain $\varepsilon_0(\equiv\varepsilon-\varepsilon_m)$ for six highly-reacted multifilamen-tary Nb$_3$Sn superconductors. The curve represents Eq. (4) evaluated with u = 1.7 and a = 900 for $\varepsilon_0 < 0$, 1250 for $\varepsilon_0 > 0$.

commercial conductors manufactured in the United States, Germany, and Japan.[1] The graph was obtained by normalizing $B_{c2}^{*}(\varepsilon)$ by its maximum value B_{c2m}^{*} and plotting versus intrinsic strain $\varepsilon_{0}(\equiv\varepsilon-\varepsilon_{m})$. These data can be fit quite accurately by the empirical expression:

$$B_{c2}^{*}(\varepsilon) = B_{c2m}^{*} \ (1 - a|\varepsilon_{0}|^{u}) \tag{4}$$

where $u \simeq 1.7$, $a \simeq 900$ for compressive strain ($\varepsilon_{0} < 0$) and $a \simeq 1250$ for tensile strain ($\varepsilon_{0} > 0$). Eq. (4) evaluated with these values of a and u is shown as a solid curve in Fig. 1. It fits the data to within $\pm$ 1% over the measured range of strain from -0.5% to 0.6%.

Equation (4) is a great simplification, for it enables the strain dependence of B_{c2}^{*} to be predicted, at least for practical multifilamentary $Nb_{3}Sn$ conductors, without lengthy high-field measurements. All that is needed is a knowledge of two sample-dependent parameters: ε_{m} and B_{c2m}^{*}. B_{c2m}^{*} varied from 20.5 T to 22.6 T for the conductors in Fig. 1. It can be taken to be about 21 T, however, with little loss of accuracy in calculating J_{c}/J_{cm} (except at fields well above normal application limits for $Nb_{3}Sn$, i.e. at B > 16 T). Substituting Eq. (4) into

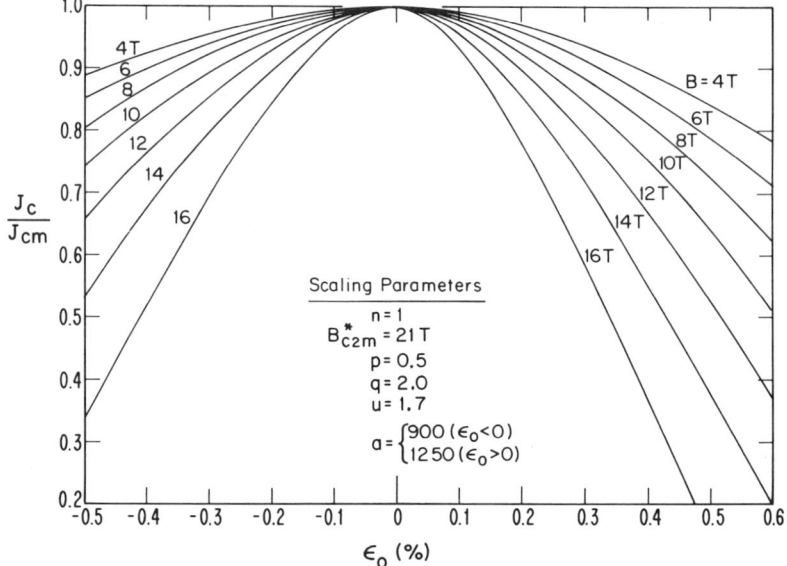

Fig. 2   Relative critical-current density J_{c}/J_{cm} as a function of intrinsic strain $\varepsilon_{0}(\equiv\varepsilon-\varepsilon_{m})$ for different magnetic fields, evaluated using Eq. (3) and the typical set of scaling parameters indicated in the figure.

Eq. (3) and using this typical set of parameters results in the
values of J_c/J_{cm} plotted in Fig. 2. These curves were found to
describe quite well the wide range of commercial conductors
studied in Ref. 1; the results of Fig. 2 were within 5% of the
measured values of J_c/J_{cm} in all the conductors at fields up to
12 T and well within 10% at fields up to 16 T.

These results should serve to give at least a reasonable
estimate of the percentage critical-current degradation that can
be expected for most practical multifilamentary Nb_3Sn conductors
as a function of magnetic field. In particular, they can be used
to determine the margin for error in applying the uniaxial-strain
criterion to the design of most multifilamentary Nb_3Sn magnets.
For example, if a 10% I_c-degradation tolerance is assumed in
applying this criterion, it is seen from the width of the I_c
vs. ε_o peak in Fig. 2 that at 10 T, ε_o must be limited to the
range $-0.29\% \leq \varepsilon_o \leq 0.24\%$. The limits on ε_o become significantly more stringent as the field is increased, becoming about half
the 10-T value at 16 T, for example. Uniaxial-strain limits corresponding to a 10% I_c-degradation criterion are given at other
fields in Table I. Because of the nearly universal nature of
strain degradation in highly-reacted bronze-process Nb_3Sn, these
uniaxial-strain limits should apply generally to most practical
multifilamentary Nb_3Sn conductors.

PREDICTION OF BENDING-STRAIN DEGRADATION FROM UNIAXIAL STRAIN MEASUREMENTS

Uniaxial strain data have the advantage that they can be
universally applied to a multitude of multifilamentary conductors
in terms of the intrinsic strain.[5] Bending strain degradation
on the other hand is very much a function of the individual conductor geometry. Fortunately it can be calculated for a particular geometry from uniaxial strain data simply by averaging over
the uniaxial strain curve.[11] The trick is taking the right

TABLE I. Uniaxial and bending strain limits for 10%
critical-current degradation

H	ε_o		ε_B
4 T	−0.47% to	+0.39%	< 0.88%
6 T	−0.40% to	+0.34%	< 0.75%
8 T	−0.34% to	+0.28%	< 0.64%
10 T	−0.29% to	+0.24%	< 0.53%
12 T	−0.24% to	+0.20%	< 0.45%
14 T	−0.20% to	+0.16%	< 0.37%
16 T	−0.15% to	+0.14%	< 0.29%

average. We consider two possible averaging schemes for multifilamentary conductors depending on whether the filament twist pitch
is long or short compared to the current transfer length.[13]

Long Twist Pitch

In a multifilamentary wire where the twist pitch is long, the
filaments in the outer half of the conductor away from the center
of curvature will experience tension, those toward the center compression. To obtain the overall current carried by the conductor,
$J_c(\varepsilon)$ is integrated over the different tensile and compressive
regions of the conductor.

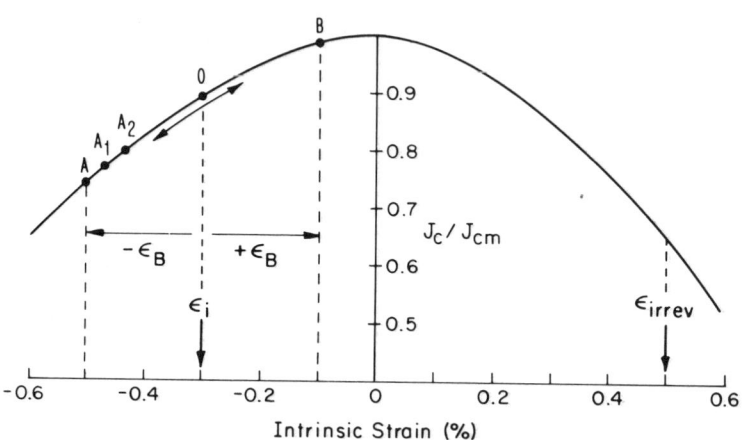

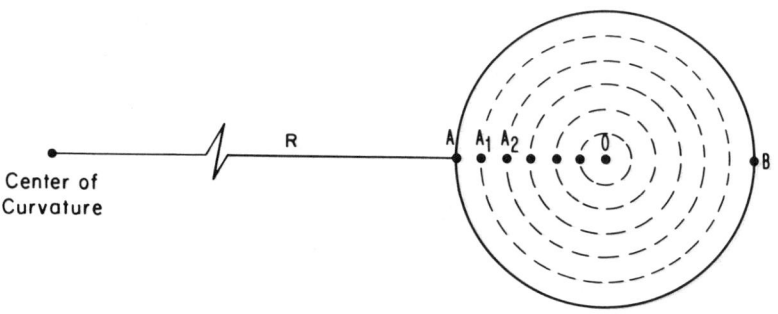

Conductor Cross Section

Fig. 3   Schematic diagram of the effect of bending strain on the
critical current of multifilamentary Nb_3Sn superconductors.

This is illustrated in Fig. 3. Assume the conductor has a level of internal compressive prestrain which places it at an initial uniaxial strain ε_i shown in Fig. 3. The overall current carried by the conductor is obtained by integrating J_c from $\varepsilon_i - \varepsilon_B$ to $\varepsilon_i + \varepsilon_B$ (i.e. from point A to point B in Fig. 3), where ε_B denotes the bending strain applied to the outermost filament. In this paper ε_B is calculated from the ratio of the bronze + niobium core diameter, d, to the bend diameter, D (i.e. $\varepsilon_B \equiv d/D$). In the case of cable conductors the strand core diameter, not the cable diameter, is used to determine ε_B (this assumption of strand independence in cabled conductors is supported by the data presented below). The values of $J_c(\varepsilon)$ are weighted by the cross-sectional area of the conductor corresponding to each strain. For a square or aspected conductor a simple average is taken:

$$\frac{I_c}{I_{cm}} = \frac{1}{2\varepsilon_B} \int_{-\varepsilon_B}^{\varepsilon_B} \frac{J_c(\varepsilon_i + x)}{J_{cm}}\, dx \tag{5}$$

For a circular conductor, the extremal strain values have a greater cross-sectional area weighting, resulting in:

$$\frac{I_c}{I_{cm}} = \frac{2}{\pi \varepsilon_B^2} \int_{-\varepsilon_B}^{\varepsilon_B} (\varepsilon_B^2 - x^2)^{1/2} \frac{J_c(\varepsilon_i + x)}{J_{cm}}\, dx \tag{6}$$

If there is a shift of the neutral axis due to yielding of the matrix during bending[12], ε_i is simply replaced by $\varepsilon_i + \delta/R$, where δ is the distance the neutral axis is shifted toward the center of curvature. The averaging procedure remains otherwise the same.

The effect of bending strain for these two shapes of conductors has been evaluated from Eqs. (5) and (6) using Eq. (3) for $J_c(\varepsilon)/J_{cm}$ (taking the typical set of scaling parameters applicable to practical multifilamentary Nb_3Sn described in the previous section). The results at 10 T are shown in Fig. 4 for different initial uniaxial strains ε_i. For both geometries, the degradation with bending strain is relatively gradual. This is because the average in Eqs. (5) and (6) is taken over the entire conductor cross-section and the changes in J_c are nearly symmetrical about point 0 in Fig. 3.

For very large bending strain the decrease is expected to become much more precipitous than indicated in Fig. 4. This occurs when the strain in the outermost filaments (point B in Fig. 3) reaches the irreversible strain limit[2] ε_{irrev} where filament breakage occurs. As described in Ref. 2 this usually occurs in practical multifilamentary Nb_3Sn conductors at intrinsic strains typically in the range from 0.4% to 0.7%; for in-situ conductors ε_{irrev} can occur at intrinsic strains as high as 1%.

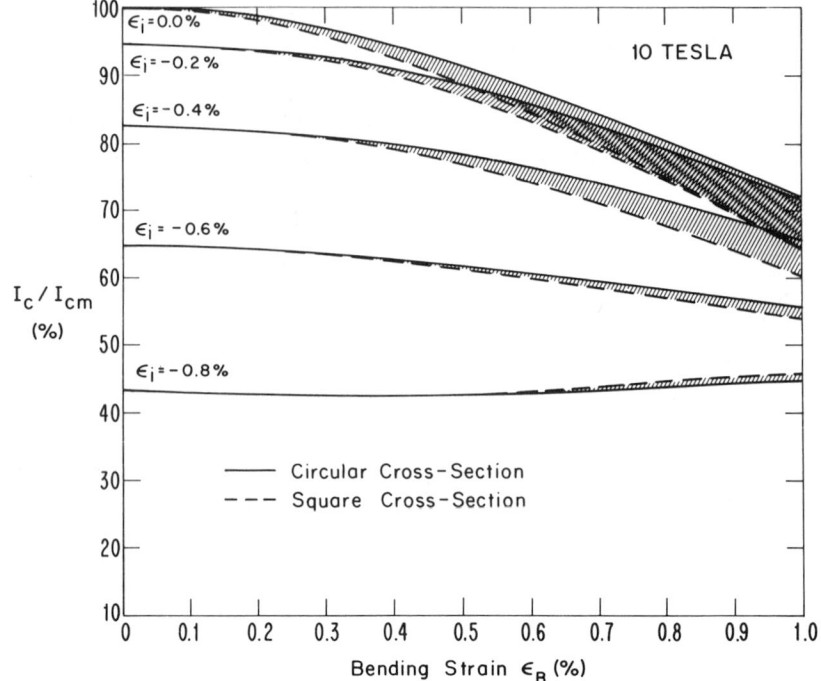

Fig. 4   Comparison of bending-strain degradation for circular and
         square cross-section multifilamentary Nb_3Sn conductors,
         assuming long-twist pitch ($\ell \gtrsim L$). ε_B represents the
         bending strain applied to the outermost filaments, calcu-
         lated by the ratio of the bronze + niobium core diameter
         to the bend diameter. ε_i represents the initial uniax-
         ial-compressive-strain state of the conductor.

Short-Twist Pitch

The effect of bending in highly-twisted multifilamentary conductors is expected to be much more severe than in conductors with
a longer twist length or in monofilamentary conductors. Each filament experiences alternate compression and tension as it spirals
through the conductor. If the twist pitch is shorter than the
current-transfer length[13], the current-carrying capacity of
the filament will be determined by J_c at its most highly
strained point. Thus in a circular conductor under compressive
prestrain, the critical-current density of the entire outer ring
of filaments will be determined by J_c at the inner radius of the
bend, i.e. point A in Fig. 3. J_c of the next ring closest to
the center of the conductor is determined by J_c at point A_1 in
Fig. 3, and so on. The result is that the overall current carried
by a highly-twisted multifilamentary conductor is obtained by

taking a weighted integral of J_c from point A to point 0 in Fig. 3 (not point A to B as for long-twist-pitch or monofilamentary conductors). The effect of a given bending strain is thus much more severe than for the long-twist-pitch averaging scheme.

For a highly-twisted circular conductor with $\varepsilon_i < 0$:

$$\frac{I_c}{I_{cm}} = \frac{2}{\varepsilon_B^2} \int_0^{\varepsilon_B} x \; \frac{J_c(\varepsilon_i - x)}{J_{cm}} \; dx. \tag{7}$$

Equation (7) has been evaluated using Eq. (3) for $J_c(\varepsilon)/J_{cm}$ and the typical set of scaling parameters. The results are compared with the long-twist-pitch averaging scheme in Fig. 5. As expected, the short-twist-pitch average leads to a much greater critical-current degradation.

Which type of averaging procedure should be used? The parameter that determines this is the ratio of the twist pitch ℓ to the current-transfer length L. For $\ell \gtrsim L$, the long-twist pitch

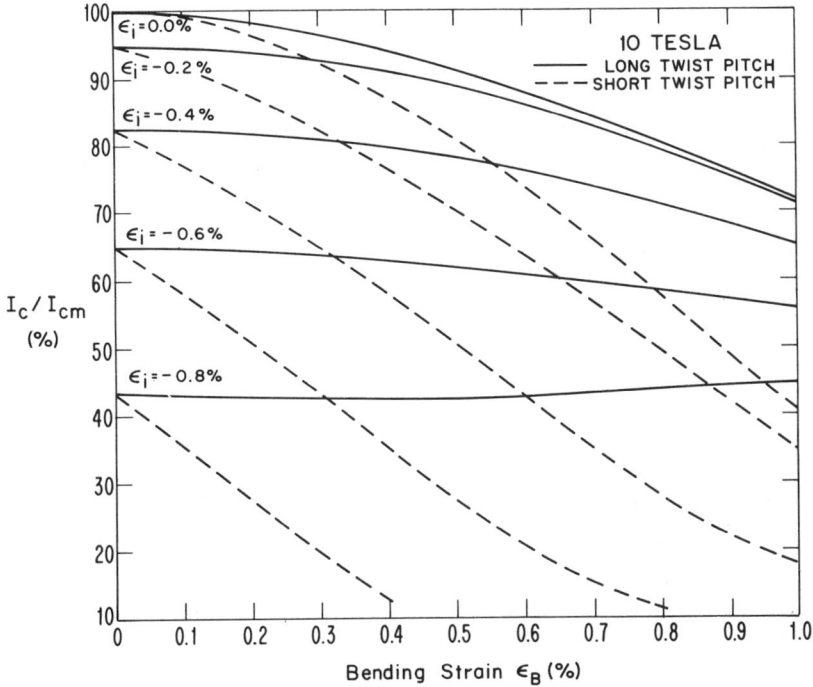

Fig. 5   Comparison of bending-strain degradation for long-twist-pitch ($\ell \gtrsim L$) and short-twist-pitch ($\ell \ll L$) multifilamentary Nb$_3$Sn superconductors.

averaging procedure applies; for $\ell \ll L$ the short-twist-pitch procedure applies. L has been calculated[13] to have the relatively simple form:

$$L \simeq \left(\frac{0.1}{n}\right)^{1/2} \left(\frac{\rho_m}{\rho^*}\right)^{1/2} d \qquad (8)$$

where ρ_m is the transverse resistivity of the matrix, d is the conductor diameter, n characterizes the resistive transition in the superconductor by the empirical formula $\rho = kJ^n$, and ρ^* is the superconductor-resistivity criterion used to define I_c. Physically, L represents the minimum distance along a wire needed for current to redistribute among the filaments without generating a resistivity in the superconducting-reaction layer greater than ρ^*. Note that the current transfer length L is defined[13] relative to the resistivity criterion ρ^* used to determine I_c, decreasing as $\rho^{*-1/2}$. For a typical superconductor-resistivity criterion of 10^{-11} Ωcm, L ranges from about 2 wire diameters for copper-stabilized NbTi to about 20 wire diameters for bronze-process Nb_3Sn.[13,14] In a typical 0.5 mm diameter strand, this corresponds to a current-transfer length of about 1 mm for NbTi and 1 cm for Nb_3Sn. This is to be compared with the filament twist pitch ℓ which is usually on the order of 1 cm for most practical conductors. Thus fortunately, for a typical superconductor-resistivity criterion of 10^{-11} Ωcm, we would expect both Nb_3Sn and NbTi multifilamentary to be described by the long-twist-length averaging scheme. For Nb_3Sn, however, it should be noted that at superconductor-resistivity levels of 10^{-12} Ωcm and lower, current-transfer voltages will start to become significant and the effects of bending will be greatly enhanced as the short-twist-pitch averaging scheme starts to apply.

Comparison with Experiment

A comparison of the results of the long-twist-length averaging scheme with bending strain data for two commercial multifilamentary Nb_3Sn conductors is shown in Figs. 6 and 7. The data in Fig. 6 were obtained by Walker et al.[15] on a conductor with an ε_i of -0.18% and an ℓ/L ratio of about 1; those in Fig. 7 were measured by Sanger et al.[16] on a conductor with an ε_i of -0.32% and an ℓ/L ratio also about equal to 1. The theoretical curves shown in these figures were obtained using Eq. (6) as described previously. Although the data in Fig. 6 are limited, the results agree quite well at 10 T where the measurements were made. Moreover, as shown in Fig. 6, the scaling relation allows the effect of bending strain to be calculated at higher fields. This represents an important predictive tool, especially for high-field magnet design. Note that the bending-strain degradation becomes significantly greater as the field is increased.

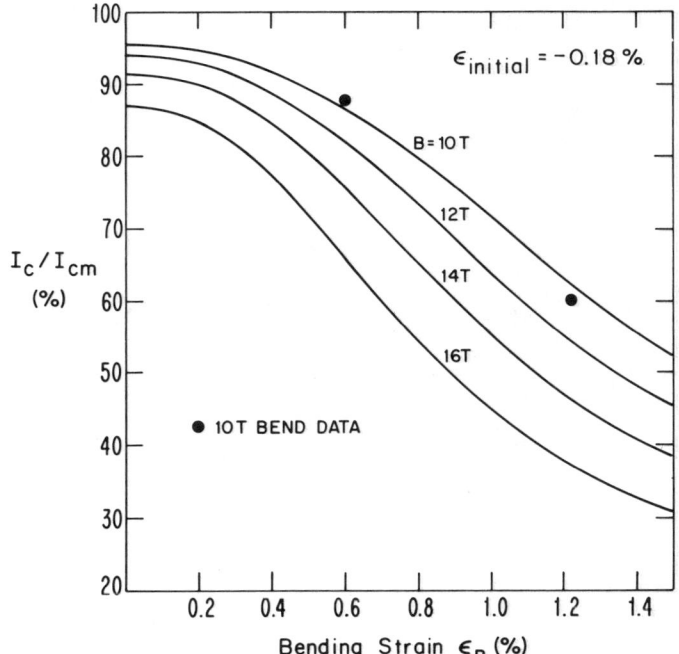

Fig. 6   Relative critical-current degradation as a function of
          bending strain ε_B for a range of magnetic field.   The
          two data points are from Ref. 15 and were measured on a
          cabled multifilamentary Nb_3Sn conductor.   The bending
          strain ε_B was calculated by the ratio of the single-
          strand Nb_3Sn core diameter to the bend diameter.

        The data in Fig. 7 are much more extensive and illustrate
very nicely the agreement with theory at bending strains below the
irreversible strain limit ε_{irrev}, as well as the rapid bend
degradation of I_c when ε_B exceeds ε_{irrev} (which for this conductor was determined from uniaxial strain measurements to be just
above 0.7%).   As seen in Fig. 7, the irreversible strain point
represents the most severe limit on bending strain at low fields.
At high fields, however, the compressive prestrain will dominate,
significantly degrading the critical current even before bending
strain is applied.

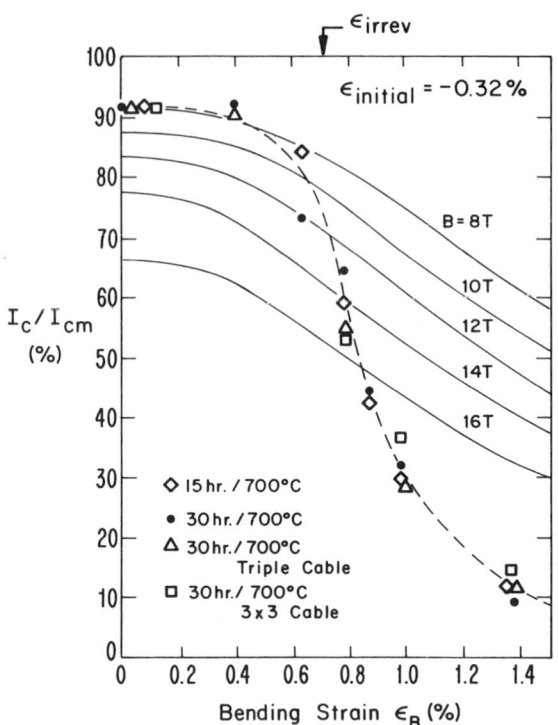

Fig. 7   Relative critical-current degradation as a function of
         bending strain ε_B for a range of magnetic field.   The
         data are from Ref. 16 and were measured on single multi-
         filamentary Nb_3Sn strands as well as cables.   For all
         the conductors, the bending strain ε_B was calculated by
         the ratio of the single-strand Nb_3Sn core diameter to
         the bend diameter.   The irreversible strain limit ε_{irrev}
         was obtained from uniaxial-strain measurements, Ref. 1.

COMBINED BENDING AND UNIAXIAL STRAIN

What happens when both uniaxial strain and bending strain are
applied to a conductor? This occurs when a reacted conductor is
wound into a magnet and then the winding is subjected to cooldown
strain and magnetic hoop strain when energized. Such is the case
for the conductor in most react-and-wind magnets. It can also be
the case in wind-and-react magnets which are not fully impregnated. If the conductor has waves or ripples reacted into it, on
initially energizing the magnet, the magnetic hoop strain will
straighten the conductor, applying bending strain as well as uniaxial strain. This is not an uncommon occurance in practical
bronze-process conductors. In fact, many of the commercial conductors that we have tested have had varying degrees of waves
inadvertently reacted into them.

An example of this "wavy-conductor effect" is shown in Fig.
8. The upper uniaxial-strain curve was obtained on a specimen
which was essentially straight. The lower curve was obtained on
another specimen of the same conductor which had a very slight
wave reacted into it. Both conductors started with about the same
critical current. But when the conductor with waves was subjected
to uniaxial strain, it straightened and the critical current initially decreased because of bending strain. Then as further
uniaxial strain was applied, the critical current increased to a
maximum and then decreased similar to the usual uniaxial-strain
effect. The resulting curve is somewhat flattened, and degraded
relative to the data on the straight sample. Note also that the
peak appears to be shifted to a lower strain. In this example,
the waves were of large radius and the degradation in the peak was
only a few percent.

The results for other specimens studied showed even more pronounced degradation and flattening of the uniaxial strain characteristics, as well as a shifting of the peak to lower and lower
strains. Several specimens tested had sufficient waves reacted
into them that, upon initially applying uniaxial strain, the
bending strain (from straightening) exceeded ε_{irrev} in the
outer filaments. In this case the critical current was severely
depressed and the uniaxial strain peak disappeared altogether.

Theoretically, what is happening in these examples is that
bending strain averages the critical current over a band of
strains, and when uniaxial strain is also applied, the mid point
of the band is shifted to higher strain, i.e. region A-B is slid
along the curve in Fig. 3. The net effect is that the peak in
Fig. 3 is averaged, producing a degraded maximum critical current
which has a more gradual dependence on strain. The results of
such a calculation at 7 Tesla are shown as solid curves in Fig. 8
and are in remarkably good agreement with the data. The upper

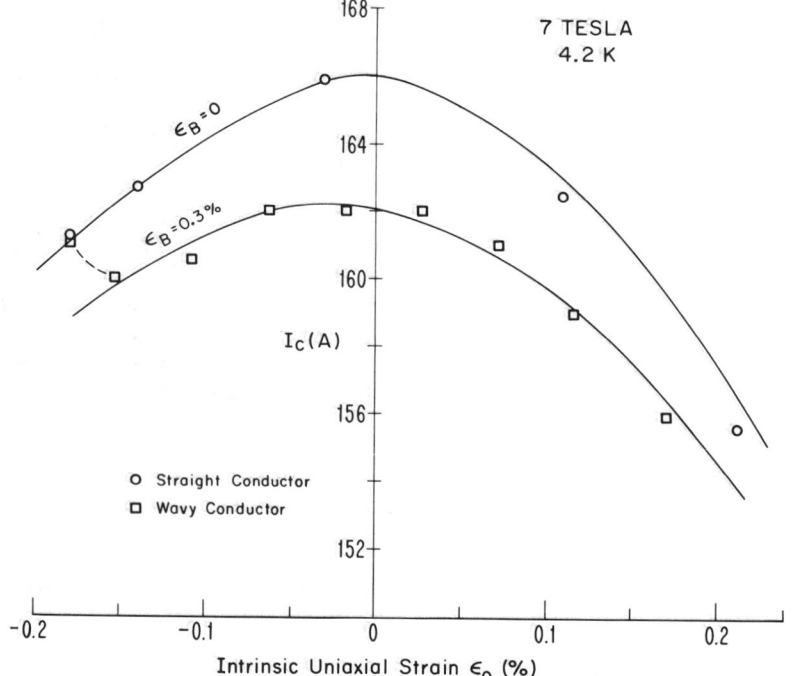

Fig. 8   Uniaxial-strain dependence of the critical current, illustrating the "wavy-conductor effect". Circles represent
data obtained on a straight conductor; squares represent
data obtained on a conductor which had very slight waves
in it. The upper solid curve was calculated at 7 T from
Eq. (3) using the typical set of scaling parameters shown
in Fig. 2 and an I_{cm} of 166 A. The lower solid curve
was calculated from Eq. (6) using the same parameters,
except a fixed bending strain of 0.3% was assumed to be
present.

curve is a direct evaluation of Eq. (3); the lower curve was
calculated from Eq. (6) assuming a fixed bending strain ε_B of
0.3% and a variable uniaxial strain ε_i.

    If different bending-strain averages are taken over the uni-
axial-strain characteristics shown in Fig. 2, the effect of bending on the uniaxial-strain characteristics of practical Nb_3Sn
conductors can be predicted over an extensive range of strain and
field. The effect has been calculated for different constant
bending strains using Eq. (6) with a variable uniaxial strain
ε_i. The results at 14 T, for example, are shown in Fig. 9.
This figure illustrates three points. First, note that the peak
value of I_c is shifted by bending to compressive strains, just

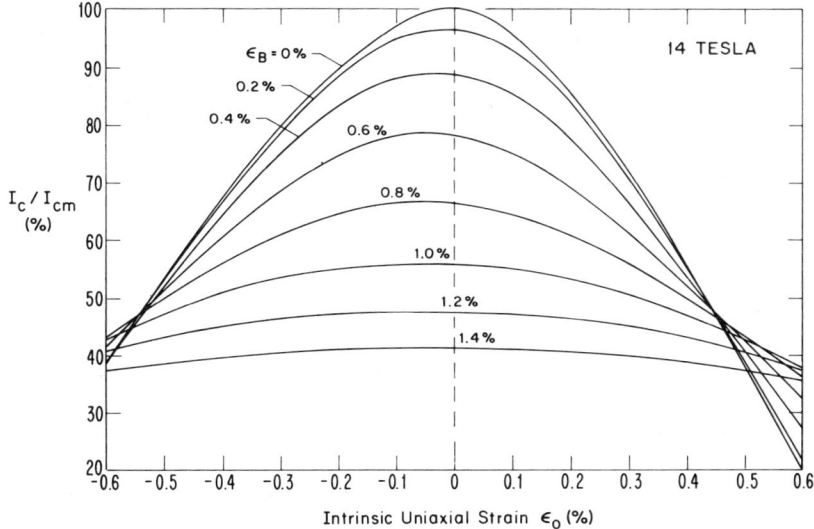

Fig. 9   Uniaxial-strain characteristics for multifilamentary
          Nb$_3$Sn at 14 Tesla for a range of bending strains ε_B.
          Uniaxial and bending strain limits at 14 Tesla may be
          obtained from such a plot for any given I_c-degradation
          criterion.

as observed in the data of Fig. 8.   This is caused by the I_c vs.
ε relationship in Fig. 2 not being symmetric.   The I_c degradation under tension is greater than for compression.   Thus a strain
average centered slightly on the compressive side of maximum produces the highest critical current.   The opposite would be true if
the I_c vs. ε relation were skewed the other way.   In general,
bending strain will shift the uniaxial-strain peak toward the side
where the uniaxial-strain degradation is the least.

The second point to note in Fig. 9 is that the relative
effect of bending strain diminishes at uniaxial strains away from
zero intrinsic strain.   Interestingly enough, at high enough uniaxial strain, bending can even enhance the critical current.   This
can be seen in the 14 T results of Fig. 9 where for $|\varepsilon_o| \gtrsim 0.5\%$,
the critical current is increased by bending.   What happens in such
case is that the critical current is so close to zero that the
strain distribution caused by bending produces significantly
higher critical currents on one side of the conductor, while only
negligible degradation on the other side.   This is only of academic interest, however, for such a bending-strain enhancement
does not occur until the critical current is degraded more than 50%.

The third and essential point for practical applications is that the maximum of the critical current can never be attained if there is bending strain present in the conductor. This places limits on the amount of bending strain that can be tolerated in magnet design. The bending-strain limits are coupled to the uni-axial-strain limits discussed previously. For example, if a 10% I_c-degradation criterion is used to determine uniaxial-strain tolerances and no bending strain is present, it was found in Table I that at 14 T the intrinsic strain ε_o must be limited to the range $-0.20\% \leq \varepsilon_o \leq +0.16\%$. As seen in Fig. 9, the presence of bending strain reduces these uniaxial strain limits, but fortu-nately the change is only a few tenths of a percent until a bend-ing strain of 0.37% is reached. Beyond this bending strain limit, the critical current degradation exceeds 10% regardless of what the uniaxial strain is. Bending-strain limits corresponding to a 10% I_c-degradation criterion have been determined for magnetic fields from 4 to 16 T and are included in Table I along with the corresponding uniaxial-strain limits. It is implicitly understood that the strain values in Table I are restricted by a conductor's irreversible-strain limit.

SUMMARY

The strain-scaling law has been used to determine uniaxial-strain characteristics which describe the critical-current degrada-tion in commercial multifilamentary Nb_3Sn conductors from 4 T to 16 T. Bending-strain degradation has then been calculated from the uniaxial-strain characteristics by taking an average appropri-ate for the twist-pitch to current-transfer-length ratio of the conductor. The results indicate that for typical applications, the long-twist-pitch averaging scheme should apply. At supercon-ductor resistivity levels of 10^{-12} Ωcm and lower, however, the calculations indicate that bending degradation will be much more severe than at higher resistivity levels.

In most practical situations, both uniaxial and bending strain degradation will be combined. These effects have been cal-culated and found to account for critical-current data in conduc-tors where both types of strain were present. Strain limits for optimizing the critical current in magnet construction were calcu-lated for both uniaxial and bending strain at fields from 4 T to 16 T. Because of the nearly universal nature of strain degrada-tion in highly-reacted bronze-process Nb_3Sn, the strain limits listed in Table I should apply generally to the design of most multifilamentary Nb_3Sn magnets.

ACKNOWLEDGMENTS

The assistance of H. Schoon in obtaining the uniaxial strain measurements and evaluating the various expressions in this paper is gratefully acknowledged. Helpful discussions with A. F. Clark during the course of this work were appreciated.   This work was supported by the Naval Ship Research and Development Center, Annapolis, Maryland.

REFERENCES

1. J. W. Ekin, submitted to Cryogenics, April 1980.
2. J. W. Ekin, IEEE Trans. on Magnetics MAG-15:197 (1979).
3. H. Hillmann, H. Kuckuck, H. Pfister, G. Rupp, E. Springer, M. Wilhelm, K. Wohlleben, and G. Ziegler, IEEE Trans. on Magnetics MAG-13:792 (1977).
4. G. Rupp, IEEE Trans. on Magnetics MAG-13:1565 (1977).
5. J. W. Ekin, Proc. Int. Cryog. Mat. Conf., Aug. 1977; Adv. Cryog. Eng. 24:306 (1978).
6. E. J. Kramer, J. Appl. Phys. 44:1360 (1973).
7. R. G. Hampshire and M. T. Taylor, J. Phys. F 2:89 (1972).
8. A. M. Campbell and J. E. Evetts, Adv. Phys. 21:199 (1972).
9. D. Dew-Hughes, Phil. Mag. 30:293 (1974).
10. E. H. Brandt, J. Low Temp. Phys. 26:709 (1977).
11. J. W. Ekin, reported at session CC of the Applied Superconductivity Conf., Aug., 1978, unpublished.
12. K. Kaiho, T. S. Luhman, M. Suenaga, and W. B. Sampson, Appl. Phys. Letters 36:223 (1980).
13. J. W. Ekin, J. Appl. Phys. 49:3406 (1978).
14. J. W. Ekin, A. F. Clark, and J. C. Ho, J. Appl. Phys. 49: 3410 (1978).
15. M. S. Walker, J. M. Cutro, B. A. Zeitlin, G. M. Ozeryansky, R. E. Schwall, C. E. Oberly, J. C. Ho, and J. A. Woolam, IEEE Trans. on Magnetics MAG-15:80 (1979).
16. P. A. Sanger, E. Ioriatti, C. Spencer, and C. Heyne, presented at the 8th Symposium on Eng. Prob. in Fusion Research, Nov., 1979.

EVIDENCE FOR MICROSTRUCTURAL EFFECTS UNDER STRAIN IN BRONZE

PROCESS Nb$_3$Sn

D. M. Kroeger, D. S. Easton, C. C. Koch, and A. DasGupta

Metals and Ceramics Division
Oak Ridge National Laboratory
Oak Ridge, TN   37830

INTRODUCTION

The possible effects of externally applied strain on the flux pinning process can be divided into two groups according to whether the significant property changes produced by strain are associated with the crystalline defects which act as flux pinners, or with the bulk superconducting material which carries the supercurrent. Changes in the latter are reflected in the equilibrium properties such as the upper critical field, B_{c_2}, and the Ginzburg–Landau parameter, κ. To first approximation, changes in the former are not seen in the equilibrium properties, but may affect the bulk pinning force, F_p, by changing the number of pinning centers or their strength. For want of a better term, we have called changes associated with the flux pins microstructural effects, even though they may or may not involve gross changes in structure such as the martensitic transformation which is known to occur in Nb$_3$Sn at low temperature.

A number of investigators[1–3] have concluded, primarily on the basis of the similarity of the dependence on strain of the critical current density, J_c, to the strain dependences of B_{c_2}, and the transition temperature, T_c, that, for strains small enough that cracks and filament breakage do not occur[4,5] equilibrium property changes are chiefly responsible for the strain dependence of J_c. However,

---

*
  Research sponsored by the Division of Material Sciences, U.S. Department of Energy under contract W–7405–eng–26 with the Union Carbide Corporation.

since this conclusion is based on comparisons of J_c and T_c or B_{c2} measurements made on different samples in different apparatuses, or on B_{c2} values obtained by extrapolation of J_c data, these results do not exclude the occurrence of microstructural change as an additional effect altering the strength or density of flux pins.

Whether even gross microstructural changes occur in the Nb_3Sn when uniaxial strain is applied to a bronze diffusion process conductor is not easily determined, because any change which may occur takes place at low temperatures, under cover of the bronze matrix, making the Nb_3Sn all but inaccessible to analytical tools. In the absence of the bronze matrix, little stress can be applied to the brittle Nb_3Sn, and removal of the matrix after straining would change the stress state of the Nb_3Sn.

Since direct evidence is not easily obtained, we chose to investigate the question of whether there is a microstructural component to the strain dependence of F_p by carefully determining whether the bulk property changes which occur can adequately account for all of the variation of F_p with strain. In addition to comparisons of the dependences on strain of J_c, T_c, and B_{c2}, determinations were made of the strain dependences of the parameters which enter the semi-empirical expression for F_p. Examination of these results in light of current theories of flux pinning and type-II superconductivity indicates that bulk property changes do not account for all of the change in F_p, suggesting that microstructural effects are also important.

COMPARISON OF STRAIN DEPENDENCES OF J_c, T_c, B_{c2}

Our first effort was to compare the dependences of J_c and B_{c2} on strain. Figure 1 shows such a comparison for a conductor from Harwell Laboratory, in England, with 1024 filaments, reacted for only 1 h at 700°C, to produce an 0.4 μm thick layer of Nb_3Sn. Conductors with such thin Nb_3Sn layers have been found to exhibit strong I_c variation with strain.[6] The peak in I_c/I_{co} occurs at a somewhat higher strain than does the peak in B_{c2}/B_{c2o}, suggesting that another factor in addition to the variation of B_{c2} is affecting I_c. However, this conclusion requires additional confirmation because, for experimental convenience, I_c and B_{c2} were determined on similar but separate samples, in different apparatuses. The values of B_{c2} at 4.2 K were estimated from measurements of $B_{c2}(T)$ near T_c. These measurements were made resistively, in vacuum to facilitate temperature control. However, the relatively high critical current of this conductor (~70 A at 4.2 K and 7 T) could not be introduced to the specimen in this apparatus without heating in the joints and possible interference with the measurement, so the I_c measurements were made on a different specimen in liquid He at 4.2 K while strain was applied by means of an Instron

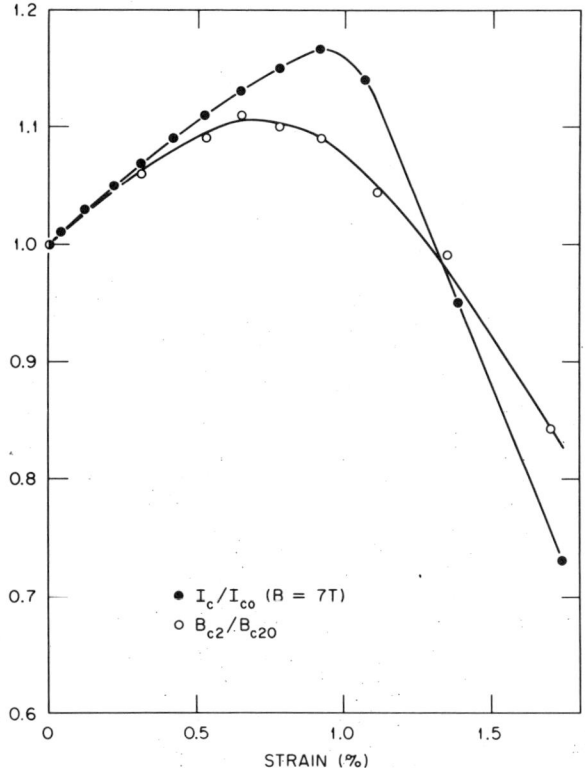

Fig. 1.   Comparison of dependences of I_c and B_{c2} on strain in a
1024 filament conductor.   Note that I_c peaks at a higher
strain than B_{c2}.

tensile testing machine (see Ref. 7,8 for description of the
procedure).   Thus, uncertainty exists exists about the comparison
of strain measurements.   To remove this difficulty, a monofilament
conductor with much smaller Nb_3Sn cross-section than the Harwell
material, and therefore smaller I_c, was made.   The bronze-to-
niobium ratio was 34/1, the specimen diameter was 0.08 cm, and the
reaction time at 700°C was 8 h, producing a Nb_3Sn layer approximately 1 μm thick.   Using the apparatus described in Ref. 9,
resistive measurements of $I_c(T)$, $B_{c2}(T)$ and T_c as functions of
strain could all be made on the same sample, in the same apparatus,
and without intervening stress or temperature cycles.   For reasons
which will be made clear in the discussion below of scaling behavior of F_p, all T_c and B_{c2} values reported here are what we have
called "finish" values, i.e., the values of T and B at which the
sample exhibits zero resistance (within limits of detection) to the
flow of a small test current.   From Fig. 2 we see that T_{cf} and
B_{c2f} peak at the same strain, but, as with the multifilament

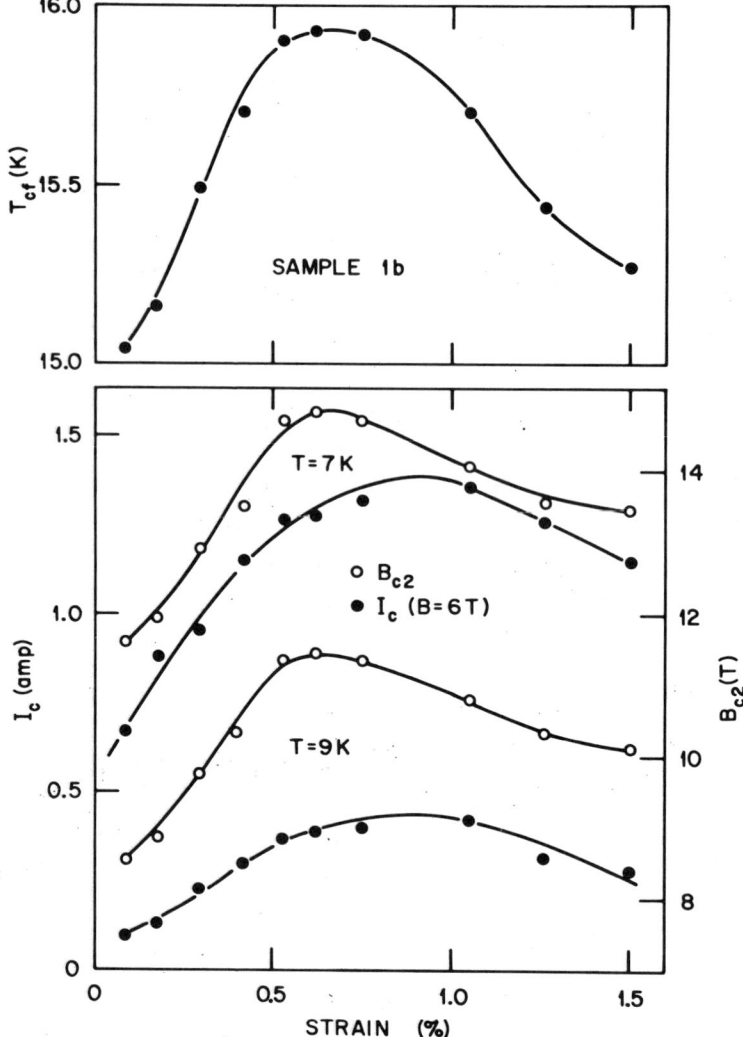

Fig. 2.   Comparison of strain dependences of I_c, T_c and B_{c2} for
          monofilament conductor. The T_c and B_{c2} peaks occur
          together, but, as for the commercial conductor of Fig. 1,
          I_c peaks at a higher strain.

conductor, I_c peaks at a somewhat higher strain.   These results
indicate that B_{c2} variation is not solely responsible for the
change in I_c.   As will be discussed below, flux pinning theory
indicates that F_p also depends on the Ginzburg–Landau parameter, κ,
so these results suggest, but do not prove definitely, that additional effects beyond equilibrium property changes must be invoked.

COMPARISON OF STRAIN DEPENDENCES OF T_c AND $(dB_{c2}/dT)_{T_c}$

Further suggestion of microstructural change is obtained by comparing plots of T_{cf}^* and $(dB_{c2}/dT)_{T_c}$ versus strain, as in Fig. 3. Here, T_{cf}^* is obtained by extrapolating the linear portion of the B_{c2} vs T curve to B_{c2} = 0, thereby ignoring the curvature which occurs very near T_c. T_{cf}^* and $(dB_{c2}/dT)_{T_c}$ behave rather similarly at low strains ($\varepsilon < \varepsilon_{maxT_c}$), but at high strains there is a region where $(dB_{c2}/dT)_{T_c}$ is increasing while T_c is decreasing. Since, in type-II superconductors $(dB_{c2}/dT)_{T_c} \propto \gamma \rho_n$ (γ = electronic specific heat coefficient and ρn = normal state resistivity), and T_c is determined by the density of states at the Fermi level and is thus a function of γ, one would expect a unique relationship between T_c and $(dB_{c2}/dT)_{T_c}$. Figure 3 indicates that strain changes this relationship.

STRAIN DEPENDENCE OF PARAMETERS IN $F_p(B,T)$

Background and Procedure

Flux pinning studies[10-16] have demonstrated that, for many type-II superconductors, the bulk pinning force density, $F_p = |\vec{J}_c \times \vec{B}|$, varies with field and temperature according to the equation

$$F_p = A \, B_{c2}^n(T) \, f(b) \qquad (1)$$

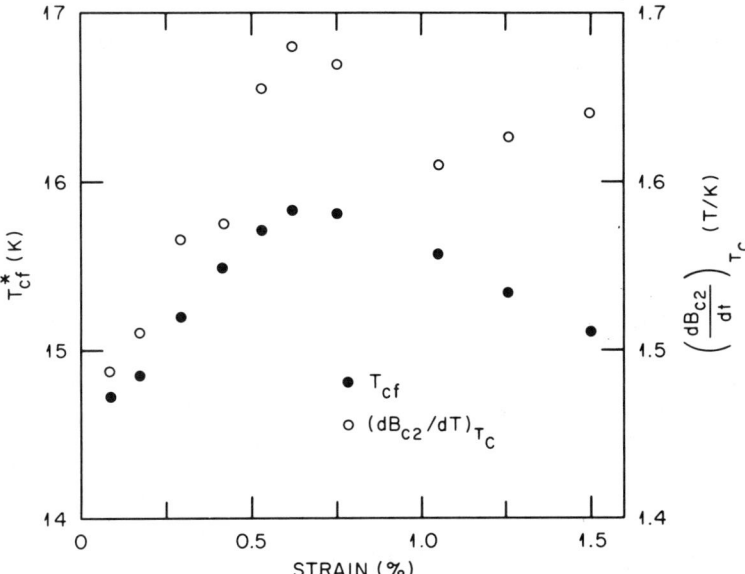

Fig. 3. Comparison of the strain dependences of $(dB_{c2}/dT)_{T_c}$ and T_c.

where A is temperature and field independent, and $b = B/B_{c_2}$. A, n, and f(b) are all sensitive to the microstructure of the specimen. Furthermore, expressions for F_p based on fluxoid interactions with the various microstructural features thought to be responsible for flux pinning indicate f(b) should have the form $f(b) = b^\ell (1 - b)^m$, where $\ell = 1/2 - 3/2$ and $m = 1 - 2$. Measurements often fit functions of this form reasonably well, although the values of the exponents, ℓ, m, and n, are sometimes not easily rationalized by theory and models. When F_p has the form of Eq. (1), the constant A and the exponents, ℓ, m, and n contain all of the influence of microstructure on pinning force, such as the volume density of pinning centers, pinning center morphology and distribution, and stress fields about microstructural features. Thus, if F_p in bronze-process Nb_3Sn conductors should have the form of Eq. (1), then any influence of stress upon F_p through microstructure should be reflected in these quantities.

On a given specimen, sufficient data were obtained, at each of several applied loads, to test for conformity of F_p to Eq. (1), and to determine values for the constants, A, n, ℓ, and m. For this purpose, I_c and B_{c_2} were measured as functions of temperature. Magnetic field limitations confined our measurements to temperatures above 60–70% of T_c. All measurements at a given load were made without cycling the temperature beyond the neighborhood of T_c and without change of load. $B_{c_2}(T)$ was determined resistively by fixing B and slowly sweeping T while passing a 3mA test current through the specimen. A voltage criterion of 0.2 µv/cm was used. The transition widths observed were about 1 K, presumably reflecting inhomogeneity within the Nb_3Sn layer. Since the transition width varied with applied strain, and tended to be smallest when T_c was a maximum, stress gradients in the Nb_3Sn may also contribute to the width. In general, F_p was found to conform well to Eq. (1) provided B_{c_2} was determined from the point in the transition at which the sample voltage went to zero, what we have called the finish point, rather than the (higher) onset point, at which full normal resistance is restored. The B_{c_2}'s corresponding to finish and onset points we call $B_{c_{2f}}$ and $B_{c_{2o}}$.

In Eq. (1), B_{c_2} refers to the upper critical field of the bulk of the superconductor. Regions of limited extent in which the superconducting properties differ from those of the bulk act as pinning centers. If the superconducting layer were perfectly homogeneous, and if the voltage criterion were arbitrarily small, $B_{c_{2f}}$ would be the field at which I_c is reduced to the constant 3mA test current, and $B_{c_{2o}}$ would be the field at which $I_c = 0$. For an inhomogeneous specimen, $B_{c_{2f}}$ is the field at which I_c for the continuous path having the highest $I_c(B)$ curve has been reduced to 3mA, and $B_{c_{2o}}$ is the field at which no bulk superconductivity remains in the specimen. The 1 K width of the transition shows that the Nb_3Sn layer is inhomogeneous, but the fact that F_p scales with $B_{c2f}(T)$ indicates the existence of a connected phase (or composition) which can be characterized by the transition curve $B_{c2f}(T)$ and which carries the transport current.

Results

Figures 4 and 5 show plots of F_p/F_{pmax} vs b for the same specimen at two different strains, $\varepsilon = 0$ and $\varepsilon = 0.64\%$. The latter is approximately the strain at which T_c and B_{c2} were maximized. In both cases, the data tend to lie on a single curve, i.e., F_p scales with B_{c2f}. In Fig. 4, normalized curves of $f(b) = b^\ell(1 - b)^m$ for various values of ℓ and m are drawn. In this way, very approximate determinations of the values of ℓ and m can be made. In Fig. 4, the data fall between the curves for $\ell = 1/2$, m = 2.5, and $\ell = 1/2$, m = 3. In Fig. 5, all except the 14.25 K data are closest to the $\ell = 1/2$, m = 3 curve. The maxima of $f(b) = b^\ell (1 - b)^m$ occur at $b_{max} = \ell /(\ell + m)$, so that increasing ℓ moves the maximum to larger b and increasing m moves it to smaller b. In all cases measured,

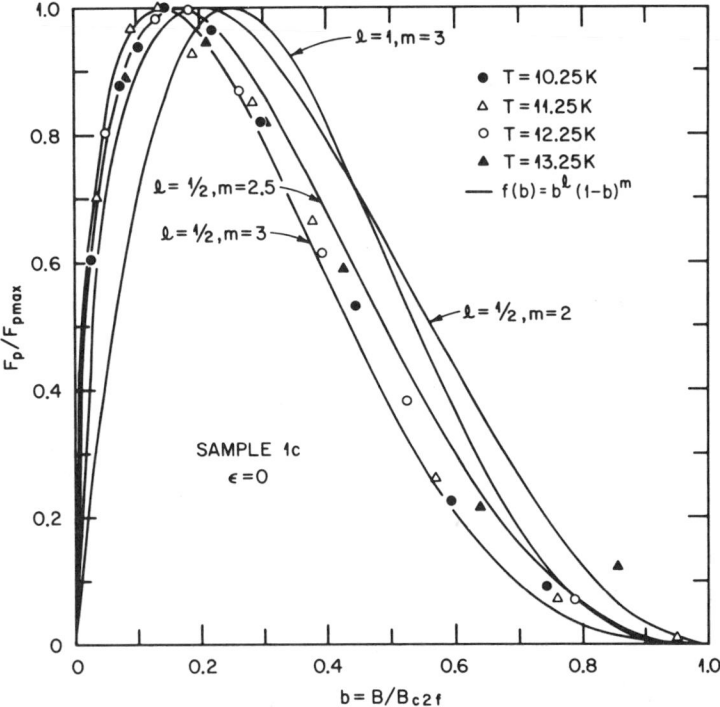

Fig. 4.   Plots of F_p/F_{pmax} vs B/B_{c2} for various temperatures tend to fall on the same curve. Plots of $f(b) = b^\ell (1 - b)^m$ are shown for various values of ℓ and m. These data, for $\varepsilon = 0$, fall between the $\ell = 1/2$, m = 2.5 and $\ell = 1/2$, m = 3 curves.

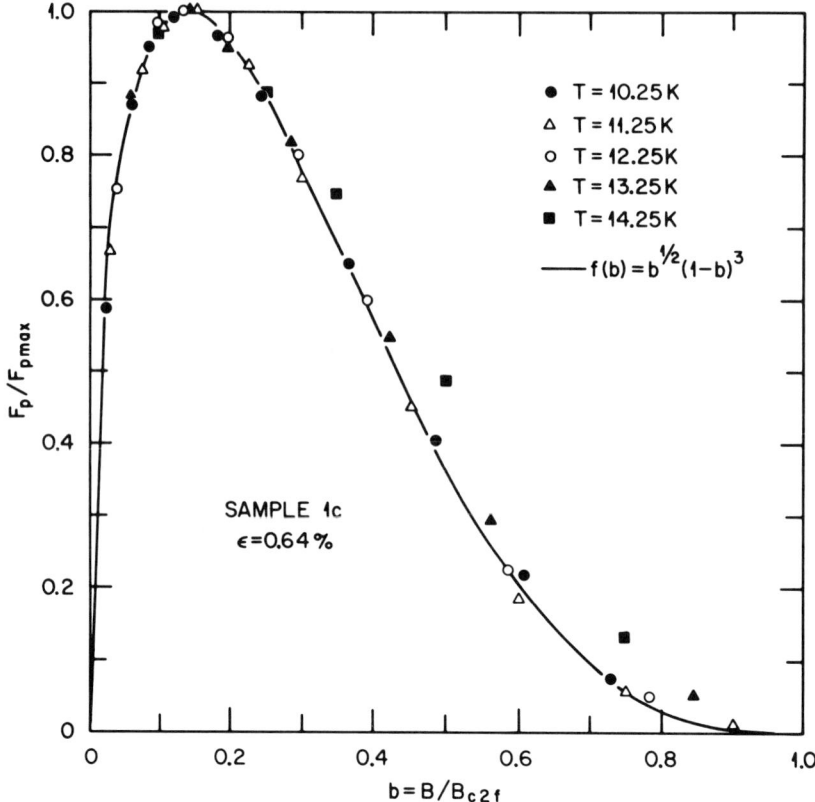

Fig. 5.   At 0.64% strain, F_p scales well with B_{c2f}, and the data
          fall approximately on the f(b) curve for $\ell = 1/2$, $m = 3$.

the position of the maximum indicates that the appropriate value of
ℓ is 1/2 if very large values of m ($\gtrsim 5$) are to be avoided.   Values
of m all lie between 2.5 and 3, with m tending to increase slightly
as the applied strain nears ε_{maxT_c}, that value at which T_c and
B_{c2} are maximized.   This trend can be seen in Figs. 4 and 5.

    Scaling was not observed in all cases.   Fig. 6 shows F_p/F_{pmax}
vs b for the same sample as in Figs. 4 and 5, but at a strain
greater than ε_{maxT_c}.   The absence of scaling is apparently not due
to the development of major cracks or break up of the Nb_3Sn layer,
since the I_c values were higher at this strain than at the lower
strains of Figs. 4 and 5.   This loss of scaling above ε_{maxT_c} was
not observed in two other specimens, results for one of which are
shown in Fig. 7 at $\varepsilon = 1.1\%$, which is well beyond ε_{maxT_c}.   The
scaling is excellent.   Also note that the best value of m is 2.5
rather than 3 for this specimen.

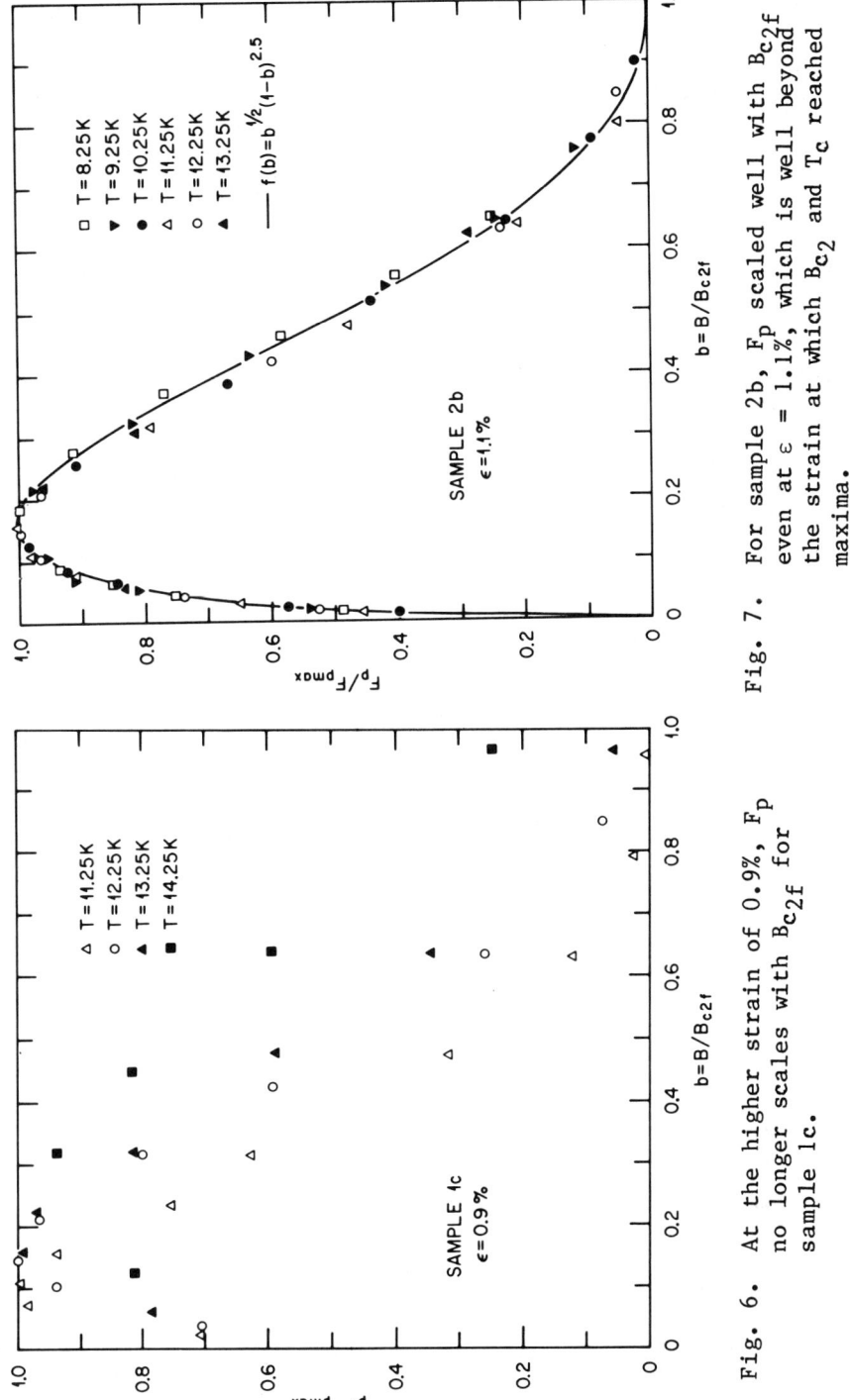

Fig. 7. For sample 2b, F_p scaled well with B_{c2f} even at $\varepsilon = 1.1\%$, which is well beyond the strain at which B_{c2} and T_c reached maxima.

Fig. 6. At the higher strain of 0.9%, F_p no longer scales with B_{c2f} for sample 1c.

The exponent n of B_{c_2} in Eq. (1) can be determined by plotting log F_p at a particular b vs log B_{c_2} for the various temperatures at which data were taken. It is convenient and customary to use F_{pmax}. Figure 8 shows such plots for the data of Figs. 4 and 5. The plots are linear, their slopes being equal to n. For sample 1c, sufficient data were obtained to determine n at five loads, and all of the values were between 2.9 and 3.2. For sample 2b (Fig. 7), n was determined at two loads, giving values of 2.5 and 2.6. Sample 2a, on which data were taken only for zero load, also gave n = 2.5. The values of n were determined by least squares fit to the data.

These results, as well as the loss of scaling for $\varepsilon > \varepsilon_{maxT_c}$ in sample 1c and the variation of m from sample to sample, illustrate the variability of behavior under stress exhibited by this material. This variability may itself be regarded as suggestive that a stress-mediated phase transformation, or other microstructural change, is occurring under stress, since electronic properties would not be expected to exhibit such irregularity.

How rapidly the sample is cooled, and whether it is under stress during cooling may also be involved in sample-to-sample

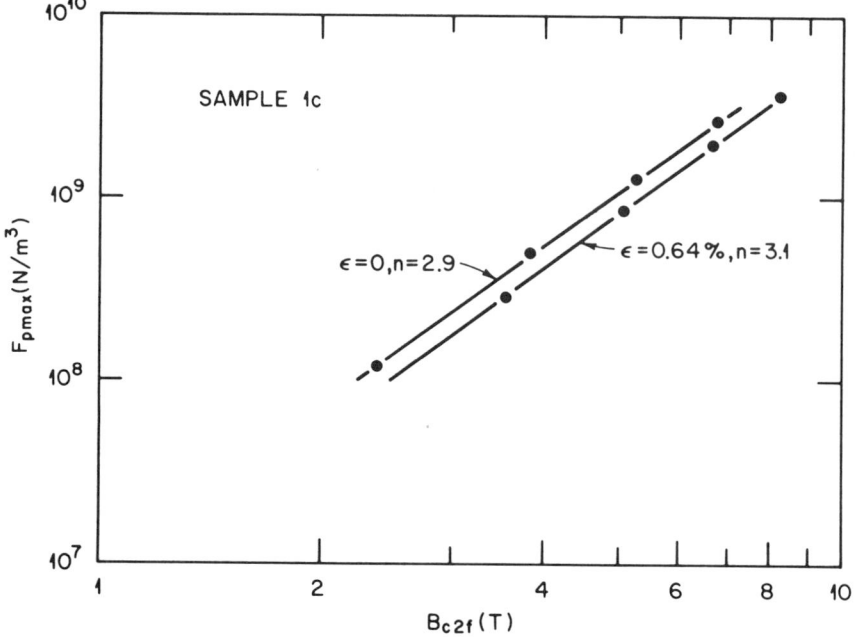

Fig. 8.   Log F_{pmax} vs Log B_{c_2f} was linear for all strains at which F_p scaled with B_{c_2f}, but the lines for different strains are displaced, indicating that A depends on strain.

variation, because of the following unusual effect. On occasion
specimens were left overnight in the cryostat, under load. During
this time, the temperature of the specimen rose to 30—35 K. The
first measurement of T_c made the next morning was invariably different from the value obtained the previous afternoon, sometimes by
as much as 0.5 K. Generally, if the load was less than that required
to produce maximum T_c, then T_c increased overnight, and if the maximum T_c had already been passed, T_c decreased. Also, some drop in
load occurred overnight. Thus, the direction of the change in T_c
was such as to suggest that additional strain occurred overnight
under constant or slightly decreasing load. A controlled experiment,
in which the sample was held under constant load at temperatures up
to about 35 K, with a strain gage attached to the specimen, confirmed that the changes in T_c were due to changes in length of the
specimen which occurred under constant load and temperature as a
function of time. This effect was studied only briefly, but one
rough determination of rate was made; at 35 K, with a load of 200 Mpa,
approximately 0.04% strain occurred in 1 h, with a consequent
increase in T_c of 0.07 K. At temperatures under about 25 K the
rate of elongation was quite small. Measurements were not made at
temperatures above 35 K. Since the components of the conductor are
always under load during cooling from room temperature, different
cooling schedules may result in different stress states for the Nb_3Sn.

Clearly, much more work is needed to understand these observations, but we may note that in these specimens the 1 μm thick layer
of Nb_3Sn represents a very small percentage of the total cross-
section, and the larger Nb core is still only about 3% of the total.
Therefore, it is reasonable to assume that this effect is due to a
process occurring in the bronze matrix. The amount of tin removed
from the bronze during reaction to form the Nb_3Sn layer is nearly
negligible in this conductor, contrary to the case of commercial
conductors, for which the matrix may have only 2—3% tin remaining
after reaction. But if this effect should be found in commercial
materials, it may have significance for operating procedures for
Nb_3Sn magnets.

The fact that the lines in Fig. 8 are not coincident indicates
that the factor A is different for the two data sets. Values of A
can be determined for each load from the constant term in a linear
fit of log F_{pmax} vs log B_{c2}. The filled circles in Fig. 9 show how
A varies with strain in Sample 1c. Values for strains larger than
approximately ε_{maxT_c} were not obtained because, as previously
stated, the F_p data did not scale.

Data are also shown in Fig. 9 for sample 1b. Measurements
sufficient for determination of all the parameters in the expression
for F_p were not obtained for this sample. Instead, at each of
several loads, $B_{c2}(T)$ was measured, along with I_c at several

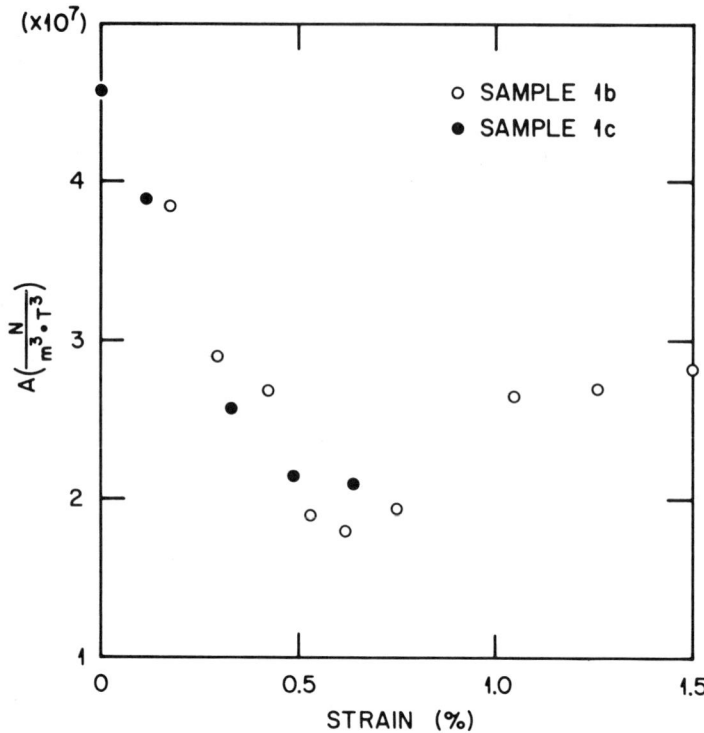

Fig. 9.   A decreased by more than a factor of two between $\varepsilon = 0$ and
           $\varepsilon \approx 0.6\%$, the latter being near that value of strain at
           which B_{c2} and T_c were maximized.

selected fields and temperatures. Plots of F_p vs $B_{c2}^3 b^{1/2} (1 - b)^3$
made from these data were found to be linear, indicating that the
behavior of this sample is similar to that of sample 1c. However,
there was no apparent change in behavior around $\varepsilon_{max} T_c$ comparable
to the loss of scaling which occurred in sample 1c. The slopes of
the F_p vs $B_{c2}^3 b^{1/2} (1 - b)^3$ plots are of course equal to the forefactor A, and the values of A so determined are plotted in Fig. 9.
It should be pointed out that if n and m are not precisely 3, as
has been assumed, plots of F_p vs $B_{c2}^3 b^{1/2} (1 - b)^3$ might still be
approximately linear, but with slopes which differ from the values
of A which would be obtained from analysis of a full set of $J_c(B,T)$
data.

      It should be stressed that the variation of A with strain
represents an effect on J_c which is independent of the change in
B_{c2} with strain. A has a minimum which approximately coincides
with the maxima in T_c and B_{c2} so that the effects of A and B_{c2} on
J_c are in opposite directions. The B_{c2} effect dominates, but for

Sample 1c, F_p calculated at ε = 0.64% for B = 3 T, assuming that A and n have the values determined at ε = 0, was ~40% higher than the measured value. Thus, if the observed B_{c_2} change were the sole effect of straining the sample, J_c at ε = 0.64% would be ~40% greater at 3 T than was measured.

Effect of Strain Dependence of κ on A

Regardless of the pinning species, or mechanism, or the manner of summation of the elementary pinning force, the forefactor A should be proportional to the number, N, of pinning centers per unit volume. It also depends on field and temperature independent parameters which determine the strength of a pin, such as its size, or in the case of pinning through the elastic energy, the stress field of a defect. Therefore, the observed variation of A may imply changes in either or both of those microstructurally determined quantities. However, A also depends inversely on the Ginzburg–Landau parameter, κ, raised to a power p, which, according to most theories, is in the range 2–4. Therefore, to discuss the question of microstructural change we must evaluate the effect of κ on A.

If we assume that A depends linearly on a parameter, a, which is related to the pin strength, then $A \propto Na/\kappa^p$. Furthermore, $\kappa \propto \gamma^{-1/2}$ $(dB_{c_2}/dT)_{T_c}$ where γ is the electronic specific heat coefficient. Therefore, the fractional changes in A, N, a, γ and $B_{c_2}' = (B_{c_2}/dT)_{T_c}$ associated with a small change in strain are related by

$$\delta A/A = \frac{\delta N}{N} + \frac{\delta a}{a} + \frac{p}{2} \frac{\delta \gamma}{\gamma} - p \frac{\delta B_{c_2}'}{B_{c_2}'} \tag{2}$$

The first two terms are the microstructural component of $\delta A/A$ and the latter two are the κ or electronic component. Note that as the sample is strained from 0 to ~0.6%, A decreases by more than a factor of two, and T_c increases by ~1 K. If this change in T_c is associated with a change in γ, then $\delta\gamma$ is small but positive, i.e., its sign is opposite to that of δA. Thus, the effect of κ on A will be overestimated if we ignore the $\delta\gamma$ term. The change in B_{c_2}' with strain was determined for both samples of Fig. 9, using the $B_{c_2}(T)$ data and ignoring the slight upward curvature which appears very near T_c. For sample 1b, when A changes from 3.8×10^7 N/m^{3}T^3 to 1.8×10^7 N/m^{3}T^3, B_{c_2}' changes from ~1.5 T/K to $\approx$1.7 T/K, giving $\delta A/\langle A \rangle$ = -0.7 and $\delta B_{c_2}'/\langle B_{c_2}'\rangle$ = 0.12. For sample 1c, the corresponding numbers are $\delta A/\langle A \rangle$ = -0.74 and $\delta B_{c_2}'/\langle B_{c_2}'\rangle$ = 0.06. Depending on the value of p, the fourth term in Eq. (2) is in the range of 15–50% of $\delta A/\langle A \rangle$, and, as was pointed out above, the third term is likely to be opposite in sign to $\delta A/A$. Therefore, it does not appear that the κ dependence of A can adequately account for its change with strain, and that the microstructural component must be responsible for a substantial portion of $\delta A/A$.

However, firmness of this conclusion must be tempered by the realization that our understanding of flux pinning may yet be so imperfect that data of this kind cannot be adequately interpreted. It may be important in this connection to note that while F_p follows the form of Eq. (1), the values of the exponents n and m are, as in some other flux pinning studies, larger than can be accounted for by theory. Also, while numerous studies have been made of the dependence of F_p on B, B_{c2}, and microstructure, direct tests of the dependence of A on κ have not, to our knowledge, been reported. Nevertheless, interpretation of our results according to current understanding of flux pinning leads to the conclusion that microstructural factors affecting F_p change when stress is applied.

MICROSTRUCTURAL EFFECTS WHICH COULD ACCOUNT FOR RESULTS

In Nb_3Sn, grain or interphase boundaries, or second phase particles are considered to be the microstructural features responsible for flux pinning. Consideration of the ways in which applied stress might alter either N or a has led to a limited number of possibilities. A brief discussion of each follows.

A change in the number of microstructural features available to act as pinning centers could be caused by strain through a stress-induced transformation, reversible twinning, and the formation of microcracks, which, in early stages of development, may be of an appropriate size to pin flux. As strain is increased, and the microcracks grow, their effect on J_c would become deleterious, in accord with the rapid decline of J_c observed at large strains. If a stress induced transformation is involved, it would not necessarily be the tetragonal transformation known to occur in unstressed Nb_3Sn at about 42 K.

External application of stress may alter the stress distribution about a pinning site, if the crystalline defect which acts to pin flux also acts as a point of stress concentration, or if the applied stress results in other microstructural change such as a phase transformation or generation of defects such as twins or dislocations. Therefore, if the crystalline defect-flux line interaction through the elastic energy is an important component of the pinning, then the strength of a pin could be altered by external stress.

If two superconducting phases are present, and pinning is due to differences in superconducting properties (κ, H_{c2}) between the two phases, then the pin strength would be altered if, as one may expect, the superconducting properties of the two phases are affected differently by strain. Similarly, if pinning is due to anisotropy of H_{c2}, stress may alter pin strength by altering the degree of anisotropy.

It is possible that more than one type of pin is operative, or that the same defect pins by more than one mechanism. If so, applied stress could alter the relative strengths of pins or pin mechanisms, thereby changing not only the pin strength but the dependences of F_p on B_{c2} and B. Bartlett, et al.[17] put forth this interpretation of changes with strain which they observed in F_p vs b curves. Additionally, if a pinning threshold exists, so that pins weaker than some threshold value are ineffective, then the population of operative pins could be changed by strain if the strength of the pins were affected.[15,16]

SUMMARY

In both mono- and multifilament conductors, we have found that the maxima in J_c and B_{c2} as functions of strain do not coincide. This indicates that variation of the bulk property, B_{c2}, does not account for all of the variation of J_c with strain, and suggests that microstructural effects, associated with the pinning centers, are involved.

Comparison of the strain dependences of T_c and $(dB_{c2}/dT)_{T_c}$ indicates that the relationship between them is not unique, i.e., that the relationship changes when strain is increased beyond that required to produce maxima in T_c and $(dB_{c2}/dT)_{T_c}$. Since $(dB_{c2}/dT)_{T_c} \propto \gamma \rho_n$ and T_c is determined by the density of states, or γ, the relationship between them should be determined uniquely by crystal structure and resistivity.

The samples exhibited variability of behavior under stress, such as the loss of scaling of F_p with B_{c2f} at large strains in sample 1c (but not in other specimens), and variation from sample to sample, as well as with strain, of the exponents n and m. The loss of scaling might result from an abrupt change in bulk properties, and the sample to sample variability may be related to the time dependent strain which was observed to develop at constant load and temperature.

From measurements of J_c and B_{c2} as functions of strain and temperature, the forefactor A in the expression for F_p was found to depend strongly on strain. A is determined by pin strength, pin density, and κ. Measurements of $(dB_{c2}/dT)T_c$ and T_c as functions of strain indicate that variation of κ with strain does not account for the change in A. Therefore, a microstructural effect, associated with the pins, is implied.

Microstructural effects which could account for these the results include a stress-induced martensitic transformation, twinning, microcracks, and changes in the stress distribution about pinning sites. If pinning involves a second superconducting phase, change in the superconducting properties of the second phase with strain could alter the pin strength.

ACKNOWLEDGMENTS

The authors would like to thank J. P. Charlesworth of A.E.R.E. Laboratory, Harwell, England for providing multifilamentary specimens, J. O. Scarbrough for performing many of the measurements, and Connie Harrison for efficient preparation of the manuscript.

REFERENCES

1. T. Luhman, M. Suenaga, and C. J. Klamut, Adv. Cryo. Eng. 24: 325 (1978).
2. G. Rupp, IEEE Trans. Magnetics MAG-15:189 (1979).
3. J. W. Ekin, IEEE Trans. Magnetics MAG-15:197 (1979).
4. D. S. Easton and D. M. Kroeger, IEEE Trans. Magnetics MAG-15: 178 (1979).
5. T. Luhman, M. Suenaga, D. O. Welch, and K. Kaiho, IEEE Trans. Magnetics MAG-15:699 (1979).
6. D. S. Easton, D. M. Kroeger, W. Specking, and C. C. Koch, to be published in J. of Appl. Phys. (1980).
7. D. S. Easton and R. E. Schwall, Appl. Phys. Letters 29:319 (1976).
8. W. Specking, D. S. Easton, D. M. Kroeger, and P. Sanger, to be published in Adv. in Cryo. Engin. 26 (1980).
9. D. M. Kroeger, D. S. Easton, A. DasGupta, C. C. Koch, and J. O. Scarbrough, to be published in J. of Appl. Phys. (1980).
10. W. A. Fietz and W. W. Webb, Phys. Rev. 178:657 (1969).
11. E. J. Kramer, J. Appl. Phys. 44:1360 (1973).
12. R. G. Hampshire and M. T. Taylor, J. Phys. F 2:89 (1972).
13. A. M. Campbell and J. E. Evetts, Adv. Phys. 21:199 (1972).
14. D. Dew-Hughes, Philos. Mag. 30:293 (1974).
15. R. Labusch, Crys. Latt. Def. 1:1 (1969).
16. H. R. Kerchner, J. Narayan, D. K. Christen, and S. T. Sekula, Phys. Rev. Letters 44:1146 (1980).
17. R. J. Bartlett, R. D. Taylor, and J. D. Thompson, IEEE Trans. Magnetics MAG-15:193 (1979).

MECHANICAL PROPERTIES OF HIGH-CURRENT MULTIFILAMENTARY Nb_3Sn

CONDUCTORS*

R. M. Scanlan, R. W. Hoard, D. N. Cornish, and
J. P. Zbasnik

University of California
Lawrence Livermore National Laboratory
Livermore, CA 94550

INTRODUCTION

Nb_3Sn is a strain-sensitive superconductor which exhibits
large changes in properties for strains of less than 1 percent.
The critical current density at 12 T undergoes a reversible degradation of a factor of two for compressive strains of about 1 percent and undergoes an irreversible degradation for tensile strains
on the Nb_3Sn greater than 0.2 percent. Consequently, the successful application of Nb3Sn in large high-field magnets requires a
complete understanding of the mechanical properties of the conductor. One conductor which is being used for many applications[1-3]
consists of filaments of Nb_3Sn in a bronze matrix, and much progress has been made in understanding the mechanical behavior of this
composite.[4] The Nb_3Sn filaments are placed in compression due to
the differential thermal contraction between Nb_3Sn and bronze which
occurs when the composite is cooled from the Nb3Sn formation temperature (typically 700°C) to the 4.2 K operating temperature. The
general behavior of the critical current when this conductor is
subjected to a tensile stress is an increase to a maximum when the
compressive strain on the Nb_3Sn is relieved, followed by a decrease
as the Nb_3Sn filaments are placed in tension. The degree of precompression is controlled largely by the ratio of bronze to Nb3Sn
in the conductor.[5]

*Work performed under the auspices of the U.S. Department of
 Energy by the Lawrence Livermore National Laboratory under
 Contract W-7405-Eng-48.

Several authors[6,7] have presented analytical methods which can
be used to estimate the residual strain on the Nb_3Sn and hence to
predict the strain-critical current behavior of the conductor.
These methods have been shown to work well for a specific type of
conductor, i.e., Nb_3Sn filaments in a bronze matrix. However,
these methods are not adequate to explain the results when other
conductor geometries are used or when additional compressive strain
occurs due to other components in the conductor. In this paper, we
present results for a composite geometry in which the triaxial
strain state is important, and we present a computer code which can
predict the conductor behavior (as well as the behavior of the
bronze-matrix-type conductor). In addition, we present results for
several practical conductors in which precompression due to components other than the bronze matrix is important.

BRONZE CORE, Nb TUBE CONDUCTOR

After the discovery that Nb_3Sn could be fabricated by reacting
Nb with Sn provided by a bronze matrix, a number of different multifilamentary configurations were proposed. One configuration, which
we designate as the internal bronze approach, consists of Nb tubes
with bronze cores in a copper matrix. This configuration has several potential advantages over the external bronze approach, namely
the Nb tubes serve as diffusion barriers so that another element
such as Ta is not necessary to prevent Sn from contaminating the
Cu, and each superconductor element is surrounded by a high conductivity Cu matrix so that current transfer lengths are much
shorter than in the external bronze case.

The fabrication and testing of several conductors based on
this approach are described in Ref. 8. A computer program was
developed which evaluates the triaxial strains on the Nb_3Sn in this
conductor, and the details are presented in Refs. 9 and 10. The
computer code, MAXIMSUPER, treats a single, cylindrical repeating
element of the entire multifilament composite. This single superconducting filament or tube represents the average geometry and
configuration (niobium-core radius, Nb_3Sn-layer thickness, bronze-
to-niobium ratio, and the amount of copper matrix separating each
filament from its nearest neighbors). Since this average filament
is repeated in a multifilamentary composite superconductor for
thousands of elements, its resulting strain fields from thermal
expansion and axial loading should be fairly representative of the
overall composite behavior. Of course, this modeling is more accurate for conductors with uniform high filament densities across
the composite cross-sectional area. The code determines the three-
dimensional strain fields by solving Hooke's elasticity equations
for an element composed of the various materials (niobium, Nb_3Sn,
bronze, and copper) in their representative proportions. The code

then iterates the elastic solutions with known plasticity data,
such as the stress-strain curves of each material, to obtain an
approximate plastic solution. Since the calculated stresses and
strains are reported as three-dimensional axial ε_z, radial ε_r, and
azimuthal ε_θ components, we calculate a geometric average strain
designated as the effective strain, where ν is the Poisson's ratio,

$$<\varepsilon> = \frac{\sqrt{2}}{2(1+\nu)}\left[\left(\varepsilon_r - \varepsilon_\theta\right)^2 + \left(\varepsilon_\theta - \varepsilon_z\right)^2 + \left(\varepsilon_z - \varepsilon_r\right)^2\right]^{1/2} \qquad (1)$$

One unique distinction is demonstrated by the MAXIMSUPER
strain field predictions for the bronze-core filament geometry as
compared to the niobium core configuration. While the two conductor designs yield approximately the same z component strains ε_z,
the radial ε_r and azimuthal ε_θ strains are much higher for the
bronze-core filament design. To illustrate this effect, we use the
geometrical averaging of the three-dimensional strain fields to
calculate the effective strain in Eq. (1) and plot this value
against the intrinsic ε_z strain component for a typical bronze-core
and a typical niobium-core filament conductor in Fig. 1. Note that
the effective strains are much higher for the bronze-core sample,

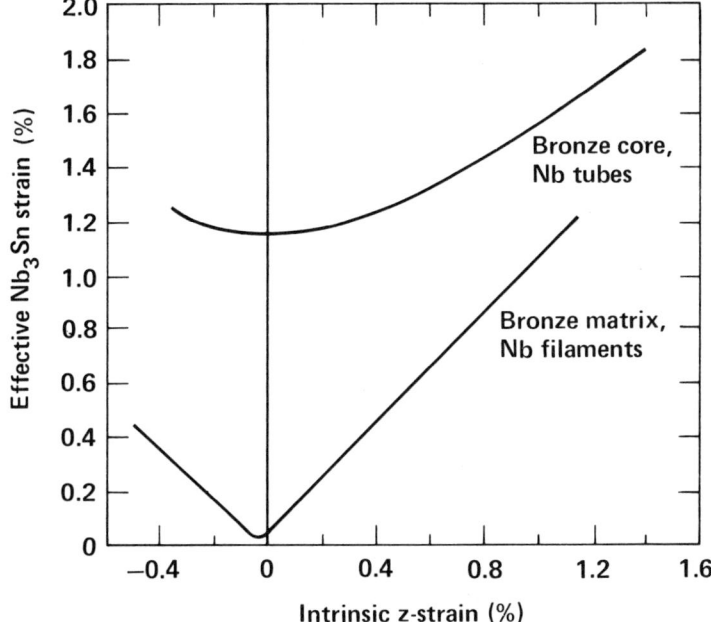

Fig. 1. The effective strain on the Nb_3Sn layers obtained from
         Eq. (1) is plotted vs the intrinsic z strain for the
         bronze core-Nb tube and the Nb filament, bronze matrix
         geometries.

because of the contributions from the larger tangential strain
components. Other studies have concluded that the peak critical
currents occur at approximately zero strain; however, for the
bronze-core geometry, the zero intrinsic ε_z strain values, which
give the peak critical current, coincide with nonnegligible tangential strain components on the Nb_3Sn zones. Hence, the maximum current displayed at zero intrinsic strain is the result of a minimum
in the effective average of the three-dimensional strain field, and
the minimum value is not necessarily zero. Analytical expressions
involving Hooke's elasticity equations in cylindrical geometry show
that the mechanical interactions and hence the stress-strain rela-
tions are not symmetric with respect to the interchange of the
material positions in a fiber composite. In fact, the effective
strain increases linearly (Fig. 1) with axial loading for the
niobium core and quadratically for the bronze-core geometry. This
prediction is experimentally verified in Fig. 2, which is a plot
of the measured critical current density of several conductors at
H = 12 Tesla and a resistivity ($\rho_r = 10^{-11}$ Ω-cm) as a function of
the MAXIMSUPER predicted residual effective strain (before tensile
loading). In general it shows that, for the same bronze-to-Nb_3Sn
ratio, we should expect the bronze-core geometry to be inferior to

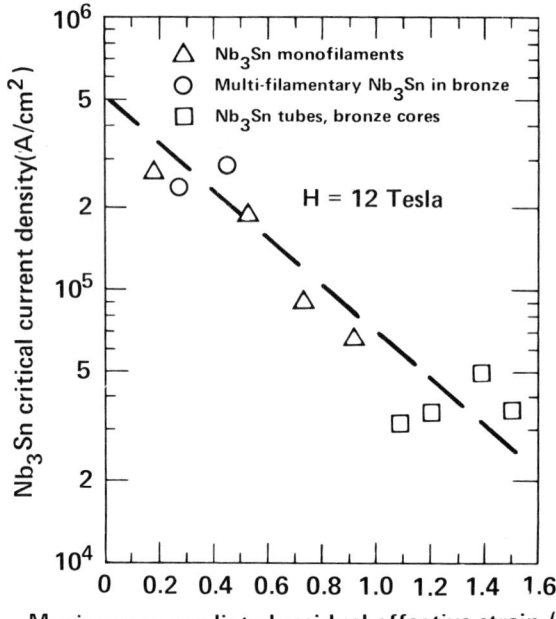

Fig. 2.   The measured critical current density as a function of
          residual strain predicted by the computer code is plotted
          for various conductor geometries.

the niobium-core design, with a decrease in critical current density at 12 T of factors from three to ten times.

Some corroborating data on this effect, obtained on these and other internal bronze samples by K. Aihara, T. Luhman, and M. Suenaga,[11,12] are shown in Table 1. For a comparable bronze-to-Nb ratio, the transition temperature depression in the internal bronze conductor is about 1.8 K, compared with about 0.2 K for an external bronze conductor.[12] Similarly, if we plot the transition temperature reduction for the internal bronze sample on a master plot[13] of T_c reduction versus strain, this plot would indicate a strain of greater than 1 percent in the Nb_3Sn layer.

HFTF Nb_3Sn SUPERCONDUCTOR

A cryostable Nb_3Sn conductor designed to operate at 5 kA and 12 T is being manufactured for the High Field Test Facility.[2] This conductor consists of a core which is clad with half-hard Cu after reaction. The core cross section (Fig. 3) consists of 58 percent Cu, 30 percent bronze, 10 percent Nb_3Sn and 2 percent Ta. The Cu in the core is annealed during the reaction step and contributes little to the precompression on the Nb_3Sn.[11] During the reaction, the bronze is depleted of Sn and is annealed. The resulting precompression of the Nb_3Sn can be predicted by graphical means and by computer modeling, when the properties of annealed, Sn-depleted bronze are used. In addition, the strain-critical current behavior can be modeled by introducing a strain dependence for the parameters in Kramer's scaling laws.[14] The experimental strain-critical current behavior of the HFTF core is shown in Fig. 4. We also present preliminary bending data for comparison with the tensile strain data. The initial increase in critical current seen in the tensile tests is not evident for the bending tests. However, the tensile data does provide adequate data with which to predict

Table 1. Transition Temperature (K) of Nb_3Sn Fabricated by the Internal Bronze Approach (Measured Inductively)

	T_c (Midpoint)	T_c (Onset)
As Fabricated	16.20	17.35
With Cu Matrix Removed	16.50	–
With Bronze Core Removed	17.97	18.10

2 mm

Fig. 3.   Cross section of the HFTF conductor core. The core
          consists of 18 strands in a Cu matrix; each strand
          contains 15,895 Nb_3Sn filaments.

general bending behavior for this conductor.

Subsequent to the reaction to form Nb_3Sn, additional Cu
(equivalent to approximately 50 percent of the total cross section) is soldered to the core. This Cu has a yield strength of
approximately 300 MPa (Fig. 5); hence, it will contribute to the
precompression on the Nb_3Sn when the conductor is cooled from the
cladding temperature to the 4.2 K operating temperature. The
calculated value of additional precompression on the Nb_3Sn due to
differential thermal contraction between 553 K and 4.2 K is 0.5
percent. This precompression occurs in two steps: during the
cooling from cladding temperature to 293 K and during cooling from
293 K to 4.2 K. The amount of compressive strain actually transmitted to the Nb_3Sn during the first step depends upon the experimental conditions, in particular, the extent that the solder
between core and cladding yields during the cladding operation.
Experiments are in progress to measure the additional precompression in the HFTF conductor due to the Cu cladding.

This precompression, due to the cladding, is useful from the
standpoint of providing more strain tolerance during coil winding
and coil operation. However, the Nb_3Sn critical current is reduced due to the compression; this must be taken into account in
designing the conductor.

Additional precompression can be expected for the JAERI design of the TMC conductor[15] and other designs employing cold-
worked copper for stabilization.

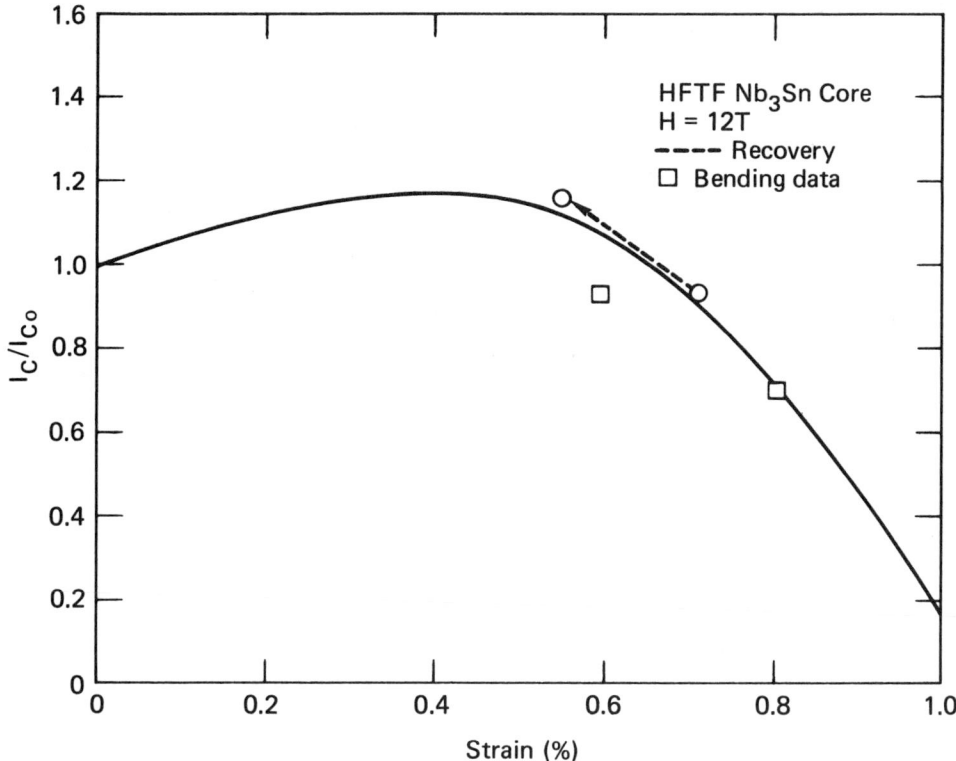

Fig. 4.   The critical current (normalized to the value for zero
applied strain) is plotted as a function of applied
tensile strain for the HFTF core. The change in critical
current with applied tensile strain is reversible up to
a strain of 0.7 percent.

LCP-TYPE CONDUCTOR

Another Nb3Sn-base conductor of practical interest is the
forced-flow conductor being utilized in the LCP.[1] Strands of
0.7-mm diameter multifilamentary Nb-bronze are insulated, cabled,
and wrapped in a stainless steel jacket prior to heat treatment to
form Nb3Sn. The strain-critical current behavior of the individual strands has been measured[16] and the results are adequately
predicted by the analytical method.[7] In order to verify the behavior of these strands in the final conductor, a series of samples were prepared[17] for testing in the LLNL tensile testing
facility. The samples consisted of 81-strand cables which were
compacted in type 304 stainless steel tubes, with approximately
33 percent of the cross sectional area available for helium

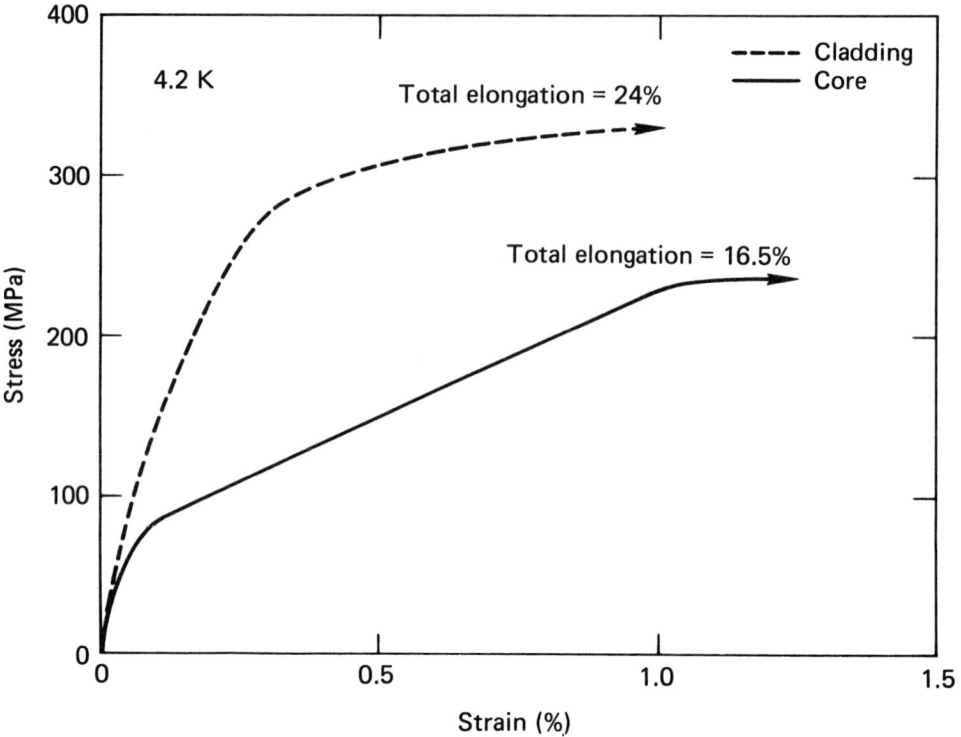

Fig. 5.   Experimental stress-strain data for the HFTF core and
HFTF cladding at 4.2 K.   The cladding is half-hard Cu.

coolant (Fig. 6).   The ends of the sample were compacted to form
current terminations, and pinned to the steel jacket, to insure
that the strain was imparted to both the cable and the jacket.
Strain gages and voltage taps were placed on the samples in the
5-cm uniform field zone of the test magnet.   The strain-critical
current behavior for these samples is illustrated in Fig. 7 and
compared with the single-strand data.   The initial critical
current values (for H = 8 T) are suppressed to 75 percent of the
single-strand values, and the peak in critical current is shifted
from ∿0.2 percent applied strain to ∿0.8 percent applied strain.
Although the strain-critical current behavior for the single
strands and the total conductor are quite different, the critical
current value at the peaks is the same for the two cases.   In
order to verify that this behavior is due to the stainless steel
jacket, identical samples with soft copper jackets were tested
(Fig. 7).   These critical current values scaled directly to the
values for the single strands.

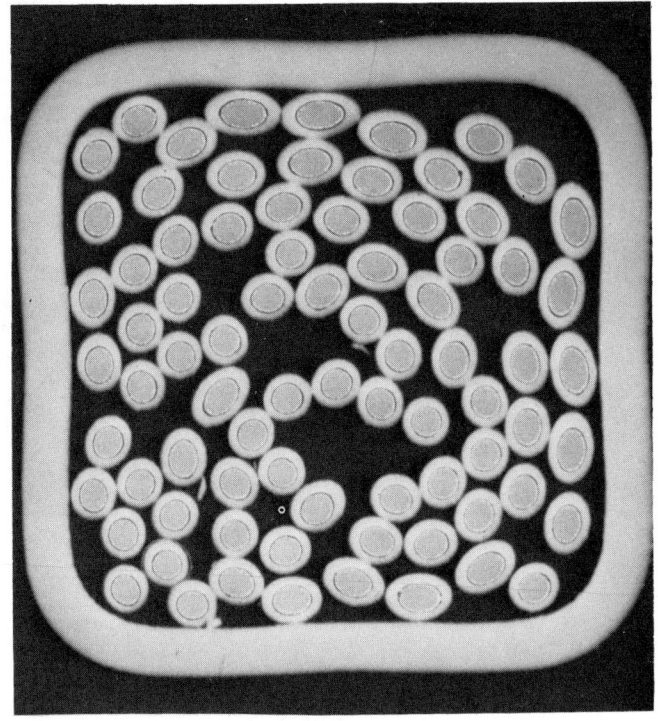

|___ 2 mm ___|

Fig. 6.   Cross section of the LCP-type conductor used in the
          strain-critical current experiments.

        The behavior of the jacketed conductor compared with that for
the single strands becomes evident when the thermal contraction of
the stainless steel jacket is considered. The stainless steel
jacket comprises about 47 percent of the conductor cross section
and has a mean thermal expansion coefficient α = 15.3 x 10^{-6}/°K.
The temperature range over which the differential thermal contraction of the stainless steel must be considered is from the reaction
temperature (973 K) to 4.2 K. When the properties of the stainless
steel are considered, two factors become important: (1) the differences in thermal expansion coefficients (stainless steel to
Nb_3Sn and bronze to Nb_3Sn) are similar; (2) the higher yield stress
of the stainless steel compared to that of bronze, especially at
higher temperatures, means that the stainless steel is much more
effective in applying a compressive strain to the Nb_3Sn. When the
data appropriate for stainless steel are used to calculate the precompression on the Nb_3Sn, a value of 0.8 percent precompressive

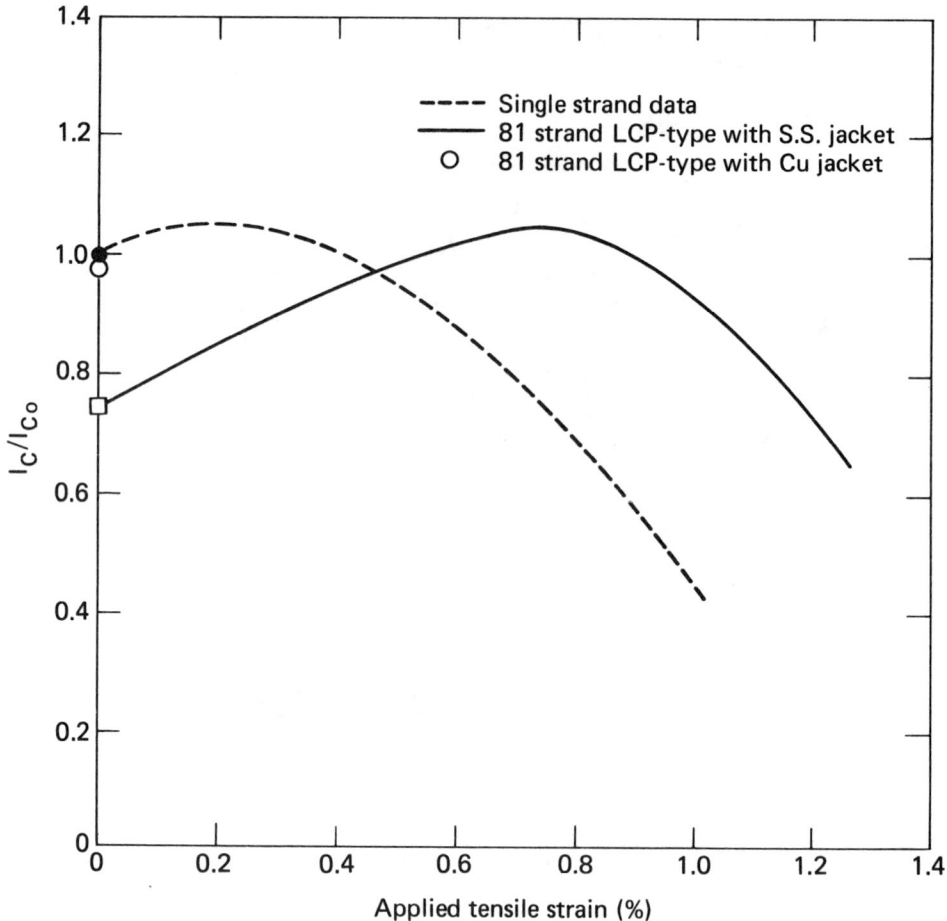

Fig. 7.   The critical current (normalized to the value for zero
           applied strain on a single strand) is plotted as a func-
           tion of applied tensile strain for a single strand and
           for 81-strand LCP-type conductors.

strain is obtained, in good agreement with the experimental results shown in Fig. 7.   This value of precompression is greater
than that necessary to protect the Nb_3Sn from damage during winding and operation.   Moreover, the decrease in critical current at
12 T, compared to the unstrained case, is a factor of two.   Consequently, this configuration presents a significant design problem to be overcome if the operation at 12 T is to be optimized.

CONCLUSIONS

The behavior expected for several configurations of Nb_3Sn conductors has been calculated and measured. The results show that the triaxial strain state of the Nb_3Sn is important for understanding the behavior of the Nb tube, bronze-core geometry.

For practical conductors in which stainless steel or work-hardened Cu are used to provide strength, the additional pre-compression on the Nb_3Sn due to these components must be taken into account.

ACKNOWLEDGMENTS

The preparation of the LCP-type samples by P. Sanger and co-workers at Airco Superconductors is gratefully acknowledged. The bronze-core, Nb-tube samples were manufactured by Supercon, Inc. Experimental assistance in sample measurements was provided by D. Hirzel, J. Johnston, and P. Waide at LLNL.

REFERENCES

1. P. A. Sanger, E. Adam, E. Gregory, W. Marancik, E. Mayer, G. Rothschild, and M. Young, IEEE Trans. on Magnetics MAG-15:789 (1979).
2. D. N. Cornish, H. L. Harrison, A. M. Jewell, R. L. Leber, A. R. Rosdahl, R. M. Scanlan, and J. P. Zbasnik, in Proc. 8th Symp. on Eng. Problems of Fusion Research, San Francisco, CA (1980).
3. M. S. Walker, J. M. Cutro, B. A. Zeitlin, G. M. Ozeranski, R. E. Schwall, C. E. Oberly, J. C. Ho, and J. A. Woollam, IEEE Trans. on Magnetics MAG-15:80 (1979).
4. Many articles on the effects of strain on critical properties of Nb_3Sn are found in the Proc. of the 1976 and 1978 Appl. Supercon. Conf. (IEEE Trans. on Magnetics MAG-13 and MAG-15) and in Adv. in Cryogenic Eng. 24 (1978).
5. T. Luhman, M. Suenaga, and C. J. Klamut, Adv. in Cryogenic Eng. 24:325 (1978).
6. G. Rupp, Cryogenics 18:663 (1978).
7. D. S. Easton, D. M. Kroeger, W. Specking, and C. C. Koch, to be published in J. Appl. Phys.
8. R. M. Scanlan, D. N. Cornish, J. P. Zbasnik, R. W. Hoard, J. Wong, and R. Randall, to be published in Adv. in Cryogenic Eng. 26.
9. R. W. Hoard, R. M. Scanlan, and D. G. Hirzel, to be published in Adv. in Cryogenic Eng. 26.

10. R. W. Hoard, Ph.D. Thesis, Univ. of Washington, to be
     published.
11. K. Aihara, M. Suenaga, and T. Luhman, in <u>Proc</u>. <u>8th</u> <u>Symp</u>. <u>on</u>
     <u>Eng</u>. <u>Problems</u> <u>of</u> <u>Fusion</u> <u>Research</u>, San Francisco, CA (1980).
12. M. Suenaga, private communication.
13. D. O. Welch, <u>Adv</u>. <u>in</u> <u>Cryogenic</u> <u>Engineering</u> 26, to be
     published.
14. E. Kramer, <u>J</u>. <u>Appl</u>. <u>Phys</u>. 44:1360 (1973).
15. S. Shimamoto, T. Ando, H. Tsuji, M. Nishi, E. Tada,
     K. Yoshida, and K. Yasukochi, in <u>Proc</u>. <u>8th</u> <u>Symp</u>. <u>on Eng</u>.
     <u>Problems</u> <u>of</u> <u>Fusion</u> <u>Research</u>, San Francisco, CA (1980).
16. W. Specking, D. S. Easton, D. M. Kroeger, and P. A. Sanger,
     <u>Adv</u>. <u>in</u> <u>Cryogenic</u> <u>Engineering</u> 26, to be published.
17. Samples were prepared by P. Sanger, AIRCO Superconductors.

STRESS EFFECTS ON W/Cu REINFORCED Nb$_3$Sn COMPOSITE CONDUCTORS

S. Murase, H. Shiraki, O. Horigami, M. Koizumi,
S. Mine, H. Takeda, and H. Baba

Toshiba Research and Development Center
4-1, Ukishima-cho, Kawasaki-ku, Kawasaki, 210 Japan

INTRODUCTION

Filamentary Nb$_3$Sn wire has become increasingly important as a high-field conductor because of its high upper critical field and critical temperature, in developing large scale magnets such as those used in a fusion reactor, energy storage and electrical machinery. Compound Nb$_3$Sn, however, has a disadvantage in that superconducting properties of Nb$_3$Sn are rapidly degraded by tensile and bend stresses[1,2,3] resulting from reeling and coiling the material and, by electromagnetic force generated when a coil is energized and by thermal contraction stresses due to cooling.

Some approaches to eliminate stresses applied to Nb$_3$Sn layers have been attempted, by keeping reinforcement materials, such as stainless steel and hard copper, in contact with Nb$_3$Sn wire by solder. Reinforcement material for Nb$_3$Sn conductor is expected to have both large Young's modulus and high conductivity, in order to protect Nb$_3$Sn wire from large stresses caused by coil construction and operation. Disadvantages for these reinforcement materials, such as stainless steel and hardened copper, are high resistivity, $\sim$5x10^{-7} $\Omega\cdot$m at 4.2 K, for stainless steel and low elastic modulus, 110 GN/m^2 at 4.2 K, for hardened copper. A W/Cu composite strip, consisting of unidirectional tungsten filaments, with larger Young's modulus (380 GN/m^2) than stainless steel (210 GN/m^2), and high conductivity copper matrix, is offered as a reinforcement material for Nb$_3$Sn conductor.

Few papers concerning stress effects on an assembled Nb$_3$Sn conductor, which consists of Nb$_3$Sn wire, reinforcement and

stabilizer, have been published. The present paper describes mechanical properties W/Cu composite strips processed by a plating
and rolling method and stress effects on W/Cu reinforced Nb₃Sn
conductors (3.4 to 8.3% of W volume fraction of the overall conductor) and Cu strip and stainless steel strip reinforced conductors in place of W/Cu for comparison. Thermal contraction of these
assembled Nb₃Sn conductors will be briefly discussed.

EXPERIMENTAL

Sample Preparation

One mm x two mm wire size 258-filament Nb₃Sn wire, which has
tubular niobium filaments filled with copper-sheathed tin inside,
all embedded in a high-conductivity copper matrix outside has been
used.[4,5] The Nb₃Sn layers were formed on the inner periphery of
niobium tubes after heat treatment at 700°C for 72 h. A W/Cu composite strip was processed by the following plating and rolling
technique. Tungsten filaments 0.1-0.2 mm in diameter were unidirectionally placed on a copper strip. They were submerged in an
electroplating solution, and the filaments and strip were consolidated by copper electroplating. Then, the composite strip was
rolled to 0.7 mm thickness and 4 mm width. The W/Cu strip, 258-
filament Nb₃Sn wire and a copper housing were assembled, using Pb-
Sn solder in a rectangular cross-section (2.5 mm x 4 mm) as a W/Cu
reinforced Nb₃Sn conductor. The composite conductors have 3.4
(Sample No. 1) and 8.3 (Sample No. 2) vol. % W of the overall conductor, as shown in Table 1 and Fig. 1. Copper strips and stainless steel strips were also prepared in place of W/Cu for reinforcement of the conductor (Sample Nos. 3 and 4, respectively).
Nb₃Sn wire without any reinforcement materials (Sample No. 5) was
used for comparison, as shown in Table 1. All the copper materials in the conductor were fully annealed, beforehand, at 600°C
for 30 minutes, to eliminate annealing effects when they were assembled by soldering.

EXPERIMENTAL PROCEDURE

Stress-strain measurements of W/Cu composite strips were performed at room temperature and 4.2 K, using a screw-driven tensile
machine with 1 ton load capacity in 250 mm gauge length and at
1×10^{-3}/minute strain rate. Stress-strain curves at 4.2 K and
change in critical current as a function of tensile stress on the
conductors were also measured, using the same tensile machine under
an applied transverse magnetic field of 4 T. Critical current was
determined at 3 μV/cm.

Table 1.   Specification of Samples

Sample No.	Reinforcement	Volume Fraction of Materials (%)						
		Cu	Sn-Cu	Pb-Sn	Nb	Nb$_3$Sn	W	Steel
I	W/Cu	87.5	2.8	1.9	2.5	1.9	3.4	—
2	W/Cu	80.2	2.9	4.0	2.6	2.0	8.3	—
3	Cu	90.3	3.0	2.0	2.7	2.0	—	—
4	Stainless Steel	66.0	2.9	2.0	2.6	2.0	—	24.5
5	—	61.5	15.0	—	13.5	10.0	—	—

RESULTS AND DISCUSSION

Ultimate tensile strengths and Young's moduli obtained from
stress-strain curves are plotted in Fig. 2 as a function of tungsten volume fraction of the W/Cu composite strips.   While the rule
of mixtures has been verified experimentally in Young's modulus of
the composite, some deviation from it has been observed in tensile
strength at higher W volume fraction.   It is considered that stress
applied to the composite to measure the tensile strength, when
higher than that for Young's modulus at larger W volume fraction,
causes a slip between copper matrix and tungsten filaments at the
W/Cu interface.

Stress-strain curves for various specimens measured at 4.2 K
are shown in Fig. 3.   Young's moduli of the specimens increase with
increasing volume fraction of the reinforcement material, which oc-
cupies the assembled conductor specimen and controls its elastic
constant.   Young's modulus of the Nb$_3$Sn wire (No. 5), which contains high volume fraction of Nb and Nb$_3$Sn with high elastic constant, is higher than that of Sample No. 2, W/Cu reinforced Nb$_3$Sn
conductor with lower W volume fraction.

Critical currents (I_c) normalized to unity at their values
under zero applied tensile stress are plotted in Fig. 4 as a function of the applied tensile stress for various specimens at 4 T.
The critical current shown in Fig. 4 somewhat increases prior to a
decrease as stress increases.   Critical stresses (σ_c), the stress
value where I_c begins to decrease as stress increases, are
170 MN/m^2, 230 MN/m^2, 125 MN/m^2, 330 MN/m^2 and 100 MN/m^2 for Sample Nos. 1, 2, 3, 4, and 5, respectively.   Volume fraction effect
of reinforcements for the overall conductors on critical stress is
shown in Fig. 5.   If a W/Cu reinforced Nb$_3$Sn conductor, with the
same 24.5 vol. % W as stainless steel volume fraction of No. 4, is
fabricated, it is expected from extrapolation of results in Fig. 5
that the critical stress will exhibit 480 MN/m^2, being 1.5 times as

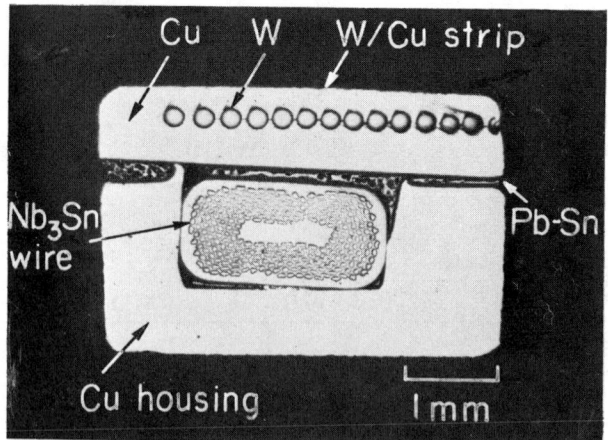

Fig. 1.   Cross-sectional view of W/Cu reinforced Nb₃Sn conductor.

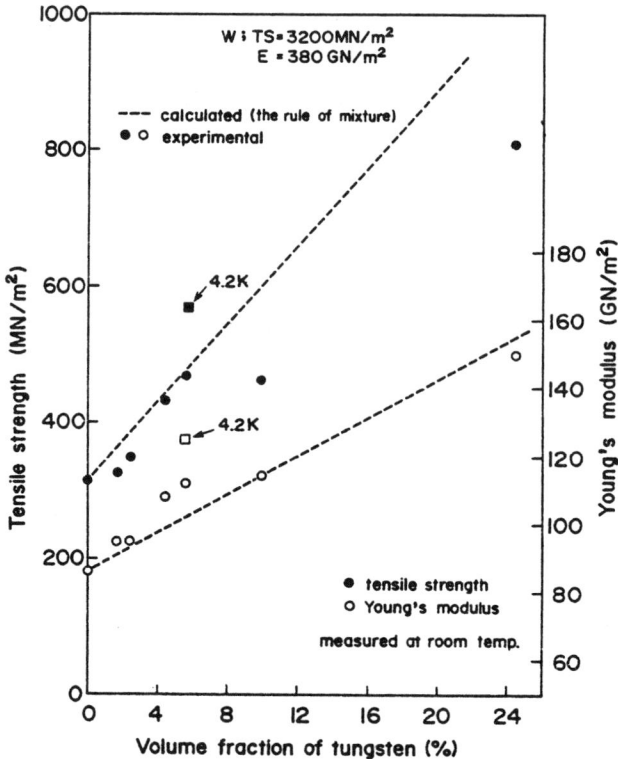

Fig. 2.   Tensile strength and Young's modulus as a function of
          tungsten volume fraction.

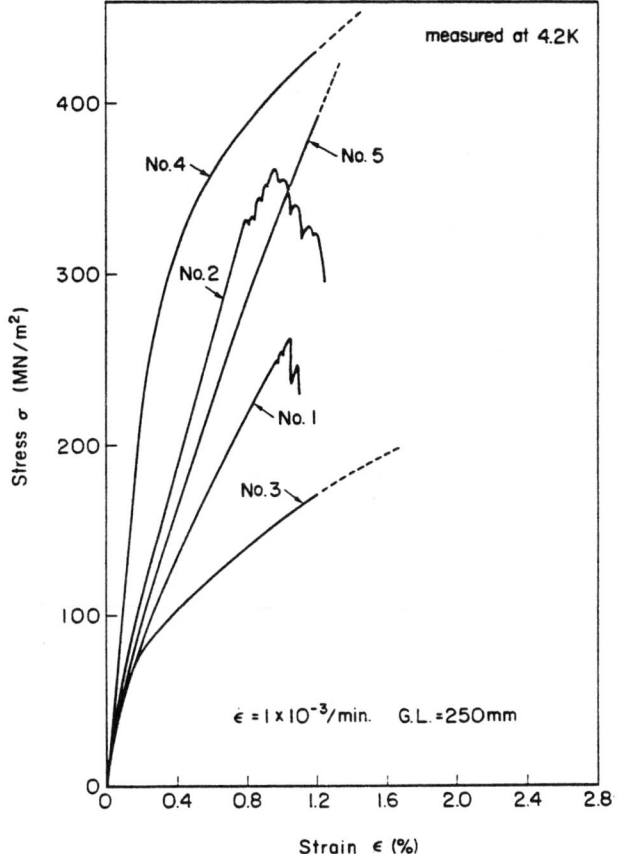

Fig. 3.   Stress-strain curves for various specimens on the uniaxial
          tensile stress at 4.2 K.

large as that of No. 4.   In this case, electrical resistivity at
4.2 K for the normal materials is assumed as being three times as
low as that of No. 4.   Figure 6 shows critical current normalized
to unity at its value under zero applied tensile strain as a function of tensile strain at 4 T.   Critical strains (ε_c), where critical current begins to degrade for various samples obtained from
Fig. 6, are 0.52%, 0.56%, 0.68%, 0.5%, and 0.23% for Sample Nos.
1, 2, 3, 4, and 5, respectively.

     I_c normalized to unity at its maximum value as a function of
intrinsic strain (ε_o), where the applied strain minus the value
when I_c is a maximum (I_{cm}), are replotted in Fig. 7.   The curves

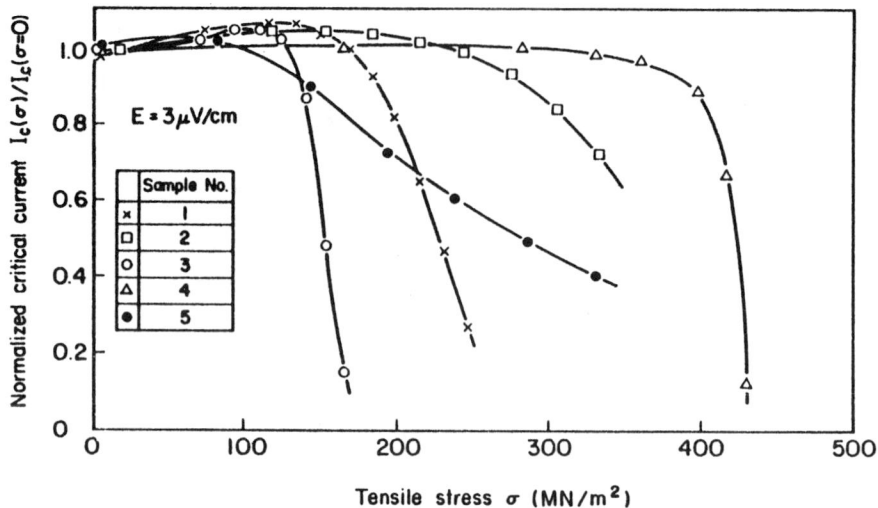

Fig. 4.   Normalized critical current as a function of applied
          tensile stress at 4 T.

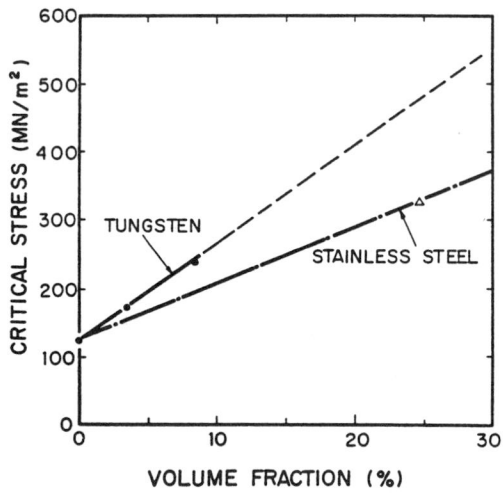

Fig. 5.   Critical stresses as a function of volume fraction of
          reinforcement materials for the overall conductors.

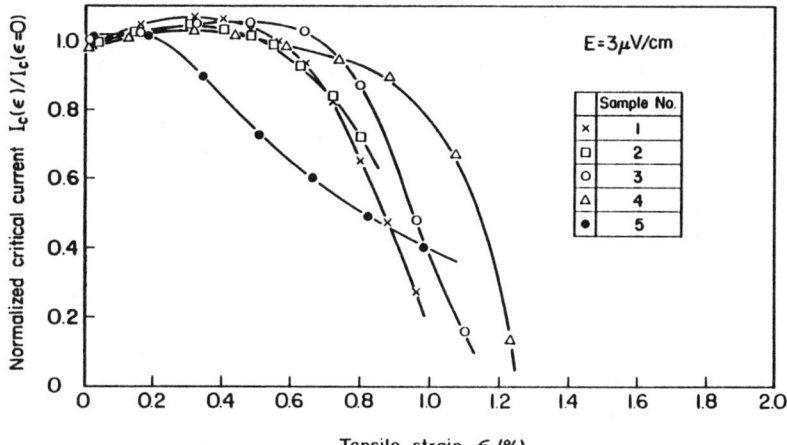

Fig. 6.   Normalized critical current as a function of applied
          tensile strain at 4 T.

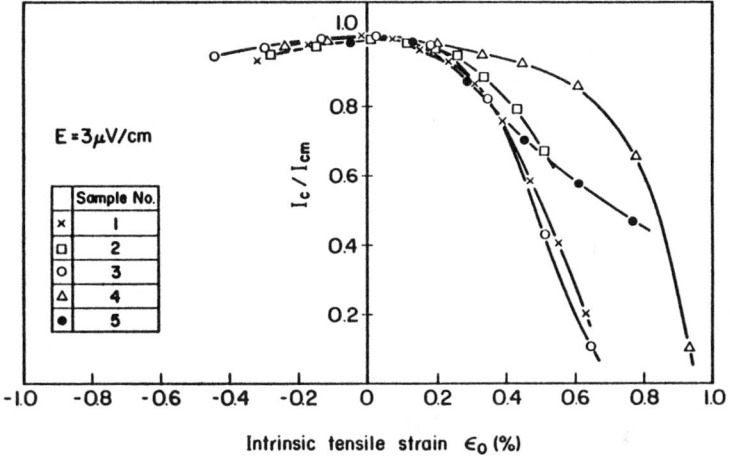

Fig. 7. I_c/I_{cm} as a function of intrinsic strain.

show asymmetry around the assumed strain-free axis ($\varepsilon_o=0$). In the tensile region ($\varepsilon_o \geq 0$), critical currents for all the samples decrease more rapidly than in the compressive region ($\varepsilon_o \leq 0$). Assuming that intrinsic compressive strains are thermally induced, they can be roughly calculated under axial loading using Young's moduli and thermal expansions of materials, such as Nb, Nb_3Sn, SnCu, Cu and stainless steel. Calculated intrinsic thermal compressive strains are 0.4% for Nos. 1, 2, 4, and 5, and 0.5% for No. 3. They approximately agree with the observed values, except 0.05% of Sample No. 5 (Nb_3Sn wire). Hoard et al.[6] showed that greater tangential strain in addition to axial strain had an effect on I_c for the bronze core geometry composite. In this case, tangential strain due to thermal contraction applied to the Nb_3Sn wire (Sample No. 5), having the bronze core geometry, can be calculated. For other samples, Nos. 1 to 4, this tangential strain effect is assumed to be small, because they are surrounded by a large amount of reinforcement materials, similarly to niobium core geometry.

SUMMARY

        A W/Cu composite strip, consisting of unidirectional tungsten filaments, having larger Young's modulus than stainless steel, and high conductivity copper matrix was developed by a plating and rolling method. The W/Cu strip, a Nb_3Sn wire and a copper housing were assembled with Pb-Sn solder into a rectangular cross-section with tungsten filaments 3.4 to 8.3 vol. % of the overall conductor. Copper and stainless steel strips were also prepared in place of W/Cu for conductor reinforcement for comparison. Stress-strain curves at 4.2 K and a change in critical current as a function of tensile stress and strain were measured. Increase in the volume fraction of tungsten in W/Cu reinforced conductor enhances critical stress. The critical stress is larger than that of stainless steel reinforced conductor for the same volume fraction.

REFERENCES

1. I. L. McDougall, IEEE Trans. Magn. MAG-11:1467 (1975).
2. D. C. Larbalestier, J. E. Magraw, and M. N. Wilson, IEEE Trans. Magn. MAG-13:462 (1977).
3. D. L. Martin, M. R. Daniel, J. M. Cutro, and R. E. Schwall, IEEE Trans. Magn. MAG-15:185 (1979).
4. Y. Koike, H. Shiraki, S. Murase, E. Suzuki, and M. Ichihara, Appl. Phys. Lett. 29:384 (1976).
5. S. Murase, M. Koizumi, O. Horigami, H. Shiraki, Y. Koike, E. Suzuki, M. Ichihara, F. Nakane, and N. Aoki, IEEE Trans. Magn. MAG-15:83 (1979).
6. R. W. Hoard, R. M. Scanlan, and D. G. Hirzel, ICMC, Madison, Wisconsin (1979).

IN SITU AND POWDER METALLURGY MULTIFILAMENTARY SUPERCONDUCTORS:

FABRICATION AND PROPERTIES*

Raymond Roberge   Simon Foner

IREQ, Institut de   Francis Bitter National
Recherche d'Hydro-   Magnet Laboratory**
Québec   and Plasma Fusion Center
Varennes, Québec   MIT
JOL 2PO Canada   Cambridge, MA 02139

INTRODUCTION

In recent years there has been an increased interest in alternate technologies for producing multifilamentary composite superconducting materials for large scale applications. We will discuss our development of both the in situ and the powder metallurgy processes which have involved a collaboration with a number of sicentists from several countries. Both processes show promise. Some highlights will be presented here; further details can be found in the references. Examples of Cu-Nb-Sn and Cu-V-Ga multifilamentary materials prepared by the in situ process and Cu-Nb-Sn using the powder metallurgy technique, are discussed here.

IN SITU

Principle

The origin of the in situ technique dates back to the early observations of superconductivity in copper containing trace amounts of niobium by Romanov et al.[1] in 1965 and by Borcherts and Silver[2] in 1968. However it was the work of Tsuei[3,4] in 1973 which stimulated efforts by several research groups, each with their own

*Supported in part by DOE and NSF.
**Supported by NSF.

particular variations.* The distinguishing feature of in situ process is that it relies on solidification of a liquid alloy to produce a superconducting element in a copper matrix.

In order to understand the technique one must consider the
phase diagram of the system under investigation. The equilibrium
Cu-Nb phase diagram, for example,[5] shows complete solubility of the
elements in the liquid state, negligible solubility of niobium in
copper at ambient temperature, and no intermediate phases. Thus
when the elements copper and niobium are melted, homogenized
(T > 1800°C) and cooled, niobium solidifies first, followed by
solidification of the nearly pure copper matrix. The structure of
the niobium phase (shape, size, ...) is controlled by the concentration, cooling rate, temperature gradients, residual elements,
environment, etc. Figure 1 shows some typical microstructures all
melted in graphite crucibles. The formation of fine homogeneous
dendrites is illustrated. The influence of the cooling rate and of
the niobium concentration on the degree of complexity of the dendrites are also visible.

Fabrication Technique

The laboratory chill casting technique[6] consists of induction
melting of ∿30 to 50 g niobium and copper in a graphite crucible.
The liquid alloy is then cast into a water cooled copper crucible
by withdrawing the graphite rod sealing the bottom of the crucible.
The solidified ingot is mechanically deformed by swaging rolling or
wire drawing to final size. The wire is electroplated with tin,
diffused and reacted to produce fine multifilamentary Nb_3Sn fibers.

Superconducting and Mechanical Properties

Overall J_c of Cu-Nb. The overall critical current density J_c
was measured by a standard four-probe technique in transverse magnetic fields. The method used to obtain the critical current was
to fix the current and to sweep the applied magnetic field until
a voltage of 1 μV (voltage leads ∿ 1 cm apart) was detected. J_c
was calculated using the total cross-sectional area of the wire.**

*The groups at Harvard University and at Iowa State University will
present results at this meeting.
**When comparing overall J_c reported in various publications care
must be taken not to attach too much importance to the absolute
values quoted because the simple definition of overall area can
vary by 30 to 60%, i.e., is the external tin deposit taken into
consideration?, is the ∿34% expansion due to the transformation of
Nb to Nb_3Sn included?

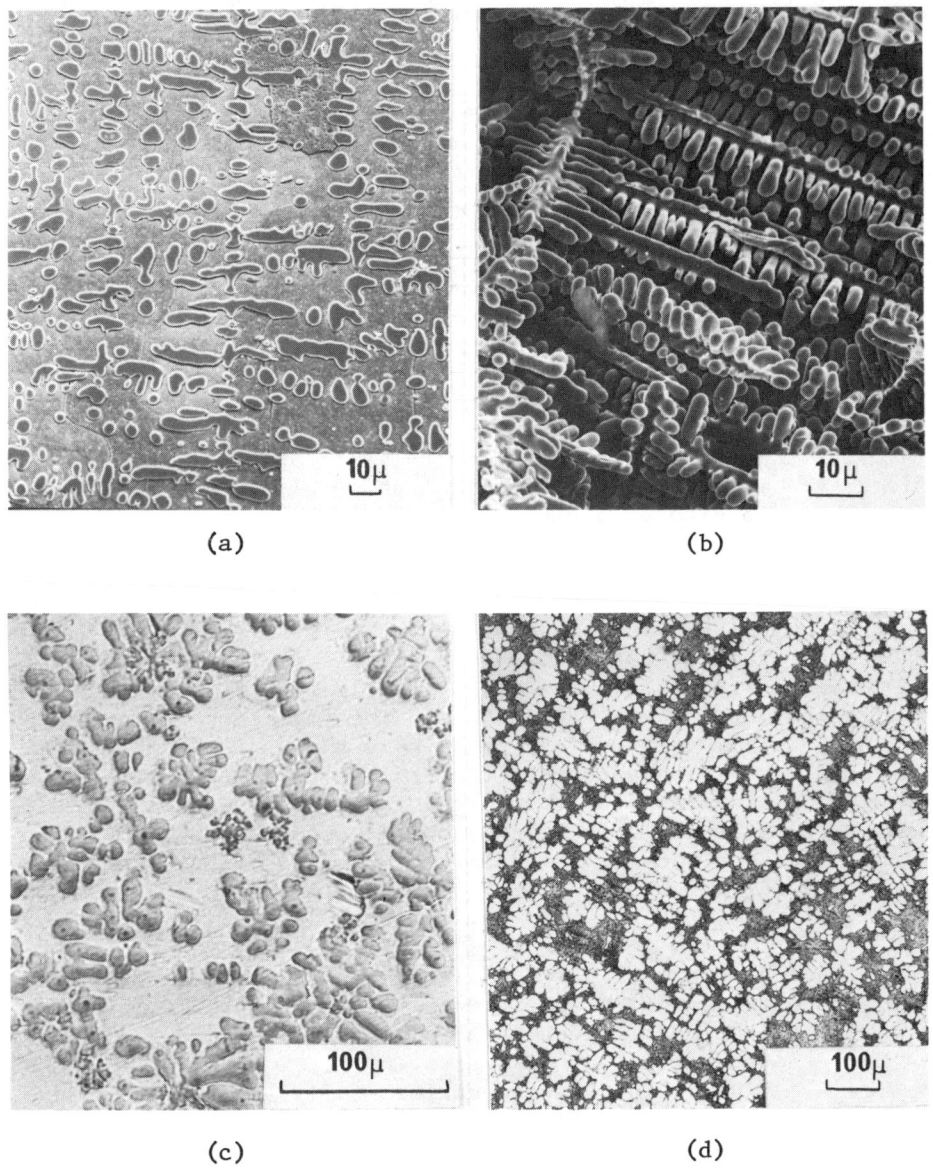

Fig. 1   Examples of as-solidified Cu-Nb microstructures;
         a) formation of fine homogeneous dendrites, Cu 30wt% Nb;
         b) similar specimen following deep etching; c) same
         concentration but lower cooling rate, note lower magnifi-
         cation; d) similar microstructure for a higher
         Cu 35wt% Nb. All alloys were melted in graphite
         crucibles.

For a fixed niobium concentration in copper the overall critical current density J_c is a strong function of areal reduction ratio
R. J_c increasing rapidly with R initially before saturating at sufficiently large values.[7] This is illustrated in Fig. 2. The value
of R for saturation depends on the initial structure of the in situ
material; large complex dendrites requiring large R values before
saturation is achieved. The main reason for saturation of J_c versus
R is that, with sufficient elongation, an homogeneous composite of
long fibers of relatively uniform cross-section is produced.

Overall J_c of Cu-Nb-Sn. The measured J_c of our earlier composites are shown together with commercial multifilamentary materials
(containing relatively thick filaments) in Fig. 3. The results for
Cu 36 wt% Nb with two different tin platings, 14.7 and 20 wt%, are
reproduced.[8] The values for Airco[9] composites were recalculated
using our definition of total area. The J_c for the MCA material[10]
is similar to Airco. The higher overall J_c for IGC[11] composite
reflects the lower bronze to niobium ratio. The low value of the
Supercon[12] composite probably may result from the large bronze to
niobium ratio.

Overall J_c of Cu-Nb-Sn as a function of dendrite size. The
initial niobium dendrite size can be varied by a controlled high
temperature gradient technique.[13] For the same values of R, one
then obtains different values of J_c; the largest J_c being measured
for wires having the smallest fibers. The variation of J_c with
fiber size is illustrated in Fig. 4 where the J_c for the smallest
fiber size materials approached that of the chill cast materials.
The lower values of J_c for the larger fiber sizes are mainly attributed to the relatively low reduction ratio ($R \simeq 500$). This
was confirmed for the larger dendrites by a further reduction in
area ($R \simeq 2000$) which showed a significant increase in J_c (see
Fig. 4).

Mechanical properties. Compared with conventional multifilamentary materials, the in situ composites show improved mechanical
properties and support larger uniaxial strain before degradation in
J_c. The general feature of the variation of J_c with applied stress
(or strain), Fig. 5, is that the maximum in J_c appears at large
strains for the in situ filaments.[14] Furthermore, the value of
the strain at the maximum in J_c is relatively independent of
field.[15] The results for the Cu-V-Ga[16] are included to illustrate
that the sensitivity to strain is different for different A15 materials. In this case, J_c is very nearly strain independent with no
degradation in J_c up to at least 750 MPa ($\varepsilon > 1.4\%$). Analysis of
the degree and influence of the fiber-matrix interactions is in
progress and will be reported at a later date. Figure 6 illustrates that the values of strain at the peak of 0.6 to 0.7% are
measured for the smallest interfilament spacings.[17]

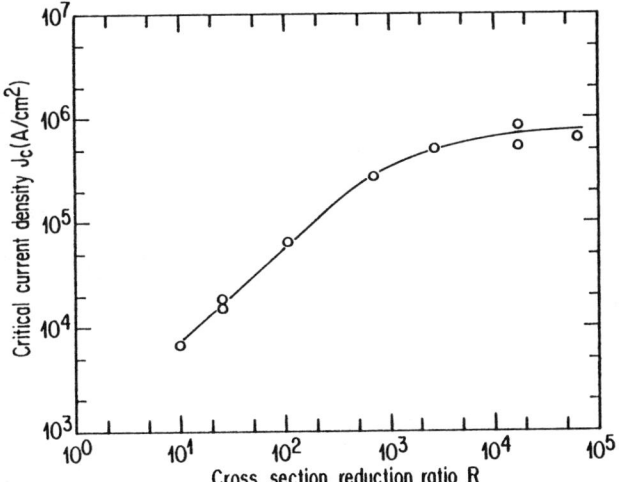

Fig. 2   Overall critical current density J_c as a function of areal reduction ratio, R at zero applied field.   Cu 17wt% Nb.

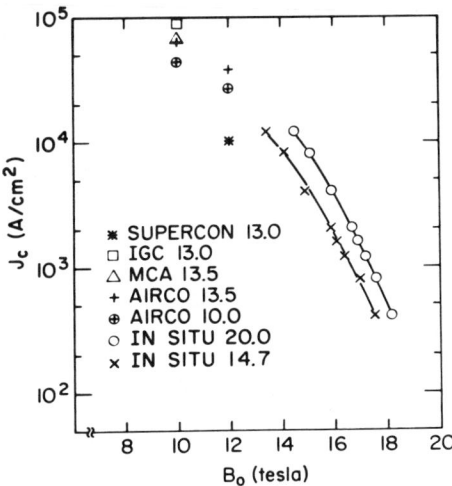

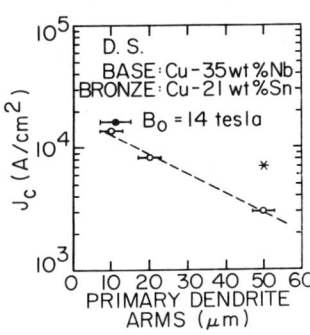

Fig. 3 Overall critical current
density J_c versus applied field
Bo for two In Situ Cu 36wt% Nb-
xwt% Sn wires compared to published results for bronze technique composites. (After Ref. 8)

Fig. 4 Overall critical current
density J_c at 14T as a function
of the primary dendrite arm size
in the alloy (open circles).
All had the same composition,
heat treatment of 1 day at $650^{\circ}C$,
and reduction ratio of 500.   The
closed circle is for a chillcast
specimen.   The added asterisk
is the measured J_c after a reduction ratio of 2000.   (After
Ref. 13)

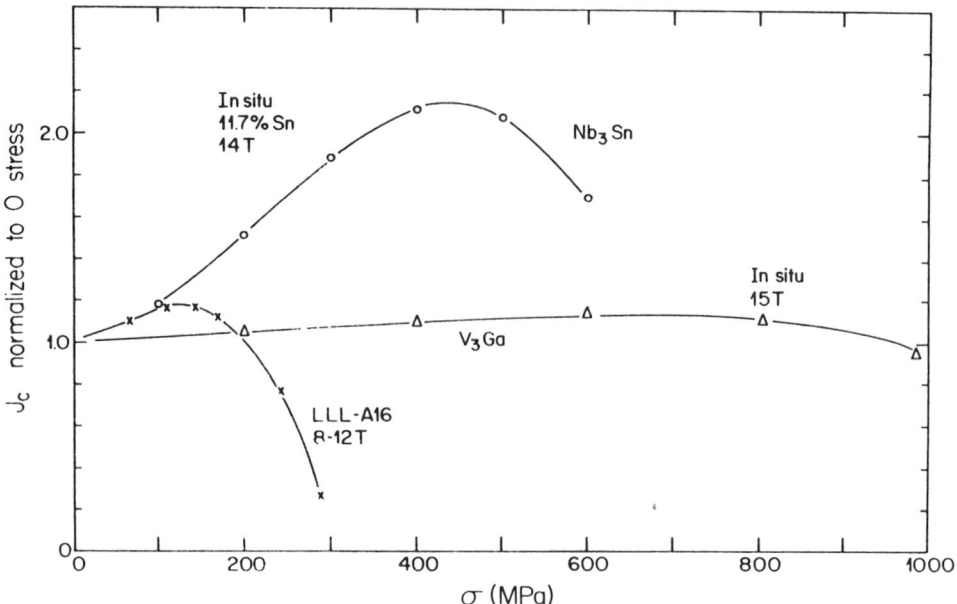

Fig. 5 J_c normalized to 0 applied stress as a function of stress
σ at indicated fields for Cu–Nb–Sn, Cu–V–Ga In Situ composites, and a commercial composite.

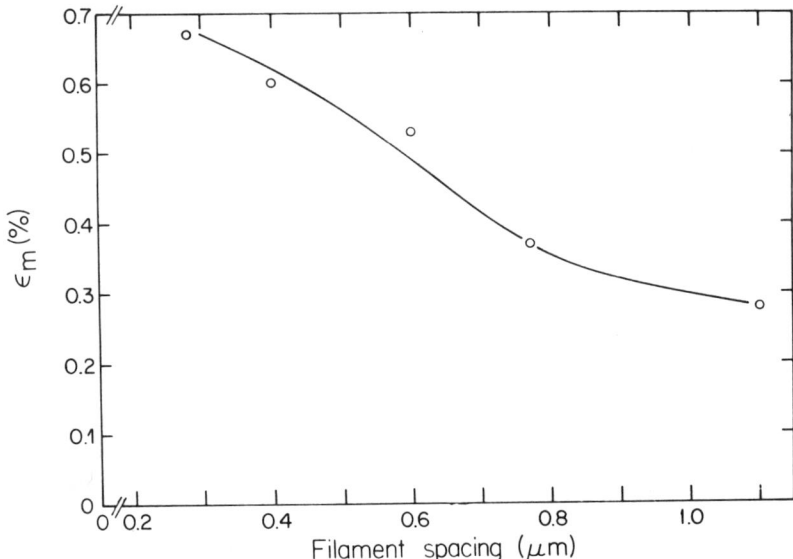

Fig. 6   Variation of ϵ_m, the strain at the peak in the Jc versus
strain curves, as a function of the submicron spacing
between the filaments. (After Ref. 17).

Overall J_c of Cu-Nb-Sn as a function of Nb concentration. The volume fraction of niobium can be varied from 0 to 100%, but for useful Nb_3Sn multifilamentary composites we have limited ourselves to ≤ 40 wt% Nb. Some of our results are given in Fig. 7. For clarity only the values at 12 T are reported. The overall J_c is plotted as a function of niobium concentration in wt%. The top scale indicates the calculated tin concentration (wt%) in copper necessary to transform all the Nb into Nb_3Sn. This concentration increases with increasing niobium concentration and it exceeds the solubility limit of tin in copper at 650°C (15 wt%) if the niobium concentration is greater than 30 wt%. The highest (recalculated) literature value of J_c for a conventional composite produced by Larbalestier,[18] is indicated by the cross. The dotted curve was drawn to show the estimated J_c assuming a linear relationship of J_c with concentration. The dashed curves in Fig. 7 indicate the range of J_c observed for different in situ materials. For in situ composites the overall J_c is a strong function of concentration in the range 5 to 25% and in the lower concentration range, J_c is particularly sensitive to the geometry of the niobium precipitate. From 25 to 40% the J_c values are highly reproducible and less sensitive to the precipitate geometry.

In order to obtain a more accurate comparison between the J_c of conventional and in situ materials, the prestress should be removed. The result for the 36 wt% Nb material at J_c max (with no prestress) is shown by the asterisk. The ultrafine fibers also allow the nearly complete reaction of Nb to Nb_3Sn which should yield a maximum J_c, whereas the conventional large fiber materials are generally not completely reacted. On the other hand J_c of the in situ material is reduced if coupling between the discontinuous fibers by either contact or proximity is not total. The high values of J_c realized by in situ type materials indicate that reduced coupling is not dominant at high Nb concentration.

The properties of the powder metallurgy process multifilamentary superconducting materials parallel those of the in situ materials and the above discussions apply as well for this process. Results for this processing will be discussed below.

Comments on Connectivity

A detailed discussion on the coupling mechanisms in in situ composites will be presented in this conference by Bevk and Tinkham.[19] However, a few interesting experimental results are presented here for consideration. Figure 8 is a summary of our results for numerous Cu-Nb specimens in the undeformed (as-cast) state, and after a reduction to final size. In the undeformed condition a sharp step in the overall critical current density is observed at the niobium concentration of ~15% (the percolation limit). Following mechanical deformation of the Cu-Nb the measured

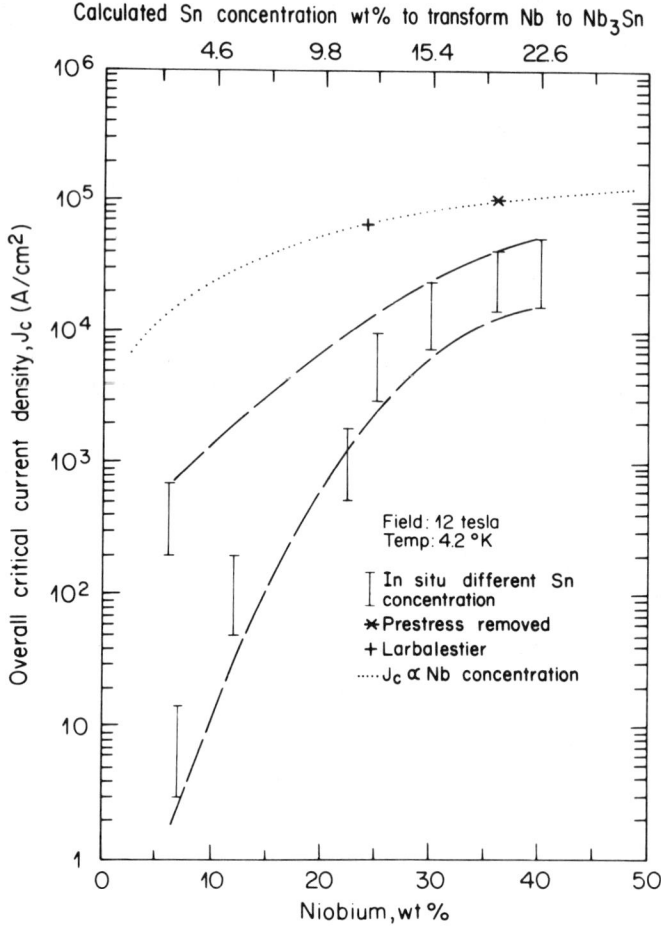

Fig. 7   Overall J_c as a function of niobium concentration in wt%
at 12T.   The top scale indicates the calculated wt% tin in copper
necessary to transform all the niobium into Nb_3Sn.   The cross
indicates the recalculated value for a conventional composite by
Larbalestier (18).   The dotted line is drawn to represent a
linear relationship between J_c and Nb concentration.

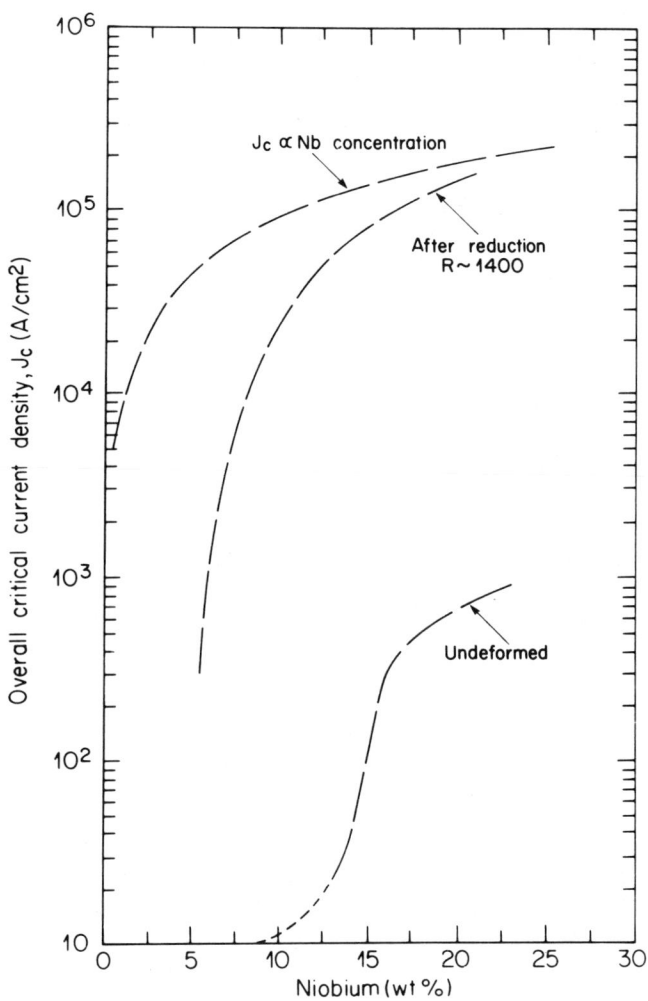

Fig. 8 Overall J_c for Cu-Nb alloys as a function of the niobium concentration in the undeformed casting and after a reduction in area of ~ 1400.

J_c is better than two orders of magnitude higher and the "critical" concentration is shifted to a lower niobium concentration. There is evidence of connectivity.

Another experimental indication of the metallurgical and electrical connectivity is shown by measuring the resistivity of in situ Cu-Nb specimen as a function of etching time. A Cu 36 wt% Nb alloy wire deformed about to the point of saturation in J_c was bent to a U shape and immersed in dilute nitric acid. The length of the wire in the acid was ∿5 cm and the etching time 4 days. The voltage was monitored as a function of time at a fixed current until all the copper was removed. Assuming a continuous niobium wire, the calculated niobium concentration agrees with the initial niobium concentration within 20%. After a reduction in area of only 170 the expected length of the initial discontinuous niobium "dendrites" would be much smaller than 5 cm.

POWDER METALLURGY PROCESSING

General Features

At MIT the powder metallurgy process for producing superconducting composites was pioneered by Flükiger. The powder metallurgy process starts by mixing the constituent elements in powder form of a selected size. The composition is determined by the starting mixture, which can be varied at will. The size of the powder particles can be chosen to match processing parameters. Many operations in the powder process are common to the conventional and in situ multifilamentary processing techniques. The important and distinguishing features involve the initial stages of fabrication.

Figure 9 illustrates the variations of the powder technique which have been experimentally evaluated to produce Cu-Nb-Sn. Of course, many others are possible. All variations begin with the selection of the powders, sieving to the desired dimensions ≃ 40 μm in most cases and finally mechanical mixing to obtain a uniform distribution. The powders are then compacted into a container (sheath) at high pressure to 80-90% of the maximum theoretical density prior to extrusion or swaging. Generally the powders are handled at a temperature low enough so that the constituents do not react until the diffusion processing. Two different routes have been investigated depending on the stage at which tin is introduced.

External tin at final size. For the external tin process where the container is a barrier,[20] e.g., monel it must be removed. If, however, Cu, Cu-Be or Cu-Zr are used as the sheath its removal is optional because these materials are "transparent"

	SELECTION, SIZING, MIXING			
COMPACTION	Cu + Nb		Cu + Nb + Sn	Cu-Nb-X-y
SHEATHING	Monel	Cu Cu - Be Cu - Zr	Cu - Be Cu - Zr	
EXTRUSION	25.4mm to 12.7 mm		Diffusion Barrier Ta or Nb	
SWAGING	"	"	"	
DRAWING	"	"	"	
REMOVAL OF SHEATHING	"	Optional	Optional	
PLATING DIFFUSION	"	"		
REACTION HEAT TREATMENT	"	"	"	

Fig. 9   Flow diagram of the laboratory scale powder metallurgy processing. The variations are indicated by the columns proceeding from top to bottom.

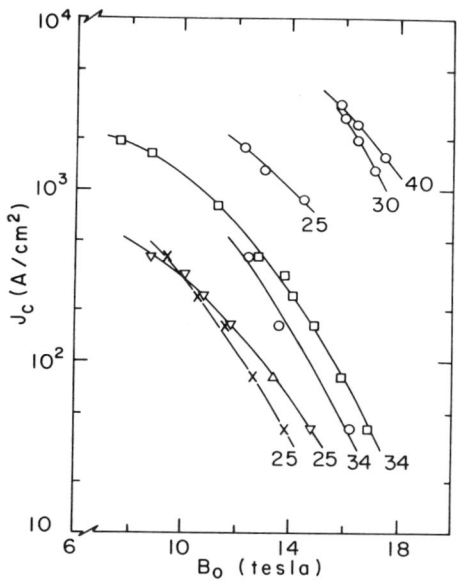

Fig. 10   Overall J_c versus applied field B_0 for cold powder metallurgy composites. The results illustrate the marked difference between the ball-milled and hydride-dehydride powders (the top 3 curves). The numbers on the curves are the wt% Nb. All were fabricated using a monel sheath which was removed before tin plating. The areal reduction ratio R ≈ 500. (After Ref. 20)

to tin and are only required in the early stages until a high density of compaction and sufficient elongation of the powders are
achieved. Prior to the electroplating at final size, two avenues
have been followed; cold extrusion of the specimen from 25.4 mm
diameter to 12.7 mm, alternatively,[21] a smaller laboratory specimen of 11 to 15 mm diameter was used to bypass the extrusion step
and to permit laboratory scale fabrication. The specimens were
swaged to ~3 mm, drawn to a final size of 0.25 or 0.125 mm and
electrolytically tin plated. Finally, the tin was diffused into
the wire and heat treated to form the Nb_3Sn.

    Tin added initially as a powder. In this case, in addition
to the sheath, a diffusion barrier of tantalum or niobium is used
to prevent the tin from diffusing into the sheath.

Materials Selection

    The success of the powder technique is greatly influenced by
the chemical and metallurgical properties of the initial powders.
The powders must be highly deformable, a property closely correlated to the interstitial content, particularly oxygen. Commercially
available niobium powders are generally produced by one of two
processes; ball-milling or hydride-dehydride. The resulting powders have very different geometry.[22] The hydride-dehydride powders show the brittle fracture characteristics whereas powders
from the ball-milling operations reveal the effect of severe mechanical deformation. The different powders have themselves a
measurable T_c difference.[22] The measured differences in T_c and
microhardness are due to the presence of oxygen and not to the
differing degrees of mechanical deformation. The presence of oxygen has also been observed by numerous authors to decrease T_c in
niobium.

    The mechanical properties of the sheath strongly influence
the effective areal reduction ratio and the deformation of the
powders. As in the in situ process, a large reduction ratio leads
to a uniform multifilamentary array with high J_c. Following the
initial results with copper and monel,[20] commercial Cu 0.2 wt% Zr
and Cu 1.8 wt% Be alloys were selected for the sheath because of
their strength, ability to withstand the severe degree of deformation without annealing, and because tin can diffuse through these
materials.

Results

    The results for the powder process have paralleled those of
the in situ processed materials in almost all cases. When suitably processed both have comparable (and high) overall critical

current densities, excellent mechanical properties, and similar submicron multifilamentary structures. We have chosen a few examples to illustrate the results of the powder process develop-ment.

   External tin diffusion. Figure 10 illustrates the marked dif-ference between the ball-milled and hydride-dehydride powder pro-cessed Cu-Nb-Sn.[20] The measured J_c realized with the more deform-able hydride-dehydride niobium powder is at least a factor of 10 larger than those resulting from the ball-milled powders. All these materials were made with a monel sheath which was removed for tin plating and diffusion. Figure 11 illustrates the effect of various sheath materials used for laboratory scale fabrication. The highest values of J_c are obtained using Cu-Be as the sheath.[21] Metallographic observations[23] indicate that the stronger sheath materials produced a finer filament size for a given reduction ratio, i.e., the effective reduction ratio is larger. Variations of J_c can also be observed for a given sheath material if the ini-tial composition, the tin content or the effective reduction ratio are altered.

   The behavior of the composite under axial loading in high magnetic field is illustrated in Fig. 12. The maximum J_c is ob-served at $\epsilon_m \simeq$ 0.5-0.6% and is independent of field in the range 14 to 18 T.[21] These results are quite similar to those for the in situ materials. The performance of the composites with the Cu-Be sheath is shown in Fig. 13 where at 800 MPa, $\epsilon \simeq$ 1%.

   Tin powder incorporated initially. In the cold process tech-nique it is expected that tin powder can be incorporated into the initial mixing stages since the reaction between the powder is avoided in the compaction and extrusion steps. Some results of J_c versus applied field for such materials are reported in Fig. 14. For the Cu-35 wt% Nb-7 wt% Sn, even though deficient in tin and with a small reduction ratio of 300, the J_c is better than $10^4 A/cm^2$ at 12 T. Also indicated are two curves for very high niobium con-centrations, when very high J_c values are observed after heat treatment for 4 days at 650°C. The small addition ($\simeq$ 2 wt%) of copper allows a reduction of the minimum formation temperature of Nb_3Sn from $\sim$900°C for Nb_3Sn to $\sim$650°C.

Potential

   Evaluation in the laboratory clearly indicates that the cold powder metallurgy techniques can produce multifilamentary compos-ites with mechanical properties and critical current densities comparable to the in situ materials. One unique aspect of the technique is that all the early operations can be carried out at low temperatures. This permits a wide latitude in preparation of new materials for evaluation prior to industrial scale-up.

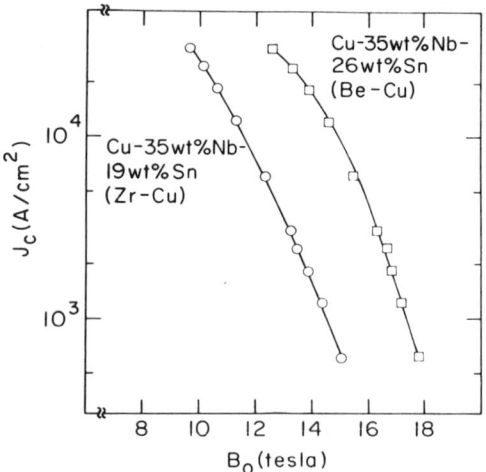

Fig. 11   Overall J_c versus applied field B_0 for cold powder metallurgy composites.   The two curves illustrate the influence of the
sheath material, Cu-Zr or Cu-Be.   The areal reduction was ~ 2000.
(After Ref. 21).

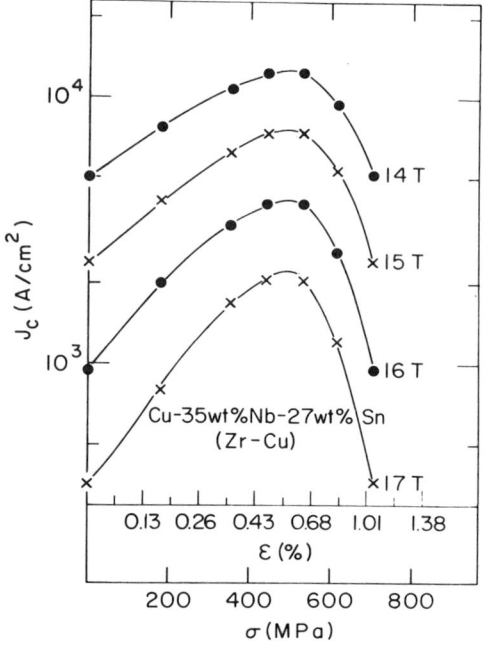

Fig. 12   Overall J_c versus stress σ at indicated fields for
Cu 35 wt% Nb - 27 wt% Sn fabricated with a Cu-Zr sheath.   The strain
ε at 4.2 K is also shown for corresponding stress (After Ref. 21).

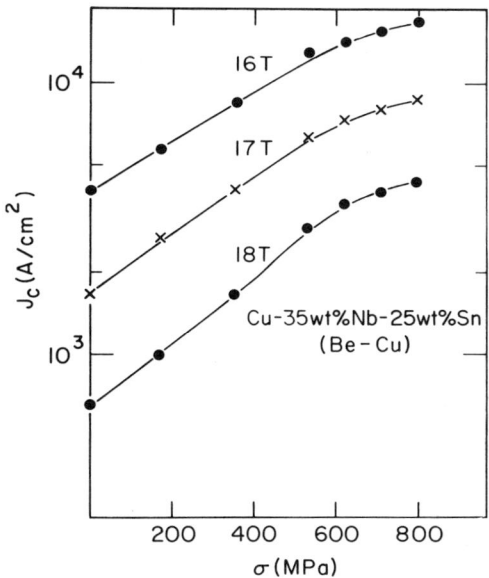

Fig. 13   Overall J_C versus stress σ at indicated fields for
Cu 35wt% Nb - 25 wt% Sn fabricated with a Cu-Be sheath.   (After
Ref. 21).

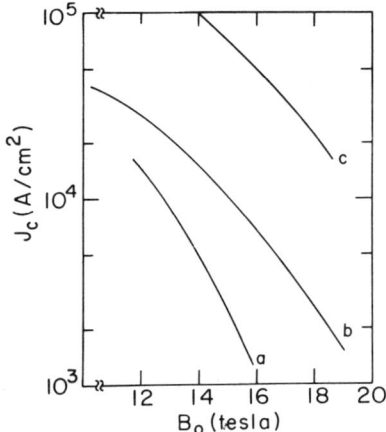

Fig. 14   Overall J_c versus applied field B_0 for cold powder metallurgy composites with tin incorporated in the initial mixing stage;
a) Cu 37 wt% Nb - 11 wt% Sn;   b) Nb 2wt% Cu - 20wt% Sn (a depar-
ture from our usual notation) after 1 day at $650^{\circ}C$;   c) after 4
days at $650^{\circ}C$. (After Ref. 23).

RESEARCH IN PROGRESS

In Situ Process

Several developments and new directions are in progress. The fibre-matrix interaction and the importance of radial and tangential stresses are being analysed using finite element analysis in order to quantify the origin of the enhanced ability of the in situ composites to support large strains before degradation of J_c. The effects of additions (Ta, Zn, Be, Al and others) to the liquid on J_c and mechanical properties are being evaluated. For scale-up to commercial production, we are constructing a small-scale continuous casting apparatus. The apparent advantage is that such casting allows a well controlled precipitation at an initial size in the range of 1 cm. The industrial wire drawing is being evaluated with IGC.

Powder Process

Several new approaches have been pursued in order to extend the powder process. These have included evaluation of high temperature processing with small scale isostatic extrusions, development of increased J_c with Nb-Ta powders replacing the niobium, using atomized Cu-Sn powders in the compact, and fabrication of new materials such as Nb_3Al which have not been feasible with the conventional bronze or in situ processes.

ACKNOWLEDGMENTS

The work summarized here has involved many scientists including, R. Flukiger, J. L. Fihey, R. Akihama, H. LeHuy, R. J. Murphy, E. J. McNiff, Jr., G. Rupp, and B. B. Schwartz who have contributed significantly to these emerging technologies.

REFERENCES

1. E. P. Romanov, L. W. Smirnow, W. D. Sadovski, and N. W. Wolkenschtein, Fiz. Metal. Metalloved. 20:3 (1965).
2. R. H. Borcherts and A. H. Silver, Bull. Am. Phys. Soc. 13:379 (1968).
3. C. C. Tsuei, Science 180:57 (1973).
4. C. C. Tsuei and L. R. Newkirk, J. Mater. Sci. 8:1307 (1973).
5. C. Allibert, J. Droile, and E. Bonnier, Compt. Rend. C 268: 2277 (1969).
6. R. Roberge and J. L. Fihey, J. Appl. Phys. 48:1327 (1977).

7. R. Roberge and J. L. Fihey, Manufacture of Superconducting Materials, American Soc. of Metals, Metals Park, OH (1977).

8. R. Roberge, S. Foner, E. J. McNiff, Jr., B. B. Schwartz, and J. L. Fihey, Appl. Phys. Letters 34:111 (1979).

9. E. Adam, E. Gregory and F. T. Ormand, IEEE Trans. Magn. MAG-13: 319 (1977).

10. Magnetic Corporation of America, private communication (1978).

11. A. Petrovich, B. A. Zeitlin, J. M. Cutro, M. S. Walker, and C. H. Rosner, IEEE Trans. Magn. MAG-13:796 (1977).

12. D. N. Cornish, D. W. Deis, R. L. Nelson, R. M. Scanlan, C. E. Taylor, R. R. Vandervoort, F. T. Wittmayer, and J. B. Zbasnik, IEEE Trans. Magn. MAG-13:454 (1977).

13. J. L. Fihey, M. Neff, R. Roberge, M. C. Flemings, S. Foner, and B. B. Schwartz, Int. Cryogenics Mat. Conf., Madison, 1979 (to be published in Advances in Cryogenic Engineering, Vol. 26).

14. S. Foner, R. Roberge, E. J. McNiff, Jr., B. B. Schwartz, and J. L. Fihey, Appl. Phys. Letters 34:241 (1979).

15. R. Roberge, S. Foner, E. J. McNiff, Jr., B. B. Schwartz, and J. L. Fihey, IEEE Trans. Magn. MAG-15:687 (1979).

16. J. L. Fihey, R. Roberge, S. Foner, E. J. McNiff, Jr., B. B. Schwartz, Int. Cryogenics Mat. Conf., Madison, 1979 (to be published in Advances in Cryogenic Engineering, Vol. 26).

17. R. Roberge, J. L. Fihey, S. Foner, E. J. McNiff, Jr., and B. B. Schwartz, Int. Cryogenics Mat. Conf., Madison, 1979 (to be published in Advances in Cryogenic Engineering, Vol. 26).

18. D. C. Larbalestier, IEEE Trans. Magn. MAG-15:209 (1979).

19. J. Bevk and M. Tinkham, this conference.

20. R. Flukiger, S. Foner, E. J. McNiff, Jr., B. B. Schwartz, J. Adams, J. Forman, T. W. Eagar and R. M. Rose, IEEE Trans. Magnetics MAG-15:689 (1979).

21. R. Flukiger, R. Akihama, S. Foner, E. J. McNiff, Jr., and B. B. Schwartz, Appl. Phys. Letters 35:810 (1979).

22. R. Flukiger, S. Foner, E. J. McNiff, Jr., and B. B. Schwartz, Appl. Phys. Letters 34:763 (1979); erratum 35:430 (1979).

23. R. Flukiger, R. Akihama, S. Foner, E. J. McNiff, Jr., B. B. Schwartz, ICMC Conference, Madison, 1979 (to be published in Advances in Cryogenic Engineering, Vol. 26).

PREPARATION AND PROPERTIES OF IN SITU PREPARED FILAMENTARY
Nb$_3$Sn-Cu SUPERCONDUCTING WIRE

D. K. Finnemore, J. D. Verhoeven, E. D. Gibson,
and J. E. Ostenson

Ames Laboratory and Physics and M.S.E. Depts.
Iowa State University
Ames, Iowa 50011

INTRODUCTION

Several years ago we began a project to make superconducting
wire by directionally solidifying Cu-Nb binary alloys so as to give
an aligned array of Nb dendrites in a copper matrix. These composites
were to be drawn to wire, coated with tin and heat treated to produce
aligned Nb$_3$Sn filaments in a bronze matrix. Initial studies, however, showed that directional freezing was not required to produce
alignment of the Nb dendrites; the wire drawing process itself produced excellent alignment of the initially random array of Nb dendrites produced by conventional casting techniques. This method of
fabricating Nb-Sn-Cu wire, which has been developed independently by
Bevk et al.[1] at Harvard and by Roberge and Foner et al.[2] at the
Hydro-Quebec Research Institute and MIT, has come to be called the
in-situ technique for preparing Cu-Nb-Sn superconductor wire. In
this paper we present a brief review of our results at the Ames Laboratory on the development of this in-situ process.

CASTING TECHNIQUES

Our initial studies[3] employed a chill casting technique in which
140 g samples of molten Cu-Nb alloy were held in a crucible at 1830-
1850°C for five minutes and then cast into a water cooled copper
mold. The liquidus for a Cu-20 wt% Nb alloy was determined to be
around 1680°C, so that a superheat of 150-170°C was employed. The
Nb dendrite arm size was around 1 to 2 μm in the chill cast ingots.
Chemical analysis showed that the composition was 20 ± 1 wt% Nb all
along the length of the resulting 1.5 x 8.2 cm ingots. Six potential
crucible materials have been examined,[4,5] BN, Al$_2$O$_3$, ZrO$_2$, ThO$_2$,

259

Y_2O_3 and graphite. Excessive oxygen pickup was found with BN, Al_2O_3 and ZrO_2. The Y_2O_3 and ThO_2 crucibles produced an oxygen pickup of only 680 and 640 parts per million atomic (ppma) and were clearly the best candidate crucible materials of the ceramics. Graphite resulted in even lower oxygen contamination (470 ppma), but produced a carbon pickup of around 2000 ppma. Surprisingly, this level of carbon contamination did not have any significant deleterious effect upon either the drawability of the Cu-Nb ingots or the resulting J_c properties.

Initial studies to scale-up the size of the ingots involved enlargement of the chill-casting design. Sound ingots were produced of dimension 5 cm x 16 cm weighing 2.8 kgms. However, a significant macrosegregation of the Nb to the bottom of these ingots was found, and this was determined to be due to inadequate mixing of the molten alloy by the natural convection currents in the crucible prior to casting. This difficulty could probably be overcome by introducing forced convection into the liquid. This was not confirmed, however, because a consumable arc casting approach appeared more promising.

Five cm diameter billets of binary Cu-Nb alloys of uniform composition have been prepared[5,6] by double pass consumable arc casting. In the first melt a Cu-Nb electrode is consumably arc melted into a graphite sleeve contained within a water cooled Cu mold. The electrode consists of a Nb rod within a Cu cylinder with the relative sizes adjusted to the desired alloy composition, generally 20 or 30 wt% Nb. The resulting billet from this first melt is rod rolled to around 3 cm and is then used as the electrode for the second melt. The second melt is required because complete homogenization is not quite achieved in the molten pool on the first melt. A typical ingot is 5 x 36 cm and weighs 6.3 kgms. Chemical analysis show that macrosegregation is small, with typical composition variations holding within 20 $\pm$ 0.5 wt% Nb both longitudinally and radially. The Nb-dendrite arm sizes are on the order of 6 μm. The carbon pickup varies from 550 to 2100 ppma in 20 wt% Nb alloys and from 1950 to 2600 ppma in 30 wt% Nb alloys.

Recent evaluation of arc cast material has revealed that the J_c values are consistently smaller than chill cast alloys by around 20 to 30%. Metallographic examination reveals that wires fabricated from arc cast material sometimes contain Nb rich particles at a volume fraction of up to 3 to 5%. These particles, which range up to around 2 μm in size, are not evident in the as-cast microstructure and are apparently carbides of Nb which are not significantly reduced in size by wire drawing. The slightly lower J_c values are probably associated with the presence of non-superconducting particles in the cross section of the wire. These particles are thought to result from detachment of the niobium carbide layer which forms at the surface of the graphite liner and could probably be eliminated by coating the graphite with a thin layer of Y_2O_3. Experiments are presently underway to examine this possibility and to attempt to

scale the arc casting process up to 10 cm diameter ingots weighing
around 30 kgms.

ADDITION OF TIN

    Adding Sn to the Cu-Nb wire by plating or dipping becomes more
difficult as the wire size increases because, (a) the diffusion
times become quite long, and (b) the thicker Sn layers required tend
to "ball-up" when the tin layer is melted. To overcome these difficulties we have developed two process in which the tin is added
within the core of the Cu-Nb alloy prior to wire drawing.

    In the first process[7] the Cu-Nb is extruded over a mandrel to
form a tube and a Sn-5 wt% Cu rod is inserted into the hole of the
tube. This composite is then drawn to wire. The J_c properties of
this single core wire is equivalent to plated wire. We have successfully prepared a few thousand feet of 0.25 mm wire with this technique, starting with 5 cm diameter billets of Cu-20 wt% Nb. However,
two difficulties have been found. After the Sn is added one cannot
anneal the wire to recrystallize the Cu matrix and, hence, the wire
drawing must be done without intermediate annealing. This has not
been a problem with Cu-20 wt% Nb alloys, but with Cu-30 wt% Nb work
hardening has limited our ability to draw the cored wire to sizes
below around 0.50 mm. In addition, there is a tendency for the Sn
metal to burst through the Cu-Nb tube wall during the heat treatment
to form Nb$_3$Sn. Typically the spacing between breaks is a few meters.
This problem is thought to be associated with local points of roughness along the inner wall of the extruded Cu-Nb tube. It is nearly
eliminated by encasing the Sn-5 wt% Cu rods in a thin walled Cu-tube
before inserting within the Cu-Nb tube, and it may be completely
eliminated by utilizing a stepped heat treatment which avoids melting
the tin matrix.[8]

    We are currently working on a process to produce a multicore
wire. In the initial work seven 0.71 cm diameter holes were drilled
into a 4.8 cm (1.875 in.) diameter Cu-20 wt% Nb billet which was then
electron-beam welded into a Cu can of 5.8 cm (2.30 in.) o.d. with a
front end cone angle of 40°. This billet was then hydrostatically
extruded at a ram speed of 2.5 mm/s through a die with an entrance
angle of 45° employing water cooling below the die exit to prevent
melting of the Sn cores. Under these conditions successful extrusions
were obtained at a reduction ratio of 7:1 (extruded rod diam. = 2.2
cm), but not at a ratio of 19:1 (extruded rod diam. = 1.3 cm),
where melting of the Sn cores occurred. If the Sn cores melt the
hydrostatic extrusion pressure is transmitted through the die causing
the Sn to burst out of the Cu-Nb jacket beyond the die exit.

    Upon drawing the extruded ingots it was found that the Cu jacket
tended to debond at wire diameters of around 5 mm. This difficulty

apparently resulted because of inadequate cleaning of the original
Cu-billet interface and it limited the final wire size to around
5 mm. This effect was overcome by removing the Cu-jacket from the
extruded rod before drawing. It was then possible to draw the rods
to 0.7 mm wire before work hardening of the Cu-Nb matrix caused breakage. Figure 1a shows the cross section of the as-extruded rod and
Fig. 1b shows the cross section of the 0.7 mm wire. It is seen that
the circular tin cross section is maintained in the hydrostatic extrusion, but that the wire drawing tends to preferentially deform the
outer Sn surfaces with the net effect of moving the Sn toward the
center of the rod. Because of the removal of the Cu jacket the
vol % Sn was 15.7, considerably above the stoichiometric value of
around 9.3 vol %. The critical currents for these wires at 8T after
reaction at 550°C for 6.8 and 10.1 days were 3.2×10^4 and 4.4×10^4
amps/cm^2, respectively. These values of J_c are similar to those
obtained with plated wire. No difficulty was experienced in the
limited initial studies due to Sn break-out with direct heating to
the 550°C reaction temperature. This multiple core technique has
obvious advantages for preparation of large diameter wire. However,
as with the single-core process it requires that the wire drawing be
done without intermediate annealing and thus we anticipate difficulty

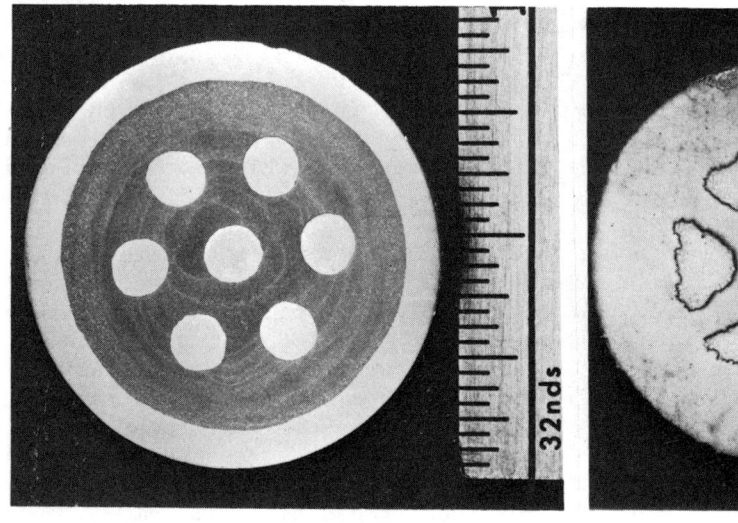

                    (a)   (b)

Fig. 1.   (a) The 7-core billet after extrusion from 5.8 cm diameter
          to 2.2 cm diameter. (b) Final wire after removal of the
          copper jacket and wire drawing. Outside diameter = 0.71mm.

in extending it to higher initial Nb contents of the Cu-Nb billet.

MICROSTRUCTURAL CHARACTERIZATION

The morphology of the as-drawn Nb filaments and the Nb_3Sn formed from these filaments has been examined using electron microscope techniques.[9] Initial work has been limited to 0.15 mm wire drawn from 12.5 mm chill cast ingots (area reduction = 7000). Figure 2a is a TEM picture of an extracted Nb filament and Fig. 2b is an SEM picture of a wire cross section showing the Nb filaments embedded in the Cu matrix. These studies indicate that the as-drawn Nb filaments are not round in cross section, but are flat filaments with a very asymmetric rectangular cross section on the order of 50-70 Å by 2000-3000 Å. The Nb filament density is generally higher than seen in Fig. 2b, but this photo shows nicely isolated filaments, illustrating their convoluted shapes. Figure 3 is an SEM picture of an extracted Nb filament from a wire heated to 550°C for only 5 hours without adding Sn. It is seen that severe coarsening of the Nb has occurred with

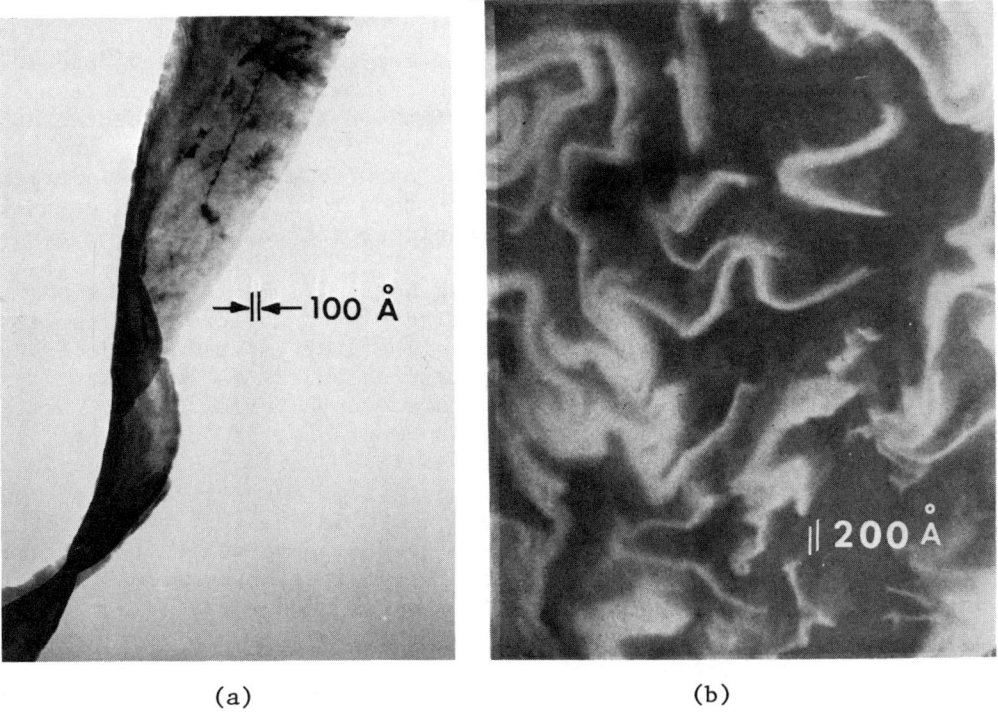

(a)   (b)

Fig. 2.   (a) Extracted as-drawn Nb filament as seen in TEM at
          100,000X.   (b) Polished cross section of Cu-20% Nb wire after
          etching away some of the Cu matrix.   SEM photo at 50,000X.

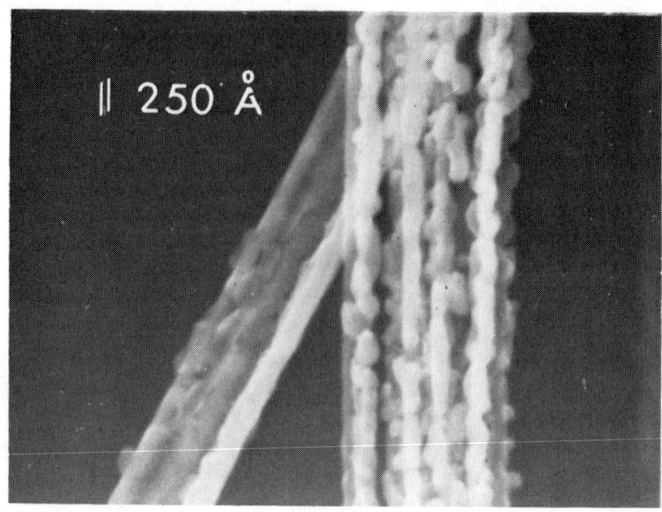

Fig. 3.   Niobium filaments extracted from wire annealed at 550°C for
5 hours – no Sn added. 40,000X.

the resulting filament consisting of a series of knobs each about
700 Å across and slightly elongated in the axial direction of the
filament. Nb$_3$Sn filaments extracted from Sn plated wire reacted at
550°C appear essentially identical to Fig. 3. Hence, it appears
that the very thin Nb filaments formed by wire drawing first coarsen,
due probably to surface diffusion, and then form Nb$_3$Sn of essentially
the geometry determined by the coarsening reaction.

From TEM studies of the extracted Nb$_3$Sn filaments[9] it was possible to determine the grain size. Figure 4 summarizes the results
and compares them to literature data. The grain size was reduced by
reacting at successively lower temperatures and it is seen that J$_c$
increased until the reaction temperature dropped below 550°C. Critical temperature measurements on wires reacted at 550°C and above
were around 17.5K but dropped to around 16.8 and 16.5 K for the 500
and 450°C reaction temperatures, respectively. Hence, it seems
likely that the drop in J$_c$ at 500 and 450°C is due to formation of
non-stoichiometric Nb$_3$Sn. The fairly good agreement with the literature data observed on Fig. 4 indicates that the Nb$_3$Sn formed by this
in-situ process behaves similarly to that of bronze processed
material.

PERFORMANCE CHARACTERISTICS

There are many parameters which can be used to tailor the
characteristics of the wire for specific uses. For those applications

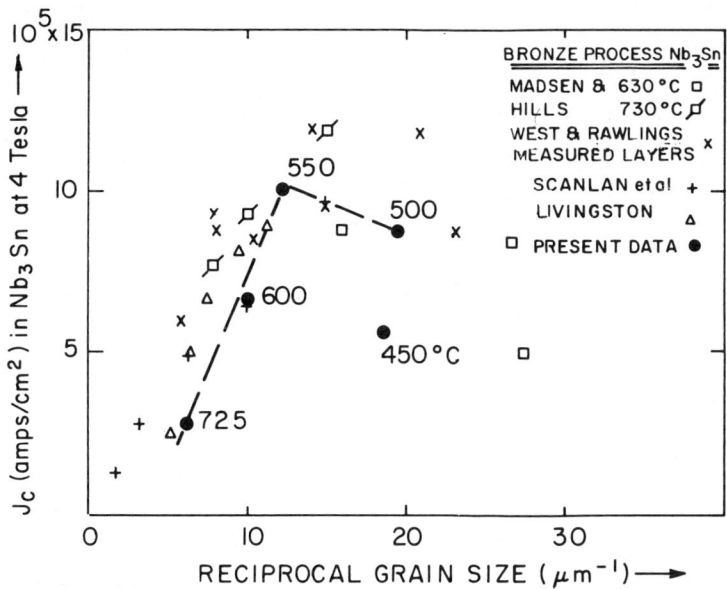

Fig. 4.   Critical current vs reciprocal grain size for wires reacted
at temperatures ranging from 725 to 450°C.   The literature
data are taken from the work of Madsen and Hills.[10]

where high J_c are required such as for dc magnets, one might want to
cast dendritic alloys with 30 to 40 wt% Nb.   As shown on Fig. 5, J_c
values for these materials are substantially above commercially
available continuous filament materials.   Because the filaments are
in close proximity with one another, however, these in-situ materials
have higher ac losses than the continuous filament materials.   For
those applications where ac losses are more of a problem, such as
ramped field magnets, one might want to cast dendritic alloys with
about 20 wt% Nb.   This reduces J_c from those values in Fig. 5 by a
factor of 2 or 3[3-7] but it improves the ac losses and the strain
tolerance.   In this 20% Nb concentration range, one typically finds
$J_c = 1.3 \times 10^4$ A/cm^2, at 10 tesla, a strain tolerance of 1.4% before
irreversible effects set in and ac losses comparable to continuous
filament wire having a 40 μm filament diameter.   We have found these
20% materials to be a good compromise and most of our recent work
has concentrated on materials having performance characteristics
in this range.

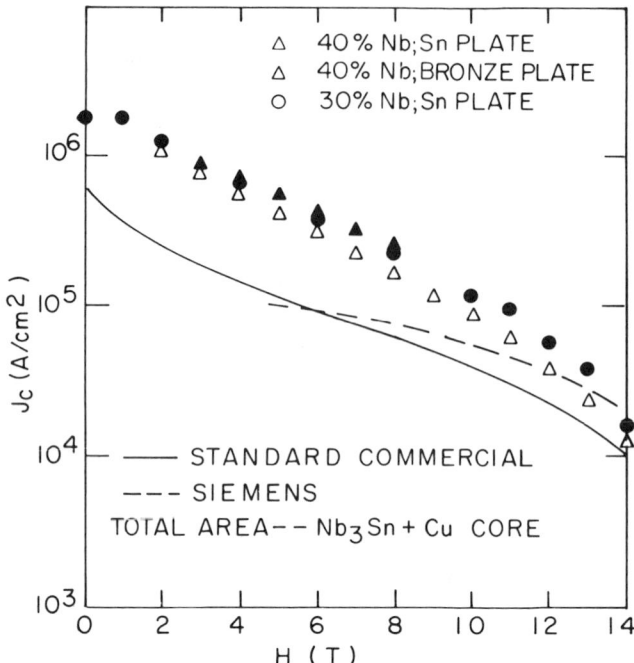

Fig. 5.   Critical current data for 30 and 40 wt% Nb samples

A series of experiments were undertaken to understand the origin of the ac losses in these materials and it was found that proximity effect coupling can play a major role. The temperature and magnetic field dependence was measured for triple layer superconductor-normal metal-superconductor, SNS, junctions having normal metal thickness d_n in the 10 to 30 μm range.[11] It was found that J_c decreases exponentially with temperature as shown by the heavy solid line in Fig. 6. In addition if one raises the magnetic field at some fixed tempera- ture, such as 4.2K, J_c decreases exponentially with magnetic field in the form $J_c = J_o e^{-H/H_o}$. This means that the fibers will decouple in high magnetic fields and the ac losses should be much lower in the 8 to 14 tesla range than they are in the 0 to 4 tesla range. In addi- tion several other techniques have been used to reduce the ac losses. Twisting the wire[12] at a pitch of 3 turns/cm reduces the ac losses by about a factor of 6 and the addition of small amounts of Ni reduces the losses somewhat. Unfortunately, there is some immiscibility in the Cu-Ni-Nb system so this cannot be pushed very far. Shen will review the ac loss characteristics in more detail elsewhere at this conference.

Several small solenoids have been wound in order to test the quality control of the wire over moderate lengths and to test the

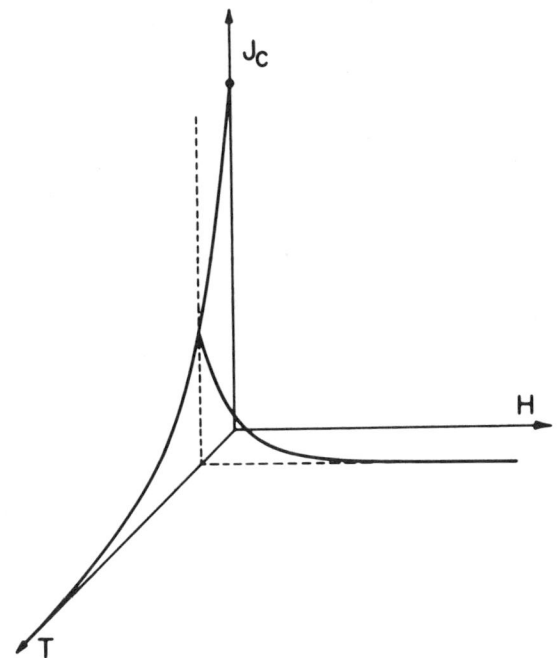

Fig. 6.   Critical supercurrents which can flow through a normal metal
          layer as a function of temperature and magnetic field.

ability to wind the wire around a small radius after reaction.
Lengths of 0.025 cm diameter wire ranging in length from 4 to 40
meters were prepared by the core process[6],[7] and reacted at 550°C on
a 3.8 cm diameter stainless steel mandrel. These lengths were then
wound spool to spool onto a NEMA grade G-10 coil form 1.27 cm in diameter with nylon thread to space the turns and nylon to separate
the layers. General Electric 7031 varnish was used to hold the wires
in place for all but one of the coils. In this coil the wires were
loose. The coils had a field to current ratio in the range of 15 to
40 mT/Amp, and typically produced a dc field of 2 to 3 tesla.

    Solenoids were tested by immersing them in an external magnetic
field, $\mu_o H_e$, and by applying a current of the form $I = I_o + I_1 \sin$
$(2\pi ft)$ to give a self field of the form $B = B_o + B_1 \sin (2\pi ft)$. The
performance varied in a regular way as variables such as twist pitch,
potting, etc. were changed. A typical case is the performance of a
six layer coil of untwisted wire with $\mu_o H_e = 8$ tesla. For $I_1 = 0$ the
coil quench current was 15 amp or $J_c = 2.4 \times 10^4$ A/cm⁴. This particular coil had a B/I_o ratio of 15 mT/A so the self field was 0.45 T
at $\mu_o H_e = 8T$. When a small ripple field of $B_1 = 11$ mT was applied

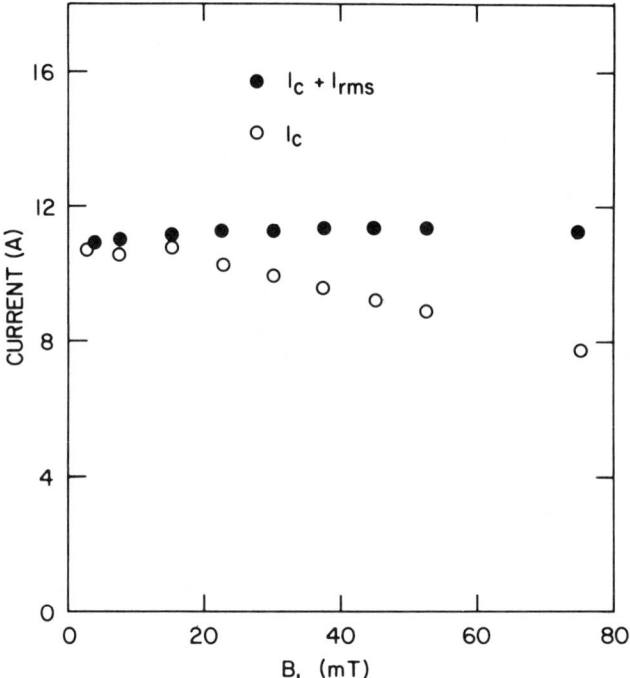

Fig. 7. Critical current versus ripple field for small coils

the value of J_c remained constant as a function of frequency out to some critical frequency, f_c, where J_c drops abruptly. At 8T, f_c = 36 Hz for this coil. As I_1 is increased, the value of I_c decreases as shown by the open circles of Fig. 7. If, however, one adds the root mean square ripple current, I_{rms} to I_c as shown by the solid circles one finds roughly a constant value implying that the quenching is a heating effect. In broad outline we find that potting of the coil in varnish and twisting the wire before reaction reduces the ac losses but leaves J_c unchanged.

CONCLUSIONS

    Consumable arc casting appears to provide a low cost and reliable method for preparing dendritic Cu-Nb alloys suitable for the large scale production of superconducting wire. Casting procedures for 5 cm diameter billets have been studied and good performance charac-teristics are found. In addition, preliminary results with 10 cm diameter castings look promising and scale up to this size seems likely to be successful. The major problem in scaling up may be to keep the dendrite size large enough to prevent excessive coarsening.

Two new methods for Sn introduction have been developed, a single core process for small diameter wire and a multicore process for large diameter wire. The overall J_c and strain performance characteristics are well suited to large scale dc magnets in the 8 to 14 tesla range.

The ac losses in these materials in the 0 to 3T range are approximately equivalent to continuous filament wire with a filament diameter of 40 μm. Twisting of the wire gives a substantial improvement in the ac losses but the addition of Ni to decouple the fibers does not look encouraging because an immiscibility occurs near 3% Ni. The proximity effect coupling of the fibers will decrease substantially with increasing magnetic fields so this contribution to the ac losses should be much lower at large $\mu_0 H$.

ACKNOWLEDGEMENTS

This work was done at the Ames Laboratory, Iowa State University, Ames, Iowa, operated for the U.S. Dept. of Energy by I.S.U. under contract No. W-7405-Eng-82. The research was supported by the Director of Energy Research, Office of Basic Energy Sciences, WPAS-KC-02-01.

We are indebted to J. J. Sue for the electron microscopy work reviewed here. The arc casting was done by F. A. Schmidt, the conventional extrusion work by C. V. Owen and L. K. Reed, the chemical analyses by R. Z. Bachman and R. J. Consemius, the E. B. welding by F. N. Linder, and the slip casting of Y_2O_3 crucibles by F. W. Calderwood. The hydrostatic extrusion was done under contract at Battelle, Columbus by R. J. Fiorentino and E. G. Smith. S. Shen of Oak Ridge National Laboratory made some of the ac loss measurements and J. Ekin of the National Bureau of Standards made the J_c vs strain measurements.

REFERENCES

1. J. P. Harbison and J. Bevk, J. Appl. Phys. 48:5180 (1977); J. Bevk and J. P. Harbison, J. Mat. Sci. 14:1457 (1979).

2. S. Foner, E. J. McNiff, Jr., B. B. Schwartz and R. Roberge, Appl. Phys. Lett. 31:853 (1977); R. Roberge, S. Foner, E. J. McNiff, Jr. and B. B. Schwartz, Appl. Phys. Lett. 34:111 (1979).

3. J. D. Verhoeven, D. K. Finnemore, E. D. Gibson, J. E. Ostenson and L. F. Goodrich, Appl. Phys. Lett. 33:101 (1978).

4. J. D. Verhoeven and E. D. Gibson, J. Mat. Sci. 13:1576 (1978).

5. J. D. Verhoeven, E. D. Gibson, F. A. Schmidt and D. K.

Finnemore, J. Mat. Sci. (1980).

6. J. D. Verhoeven, F. A. Schmidt, E. D. Gibson, J. E. Ostenson and D. K. Finnemore, Appl. Phys. Lett. 35:555 (1979).

7. J. D. Verhoeven, E. D. Gibson, C. V. Owen, J. E. Ostenson and D. K. Finnemore, Appl. Phys. Lett. 35:270 (1979).

8. H. Benz, I. Horvath, K. Kwasnitza, R. K. Maix and G. Meyer, Cryogenics 19:442 (1979).

9. J. D. Verhoeven, J. J. Sue, D. K. Finnemore, E. D. Gibson and J. E. Ostenson, J. Mat. Sci. (1980).

10. P. E. Madsen and R. F. Hills, IEEE Trans. Mag., MAG–15, 181.

11. T. Y. Hsiang and D. K. Finnemore, Sol. State Comm. 33:847 (1979); T. Y. Hsiang and D. K. Finnemore, Phys. Rev. (accepted).

12. J. E. Ostenson, D. K. Finnemore, J. D. Verhoeven and E. D. Gibson, Appl. Phys. Lett. (submitted).

SUPERCONDUCTING PROPERTIES AND COUPLING MECHANISMS IN IN SITU

FILAMENTARY COMPOSITES

J. Bevk and M. Tinkham

Harvard University, Division of Applied Sciences

Cambridge, Ma 02138

INTRODUCTION

In situ formed superconducting filamentary composites are of considerable interest from both theoretical and practical points of view. A number of authors[1] have demonstrated that these composites can carry critical currents comparable to those of the best conventional conductors with continuous Nb_3Sn or V_3Ga filaments and they are also remarkably insensitive to mechanical stress or strain. Along with the powder-metallurgically processed composites,[2] these new materials show realistic promise for potential practical applications and can now be produced also in large quantities.[3]

The physical mechanisms which govern the properties of in situ composites with very small, closely spaced and discontinuous filaments can be quite different from those in conventional composites and are not yet fully understood. It is clear, however, that size effects play an important role in both transport and mechanical behavior of these materials. Some of these effects can be directly related to the composite microgeometry (filament size, spacing, distribution, volume fraction) and are reflected in the experimentally determined flux-pinning and coupling mechanisms. Moreover, the strength of some highly reduced composites has been found to be as high as that of metallic whiskers and approaches the estimated theoretical strength of the material.[4]

The purpose of this paper is to review some of the work done at Harvard over the past few years which is pertinent to the understanding and optimization of the practical properties of in situ composites. In particular, we focus on the inter-filament

271

coupling mechanisms and the interactions between the flux-line
lattice and filament-matrix interfaces. The former are of special
importance when considering composites in which the volume fraction
of superconducting filaments is below the percolation threshold.
Since many properties of the practical composites (e.g., stability,
ac losses, bending characteristics) depend sensitively on the matrix-
to-filament volume ratio, it is clear that quantitative understanding of these mechanisms is necessary in order to optimize the
properties of in situ conductors.

SAMPLE PREPARATION

Cu-Nb and Cu-V ingots were prepared by rf melting in a water-
cooled copper crucible.[4]   The two-phase as-cast alloys with Nb
(or V) concentrations ranging from 7.5 to 20 vol.% were next swaged
and drawn to wires of various diameters (0.05 to 1.0 mm) without
any intermediate annealing. Some of the wires were also rolled to
tape with aspect ratio between 8:1 and 27:1. The samples were
then plated with tin (or gallium) and annealed in a vacuum for
various lengths of time at temperatures ranging from 450 to 600°C.

The optimum amount of tin (or gallium) necessary to convert
niobium (or vanadium) filaments into A15 compounds by external diffusion is considerably higher than the stoichiometric amount.
Consequently, the excess tin (gallium) remains in the matrix which
has typically resistivity ratio of 2 or less. This high matrix
resistivity is, in fact, desirable in some cases, such as ac conditions, since it reduces interfilament coupling and in turn eddy
current losses. For high matrix conductivity and improved stability
under dc conditions the bronze route appears to be more advantageous.
In this latter case, the appropriate amount of tin is added already
during the initial melting process, resulting in a two-phase Cu-Sn-Nb
alloy. The drawback of this approach is the need for frequent
intermediate anneals due to rapid matrix work-hardening, particularly
at high reductions of the cross-sectional area. However, since tin
is uniformly distributed in the matrix and diffusion distances are
very short (on the order of the interfilament spacing) less tin is
needed and the average tin concentration in the matrix after
annealing may be as low as a few hundredths of a percent. In fact,
in the cleanest matrices the electron scattering from the filament-
matrix interfaces becomes the dominant scattering mechanism.

A characteristic feature of all composites after mechanical
reduction is a dense distribution of <u>ribbon</u> like filaments. Their
shape can be related directly to deformation mode during composite
formation.[4]   Bcc crystals are known to develop a <110> fiber
texture in which only two of the four <111> directions are oriented
favorably to accommodate extension parallel to the fiber axis.
Consequently, further deformation produces plane strain rather than
axially symmetric flow, resulting in a band-like or elliptical cross

section of the filaments.[5]   In wire composites, the filaments
are forced to curl and fold due to the constraint of the surrounding
fcc matrix which is able to accommodate axially symmetric flow.   In
composites rolled into tape, however, the ribbon-like filaments are
flat and aligned in the rolling plane.[6]   As discussed later, this
microgeometry leads to some interesting flux-flow properties.

   The volume fraction of niobium (or vanadium) in the initial
two-phase alloys ranged from 0.075 to 0.20.   Although increasing
the volume fraction of the superconducting phase would obviously
result in higher overall critical-current densities, other practical
considerations (stability, tensile and bending strain tolerance,
ac loss behavior) dictate keeping it below ~25%.

   Most samples reported in this paper were prepared in small
quantities (a few m in length).   In one instance a 100 m long
single piece of $Cu-Nb_3Sn$ wire was prepared in order to permit the
calorimetric measurement of ac losses[7]   and to explore the feasibility of scaling up the process to larger quantities.

COUPLING MECHANISMS

   The a priori consequence of discontinuous filaments is that
current must flow through a nominally normal copper matrix to get
from one filament to the next.   The coupling mechanisms, which can
lead to a vanishingly small resistance in these composites, are not
yet fully understood and they continue to stimulate further theoretical and experimental studies.   The emphasis of this paper, however,
is on composites with large area reduction ratio (Re $\geq$ 1000), and
for this limiting case there is sufficient experimental evidence
that the critical-current density J_c is not limited by the interfilamentary coupling strength, but rather by the flux-pinning
strength of the filamentary material itself.   In the short-filament
conductors, the relative strength of the coupling is less, and the
critical current may indeed be limited by the onset of superconductive phase slippage between adjacent filaments.   Further studies
of this regime are certainly warranted, since our recent measurements[8]   and qualitative theoretical arguments[9]   both suggest that
ac losses are reduced in composites with shorter filaments and
lower filament volume fractions, which reduce the interfilamentary
coupling.   This observation indicates that wire structure might be
optimized not by maximizing interfilamentary coupling but rather
by making it no stronger than necessary to give satisfactory J_c
under dc conditions.

   Turning back to the long-filament in situ composites and their
potential usefulness, one is particularly interested in whether
their resistance will be zero or at least close enough so that they
could be used as a magnet wire.   Given the notion of a percolation
threshold[10]   at some volume fraction (~15% in 3-dimensions, ~50%

in 2-dimensions) one might expect strictly zero resistance above
that fraction, and finite (but small) resistance at smaller volume
fractions. But the effective superconducting volume fraction will
be larger than the nominal one, and will depend on temperature,
current density, and magnetic field because of the "proximity
effect", which allows superconducting electrons to diffuse roughly
a normal metal coherence length ξ_N out from the superconducting
filaments themselves. This length will be of the order of a fraction of a micron under typical conditions, and hence will become
important when filament diameters are reduced to approach a similar
scale. But because the coherence energy of the wavefunction only
falls exponentially, as e^{-x/ξ_N} , some superconductive coupling
will continue to larger distances, until this energy falls below
the thermal noise level. Thus, at low temperatures and with small
measuring currents, one expects to find true perfect conductivity,
as in a bulk superconductor, and one does, even for nominal superconducting volume fractions   well below the percolation threshold.

But the technically interesting case is one of high current
densities, $T \gtrsim 4$ K , and in the presence of strong magnetic fields.
All these influences reduce the proximity-effect coupling, shrinking
the effective superconducting volume fraction closer to the nominal
one. Thus, it is useful to consider first a simple "worst-case"
limit, in which one assumes that the proximity effect is completely
suppressed, and that the superconductive volume fraction f_s consists of a collection of long thin cylinders of length L and
diameter d , formed from individual precipitate grains by the
drawing process, and hence aligned along the wire. The rest of the
volume is assumed filled with a completely normal Cu matrix.

For a qualitative understanding, it is convenient to idealize
the deformation in the drawing process so that, if the area reduction ratio is Re , all transverse dimensions are reduced by
$(Re)^{-1/2}$, while lengths increase by a factor of Re to conserve
volume. Thus the filamentary aspect ratio d/L scales as $(Re)^{-3/2}$,
while the filamentary separation scales as $(Re)^{-1/2}$. (If the deformation is not ideal, one can define an effective Re which would
produce the actual d/L ratio.) For large deformations, it is a
reasonable approximation to model the current flow as purely axial
in the filaments and purely radial in the copper matrix. It then
follows that $J_{\parallel}$ in the filaments is J/f_s , where J is the
overall average current density carried by the wire, while the
transverse current density in the copper matrix (evaluated midway
between the filaments) is $J_{\perp} \sim J(f_s)^{-1/2}(Re)^{-3/2} \sim 10^{-4}J$ for
$f_s = 0.1$ and $Re = 10^3$. This shows how extremely effective the
long thin filamentary segments are in reducing the current density
in the copper. This has two crucial implications:

First, the energy dissipation in the copper is reduced by a
factor of $(J_{\perp}/J)^2$, leaving an effective remnant resistivity given

by[11]

$$\frac{\rho_{rem}}{\rho} \simeq \frac{1}{f_s} \frac{1}{Re^3} = \frac{1}{f_s} \frac{d^2}{L^2} \tag{1}$$

where ρ is the resistivity of the matrix. This resistance
reduction ratio would be $\sim 10^{-8}$ for the typical values of
$f_s = 0.1$ and $Re = 10^3$, reducing the resistance below the level
of detectability except with a SQUID voltmeter. A result similar
to (1) was found by Callaghan and Toth,[12] apart from the numerical
factor. By contrast with this physically reasonable result, conventional effective medium theory is not useful in dealing with
such highly anisotropic inclusions of superconducting material,
since it predicts zero resistance for any $f_s > d^2/L^2 \sim 10^{-8}$!

The second crucial consequence of the small $J_\perp/J$ ratio is
that $J_\perp$ can easily be so small that it can be carried as a supercurrent by proximity effect in the copper matrix. Again taking
$f_s = 0.1$ and $Re = 10^3$, a J of 10^5 amps/cm^2 translates into
$J_\perp \sim J f_s^{-1/2} (Re)^{-3/2} \sim 10$ amps/cm^2, small enough to be carried by
even a weak residual proximity effect superconductivity in the
copper. It is this geometrical reduction of $J_\perp/J$ that causes the
effective J_c to be limited by the filaments rather than the
matrix coupling if Re is large enough.

Since the proximity-effect critical current density $J_{c\perp}$ rises
continuously from zero at T_c of the superconducting inclusions,
there may be an intermediate temperature region between T_c and some
lower $T_{c1}(J)$, at which $J_{c\perp}$ has become large enough to carry the
$J_\perp$ for a given applied J, causing the resistance to fall all the
way to truly zero. In this intermediate regime, the continuous
T-dependence of the resistance is controlled by two distinct influences: the geometrically random array of coupling strengths between filamentary segments (described as a percolation problem) and
the thermodynamic fluctuations which would cause a gradual onset of
phase locking even in a geometrically regular array. If percolation is dominant, one would expect the remnant resistance to vanish
when the effective superconductive fraction f_s reaches a critical
value f_c. Making a phenomenological generalization of (1), we
have

$$\frac{\rho_{rem}}{\rho} = \frac{1}{f_s} \frac{d^2}{L^2} (1 - \frac{f_s}{f_c})^s \qquad f_s < f_c \tag{2}$$

$$= 0 \qquad f_s > f_c$$

where s is a critical exponent which is expected to depend on dimensionality, and f_s is assumed to decrease smoothly with increasing current, temperature, and magnetic field. Since the stiffness of the coupling against thermodynamic fluctuations would also be smoothly controlled by these same parameters, the observable behavior of $R(T)$ would be qualitatively similar, whichever rounding mechanism is dominant.

Measurements of the remnant resistivity as a function of temperature and current[11,13,14] (Fig. 1) and of applied magnetic field[15] (Fig. 2) on samples containing Nb and Nb_3Sn filaments are in good qualitative agreement with the above model. An important idea here is that, by varying $T, H,$ or I one can vary f_s, and therefore interfilament coupling, without introducing in the process undesired microstructural changes and secondary effects.

The crucial test of the model, however, and of the usefulness of the in situ material are the measurements of the critical currents. A logical extention of (1) is that in the long-filament limit, the composite critical-current density, J_c, should scale with f_s even in composites below the microstructural percolation threshold. However, early experimental data, reported by various investigators, did not confirm this predicted behavior. At high fields in particular, a decrease of f_s by a factor of two resulted typically in a decrease of J_c by an order of magnitude.

Becuase composites with low f_s are particularly useful for those applications where good stability, low ac losses and high tolerance for bending strain are required, and because they provide at the same time a sensitive test of our model, we have recently reexamined the superconducting behavior of $Cu-Nb_3Sn$ composites with only 7.5 vol.% Nb (volume fraction of Nb_3Sn filaments in the reacted composites was 0.098). Our results suggest that some of the conclusions stemming from earlier experimental tests are not valid, partly because of the difficulties in preparing uniform Cu-Nb samples with low Nb concentration[16] and partly because of the secondary or side effects introduced by varying the volume fractions of the superconducting filaments.

In the case of $Cu-Nb_3Sn$ composites, it is now known that the intrinsic properties (T_c, H_{c2}) of Nb_3Sn filaments depend rather sensitively on the matrix-to-filament volume ratio.[17] This ratio determines the magnitude of the compressive stress exerted on the filaments by the matrix due to differential thermal contraction. Reducing f_s will therefore not only reduce J_c in a direct proportion but will also lower H_{c2} of the filamentary material. A more meaningful way to compare J_c's of two composites with different f_s is therefore to plot normalized critical-current density $j_c \equiv J_c/f_s$ versus reduced critical field $h \equiv H/H_{c2}$. This is done in Fig. 3 for two composites containing 0.23 and 0.10 volume fraction

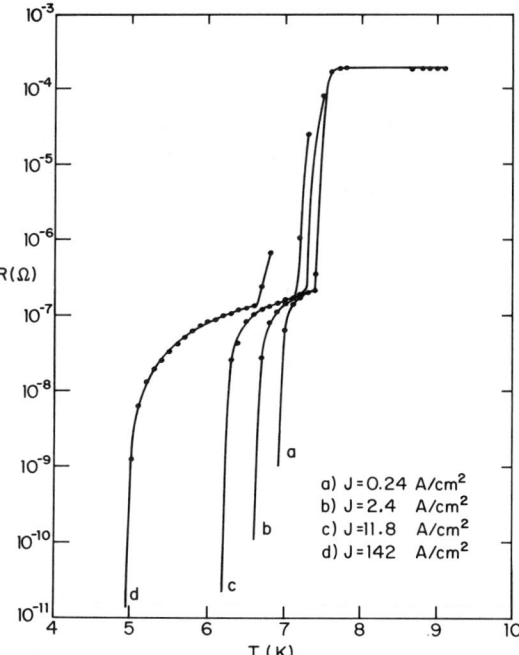

Fig. 1   Resistive transitions of an in situ sample (Re=16) consisting of 7.5 vol.% Nb in a Cu-3 at.% Ni matrix (from Ref. 14).

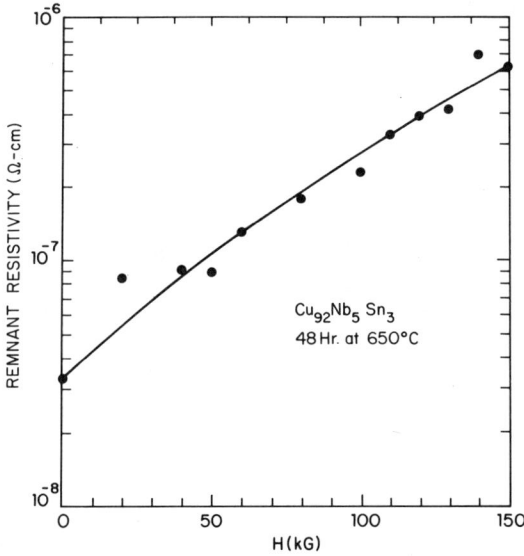

Fig. 2   Remnant resistivity as a function of the applied magnetic field in a Cu-Nb-Sn composite (Re=40) containing 9.8 vol.% Nb_3Sn (from Ref. 15).

of Nb_3Sn filaments (original Nb volume fraction before plating and annealing was 0.182 and 0.075, respectively). Furthermore, care was taken that the average matrix composition after annealing was the same in both composites. Although J_C-vs-H plot (Fig. 3a) shows almost an order of magnitude discrepancy between the two samples at high fields, the experimental points fall on a single master curve when plotted as j_c vs h (Fig. 3b).

At lower fields, proximity-effect induced coupling becomes increasingly more important and magnetization measurements in fact show that the samples can become completely superconducting.[11,19] Consequently, high critical-current densities can be achieved even in composites with low filament volume fraction. The self-field J_C of the Cu-Nb_3Sn tape with f_s = 0.098, discussed earlier, was found to be 8×10^5 A cm^{-2} at 1 $\mu V/cm$ voltage criterion. Increasing the sensitivity by two orders of magnitude to 0.01 $\mu V/cm$ (corresponding to resistivity criterion of 2×10^{-14} Ω-cm) resulted in a decrease of the measured J_C by only 10%. This indicates that the sample either behaves as a true superconductor or at the very least that its remnant resistivity is low enough to allow technological applications. An added bonus is exceptional conductor stability; samples of this composition could be repeatedly quenched, without a shunt protection, even in zero applied field with no apparent damage.

FLUX-PINNING MECHANISMS

The primary sources of flux-pinning in bronze-processed A-15 materials are known to be grain boundaries. Dependence of the pinning force $F_p(h)$ on the reduced magnetic field h can be described adequately[20] by $F_p \propto h^{1/2}(1-h)^2$ in the range of $H > H(F_p^{max})$. In materials with very small grain size, this range is further restricted to fields close to H_{c2}, where the condition that the flux-line lattice parameter is much smaller than the pinning-site spacing is satisfied. (This restriction is particularly applicable to in situ composites where relatively low annealing temperatures and short annealing times result in very small grain size of the A-15 material (Fig. 4).) At lower fields, the field dependence of the flux-line lattice parameter leads to a more complicated F_p-vs-h relationship[21] and F_p^{max} can sharply increase. Moreover, in in situ composites, where the filament thickness becomes comparable to grain size, pinning by interfaces is expected to contribute significantly to the total pinning, and therefore the orientation of the ribbon like filaments with respect to the applied field becomes important. In multifilamentary tapes, where the filaments are aligned in the rolling plane, this leads to anisotropy in critical properties and flux-flow behavior,[6] which is particularly pronounced at high fields. (It should be noted that the aspect ratio of many filaments is greater than that of the tape itself due to the fact that the filaments deform in the

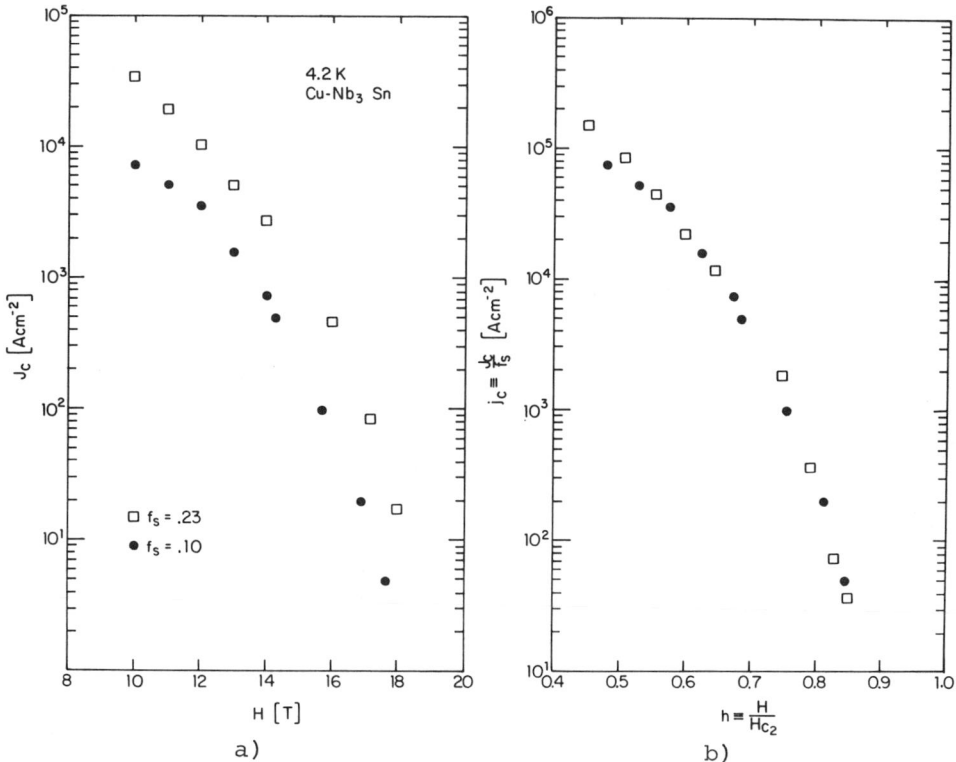

Fig. 3 a) overall and b) normalized critical-current densities
for two in situ Cu-Nb$_3$Sn tapes, containing different volume
fractions of superconducting filaments, plotted versus
applied and reduced field, respectively

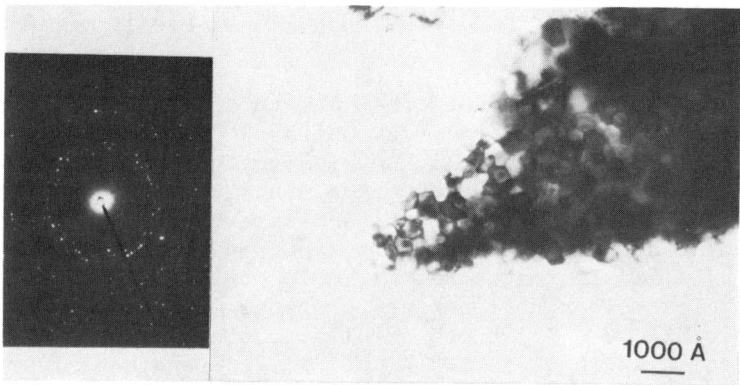

Fig. 4 Transmission electron micrograph and diffraction pattern from
a Cu-Nb$_3$Sn composite showing the fine ($\bar{d} \cong 400$Å), equi-axed
grain structure of the filamentary material (from Ref. 15).

plane strain already during the wire drawing process.)

Figure 5 shows the angular dependence of the normalized critical current measured at 3 and 16 tesla in a Cu-Nb$_3$Sn multifilamentary tape with 23 vol.% of Nb$_3$Sn filaments. θ is the angle between
the applied magnetic field and the tape (or filament) surface. As
θ increases from 0 to 90° the initial depinning current decreases
and the flux-flow characteristics gradually change.[6] The strongest
pinning occurs when the fluxoids are parallel to the filament sur-
faces; furthermore, the anisotropy factor $\Delta(H) \equiv [J_c(\|)/J_c(\perp)]_H$
depends strongly on the magnitude of the applied field (Fig. 5) and
can reach almost an order of magnitude in tapes with high aspect
ratio, measured at high fields $(H \simeq 0.9\ H_{c2})$.

In Cu-Nb filamentary tapes with single-crystalline Nb filaments,one would expect the relative contribution of the surface
pinning to the bulk pinning force to be even more important. Indeed,
our recent study[22] of transport behavior of Cu-Nb composite tapes
reveals a pronounced peak in $\Delta(H)$ as a function of H with
$\Delta(H)^{max}$ exceeding two orders of magnitude.

Turning back to the high-field Cu-Nb$_3$Sn composites, the combined effect of small grain size, surface pinning and proximity
effect results in very high critical-current densities (Figs. 6,7),
particularly in highly reduced in situ tapes.[7] The self-field
critical-current density of tape No. 2 (Fig. 7), normalized to the
filament volume fraction was 1.4×10^7 A cm^{-2}, and the maximum flux
pinning force F_p at 3T exceeded 7×10^{10} N m^{-3}. These values
are comparable to the highest values of J_c and F_p in Nb$_3$Sn
thin films and layered composites. It should be again stressed
that J_c behavior in the long-filament limit is dictated by the
properties of the filamentary material, including filament-matrix
interfaces. While proximity effect coupling is not crucial to keep
ρ_{REM} low it nevertheless enhances J_c, particularly at low fields,
through increased f_s.

We have recently extended the in situ approach to preparation
of Cu-V$_3$Ga composites.[23,24] In contrast to earlier published
work, we have demonstrated that Cu-V$_3$Ga conductors with excellent
critical properties can be produced by in situ techniques. Transition temperature (15.5 K midpoint) and upper critical field (22.4 T)
in particular are among the highest reported in the literature even
for bulk V$_3$Ga and indicate that stoichiometric V$_3$Ga can be grown at
relatively low temperatures in the presence of the copper matrix.
Typical overall J_c for these samples, containing only 20 vol.% V,
was 10^4 A cm^{-2} at 18 T and 10^3 A cm^{-2} at 21 T and clearly exceeded
the J_c of Cu-Nb$_3$Sn composites with comparable f_s and diameter
at fields above ~10 T (Fig. 6).

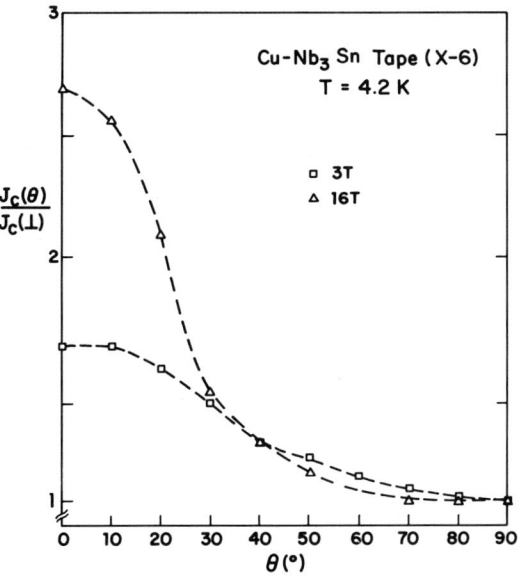

Fig. 5   Normalized J_c versus angle θ for a Cu-Nb$_3$Sn multifilamentary tape with 23 vol.% of Nb$_3$Sn filaments (from Ref. 6).

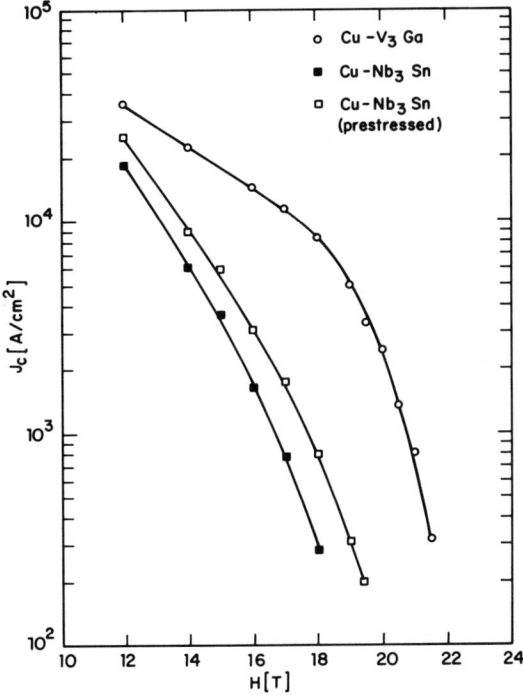

Fig. 6 J_c of Cu-V-Ga and Cu-Nb-Sn *in situ* composites with comparable cross-sectional area ($\sim 4.4 \times 10^{-4}$cm^2) and filament volume fraction (23-25 vol.%) (from Ref. 24).

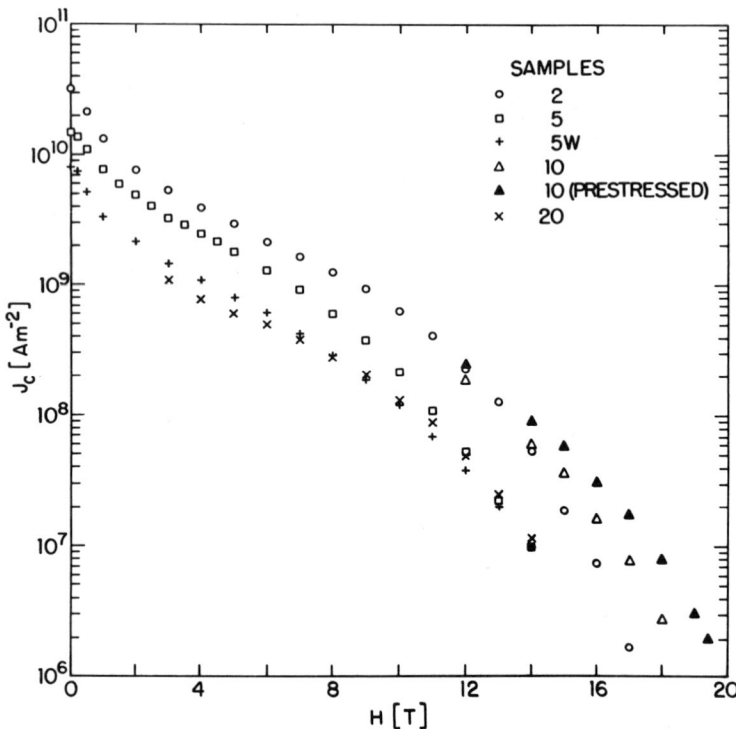

Fig. 7 Overall critical-current density versus applied transverse
        magnetic field for Cu-Nb$_3$Sn in situ tapes and wire (5 W)
        fabricated from two-phase Cu-Nb alloys containing only
        18.2 vol.% (17.3 wt.%) Nb. Sample numbers indicate the
        diameter (in mils) of the Cu-Nb wires before rolling
        to tape (from Ref. 7).

MECHANICAL PROPERTIES AND STRENGTHENING MECHANISMS

One of the most striking characteristics of the ultrafine filamentary composites prepared by in situ techniques is their exceptional mechanical strength. Our initial observations of enhanced yield stress and ultimate tensile strength of Cu-Nb$_3$Sn composites,[25] and of enhancement in J_c in composites under bending strain,[25] stimulated a more detailed study of strengthening mechanisms in the Cu-Nb system.[4]   We found that in composites with very small filaments (d $\lesssim$ 1000 Å) the ultimate tensile strength increases anomalously even at very low volume fractions (10-18%) of filaments. The highest values (2.2 GPa at R.T.; 2.9 GPa at 77K) were found to approach the estimated theoretical strength of the material and equal the strength of the best copper whiskers (Fig. 8). This remarkable behavior is linked to the presence of densely spaced interfaces which inhibit dynamic recovery in both the matrix and the filaments and result in accelerated work hardening and extremely high dislocation densities (N_{disl}). Since the flow stress in workhardened materials scales with $N_{disl}^{1/2}$, it is clear that increasing N_{disl} by a factor of 40-50 over the values attainable in a single-phase material should result in a dramatic increase in strength. In the smallest composites (D = 25 µm), however, one finds an interesting coexistence of two high-strength components with diametrically opposite defect structure: Nb filaments with virtually no dislocations and therefore whisker-like behavior, and the matrix, packed with so many dislocations and other defects that the stresses, required to move dislocations from one interface to another, approach theoretical strength.

Although Cu-Nb composites cannot be used as high-field super-conductors because of the relatively low H_{c2} of niobium filaments, their development may nevertheless be of crucial importance in the current efforts to achieve higher magnetic fields. The pulsed normal-state magnets used for this purpose are limited in their performance by the combination of strength and conductivity of the material from which they are constructed. In terms of their strength and conductivity, the Cu-Nb conductors (and other similar composites currently studied at Harvard) are far superior to the best presently known conventional materials.[26]   Moreover, in contrast to other high-strength and high-conductivity materials, whose electrical conductivity is limited by the impurity scattering and therefore virtually independent on the temperature, the main sources of scattering in Cu-Nb composites are dislocations and interfaces. Lowering the temperature to, say, 77K will therefore result not only in 25-30% increase in strength but also in 2-3 fold increase of electrical conductivity.

The matrix dislocation density in Cu-Nb$_3$Sn and Cu-V$_3$Ga composites is of course much lower than in Cu-Nb due to the necessary high-temperature anneal. The resistance to plastic flow and the

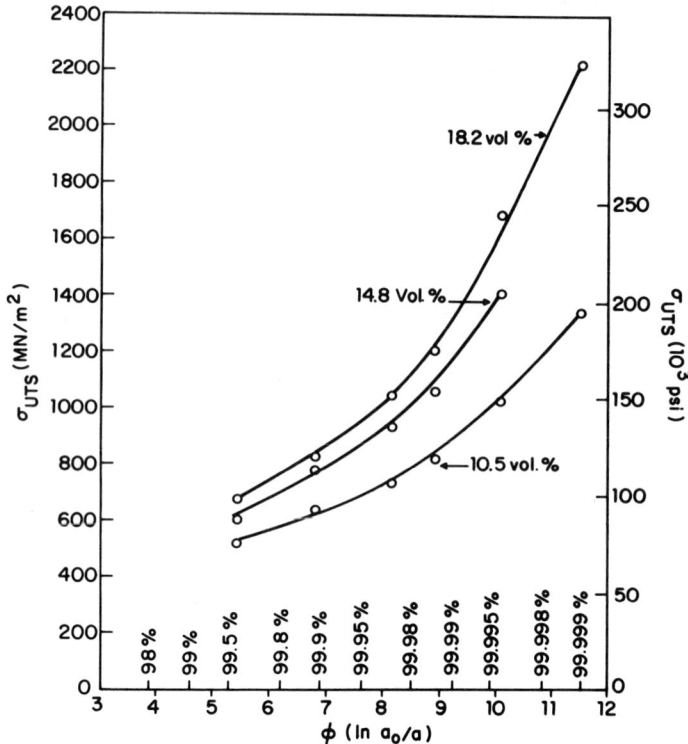

Fig. 8   Ultimate tensile strength of in situ Cu-Nb composites as a
function of true strain ϕ (from Ref. 4).

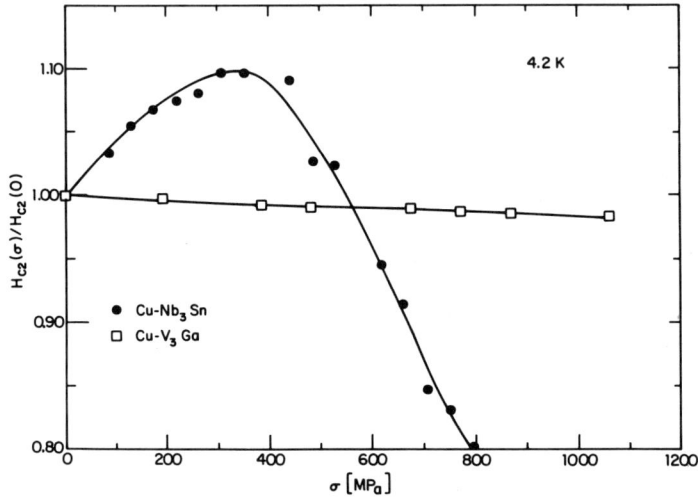

Fig. 9 H_{c2} dependence on the applied uniaxial stress for $Cu-Nb_3Sn$
and $Cu-V_3Ga$ in situ composites with comparable diameter
(~ 0.25mm) and filament volume fraction ($\sim 0.23 - 0.25$).

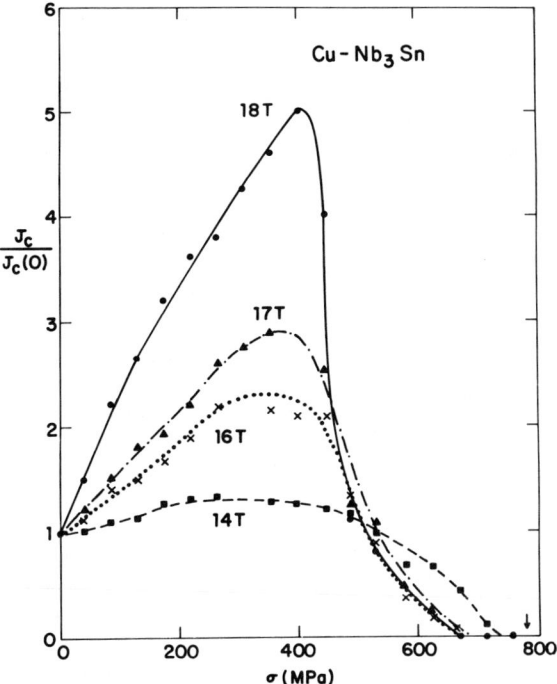

Fig. 10   Normalized J_c vs σ for a Cu-Nb$_3$Sn in situ tape with 23 vol.% of Nb$_3$Sn filaments.

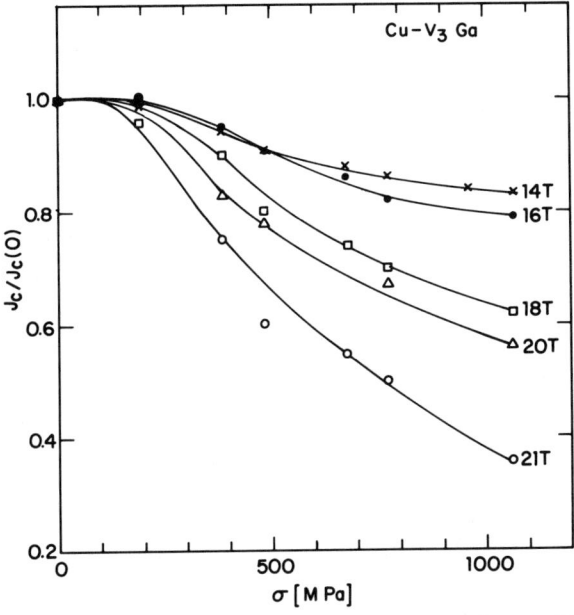

Fig. 11   Normalized J_c vs σ for Cu-V$_3$Ga composites (from Ref. 27).

ultimate tensile stress are nevertheless very high due to small
interfilament spacing and strong filament-to-matrix bonding.[25]
The most striking difference between in situ and conventional composites, however, is not in their strength but rather in the stress/
strain tolerance of their critical properties. Our experiments show
that both $Cu-Nb_3Sn$ and $Cu-V_3Ga$ composites can be subjected to
stresses close to their respective σ_{UTS} without any permanent degradation in J_c or H_{c2}. In fact, in $Cu-Nb_3Sn$ composites, whose
H_{c2} depends sensitively on the applied stress (Fig. 9), one can
substantially reduce the negative effect of the compressive stress
exerted on the filaments by the matrix by simply prestressing composites to $\sim$ 60-70% of their ultimate tensile strength.[7]   The
net effect of this procedure is a permanent increase of $J_c(H)$ to
the peak values of $J_c(H)$-σ curves (Fig. 10, also Fig. 6). In
contrast, H_{c2} of $Cu-V_3Ga$ composites depends only weakly on the
applied stress (Fig. 9) and, consequently, stress effects on J_c
are much smaller (Fig. 11) but, again, fully reversible.[27]

CONCLUSIONS

     The superconducting and mechanical behavior of in situ fila-
mentary composites is in many respects markedly different from that
of continuous filamentary conductors. In order to optimize their
practical properties, one should, therefore, not limit oneself
only to conventional approaches but also exploit the mechanisms which
come into play because of their unique microstructural features,
particularly high interfacial density. Interaction of matrix dislocations and of the flux-line lattice with interfaces are only two
examples which should be further explored.

     One should also bear in mind that the main attractiveness of
in situ composites is their unique combination of properties. Although composites with higher volume fraction of superconducting
filaments generally carry higher critical currents, one should be
cautious when using this approach to improve their J_c performance.
It may come at the expense of lower stability, lower bending strain
tolerance, higher ac losses,[28]   etc., the very factors which stimulated the development of in situ and other alternative techniques
in the first place. At this stage, it appears that possibilities
for further improvement of high-field properties by proper alloying,
by manipulating the microgeometry and microstructure, and by prestressing, have not yet been exhausted.

ACKNOWLEDGMENTS

     The authors acknowledge the contributions during various stages
of this work by M.R. Beasley, W.J. Skocpol, A. Davidson, J.P.
Harbison, C.J. Lobb, F. Habbal and K.R. Karasek. This work has
been supported by NSF Grants DMR79-05011 and DMR-76-01111. All
high-field measurements reported in this paper have been performed

at the Francis Bitter National Magnet Laboratory.

REFERENCES

1. The work of other groups active in this field is reviewed elsewhere in these Proceedings.
2. H. C. Freyhardt and R. Bormann, these Proceedings; R. Roberge and S. Foner, these Proceedings.
3. D. K. Finnemore and J. D. Verhoeven, these Proceedings.
4. J. Bevk, J. P. Harbison, and J. L. Bell, J. Appl. Phys. 49:6031 (1978).
5. W. F. Hosford, Jr., Trans. Metall. Soc. AIME 230:12 (1964).
6. F. Habbal and J. Bevk, AIP Conf. Proc. 58:299 (1980).
7. J. Bevk, J. P. Harbison, F. Habbal, G. R. Wagner, and A. I. Braginski, Appl. Phys. Letters 36:85 (1980).
8. A. I. Braginski, G. R. Wagner, and J. Bevk, Adv. Cryo. Eng. 26, in press.
9. M. Tinkham, unpublished.
10. See, for example, V. K. S. Shante and S. Kirkpatrick, Adv. Phys. 20:325 (1971); H. Scher and R. Zallen, J. Chem. Phys. 53:3759 (1970); R. Zallen and H. Scher, Phys. Rev. B4:4471 (1971).
11. A. Davidson, M. R. Beasley, and M. Tinkham, IEEE Trans. on Magnetics MAG-11:276 (1975).
12. T. J. Callaghan and L. E. Toth, J. Appl. Phys. 46:4013 (1975).
13. A. Davidson and M. Tinkham, Phys. Rev. B13:3261 (1976).
14. C. J. Lobb, M. Tinkham, and W. J. Skocpol, Solid State Commun. 27:1273 (1978).
15. J. Bevk and J. P. Harbison, J. Mat. Sci. 14:1457 (1979).
16. R. Roberge, private communication.
17. T. Luhman and M. Suenaga, Appl. Phys. Letters 29:61 (1976).
18. J. W. Ekin, IEEE Trans. on Magnetics MAG-15:197 (1979).
19. A. I. Braginski and J. Bevk, Bull. Am. Phys. Soc. 25:385 (1980).
20. E. J. Kramer, J. Appl. Phys. 44:1360 (1973).
21. T. Luhman, C. S. Pande, and D. Dew-Hughes, J. Appl. Phys. 47:1459 (1976).
22. F. Habbal, K. R. Karasek, and J. Bevk, to be published.
23. J. Bevk, F. Habbal, C. J. Lobb, and J. P. Harbison, Appl. Phys. Letters 35:93 (1979).
24. J. Bevk, F. Habbal, C. J. Lobb, and G. Dublon, Adv. Cryo. Eng. 26, in press.
25. J. P. Harbison and J. Bevk, J. Appl. Phys. 48:5180 (1977).
26. K. R. Karasek and J. Bevk, Scripta Met. 13:259 (1979).
27. J. Bevk and F. Habbal, Appl. Phys. Letters 36:336 (1980).
28. ac losses in in situ composites are discussed elsewhere in these Proceedings.

POWDER METALLURGICALLY PREPARED A15 MICROCOMPOSITE SUPERCONDUCTORS

H. C. Freyhardt, R. Bormann, and K. Mroviec

Institut für Metallphysik, Universität Göttingen, and
Sonderforschungsbereich 126
Göttingen/Clausthal, Germany

INTRODUCTION

Superconducting composite materials[1] became of technological interest because they combine the excellent performance of brittle A15 superconductors with the good mechanical properties of ductile copper. Commercially they are produced by utilizing the bronze technique and are available as multicore wires, where superconducting filaments are embedded in a mechanically and electrically stabilizing copper or copper-bronze matrix.

The A15 phase, e.g., Nb_3Sn or V_3Ga, is formed in a diffusion and reaction treatment at temperatures between 600–700°C at the interface between the transition metal (Nb or V) and the bronze matrix. Its microstructure and stoichiometry are determined by the reaction kinetics and the ternary equilibrium phase diagrams (Cu-Nb-Sn, Cu-V-Ga). Stoichiometric A15 phases form only from a concentrated Cu bronze; the stoichiometry is adjusted according to the tie lines in the α-bronze/A15 two phase region.[2]

Although this process is used to fabricate commercial A15 conductors, the generally required large filament densities and small filament diameters (≤ 1 μm) are not readily obtainable, in particular not without repeated bundling or without intermediate annealing treatments. Moreover, temperatures above 650°C are necessary, if the Nb or V filaments were to be transformed completely into the A15 phase. Larger reaction temperatures, however, tend to result in coarser grains and deteriorate the critical current densities, in particular those of V_3Ga.[3]

Recent investigations have searched for alternative processes to produce A15 superconductors. Mainly there are three different approaches involved: (1) In the in-situ technique[4-8] finely dispersed Nb particles or dendrites form during a (continuous) rapid casting or solidification of a Cu-Nb (sometimes a Cu-Nb-Sn) melt. By rolling and wire drawing, these Nb particles (1-20 µm) are deformed into long filaments of diameters $\leq$ 100 nm. Nb is converted into Nb_3Sn during a reaction treatment of the wires after they have been plated with Sn. (2) In the powder-metallurgical process[9-11] a two- (CuNb) or multi- (CuNbSn) phase microstructure is obtained by mixing and compacting powders of the constituents. The subsequent deformation into wires or ribbons and the reaction treatment is analogous to the in-situ technique. (3) The infiltration[12] method involves the preparation of a ductile Nb matrix containing a controlled network of pores which are infiltrated with liquid metals (Sn) or low-melting-point eutectics (e.g., Al-Ge).

In this contribution we discuss the hot-powder-metallurgical process, developed at the Universities of Göttingen and Clausthal.

(HOT) POWDER METALLURGICAL PREPARATION OF A15 COMPOSITE SUPER-CONDUCTORS

Principle

A mixture of Cu and Nb (or V) powder (particle size $\leq$ 50 µm, sifted or unsifted) is hot extruded at temperatures between 900-1000°C subsequently drawn into wire, plated with Sn (or Ga), and reaction treated, generally at 550 to 650°C.

Initially the desired filamentary structure of the Cu-Nb wires was only poorly developed because of a severe solution hardening of the bcc Nb mainly due to interstitial oxygen. After the hot extrusion the oxygen content in the Nb reached $\sim$2-3 at.%, which is about the solubility limit at the extrusion temperatures. Obviously the oxygen adsorbed at the surface of the powder particles must have diffused into the Nb, increasing its microhardness to $\sim$3500 MN/m^2. A successful co-deformation of Nb and Cu particles requires, therefore, a sufficient reduction of the oxygen content in the Nb. Three alternative purification methods were tested. (1) Because the oxygen solubility in Nb decreases with temperature, part of the oxygen could be precipitated in the form of oxides during an annealing treatment at 600-700°C subsequent to the hot extrusion at $\sim$1000°C. This treatment indeed proved to be beneficial, however, the deformation of the Nb particles was still insufficient. (2) Similarly, a reduction treatment of the composite in an H_2 or CH_4 atmosphere (below the melting point of the copper matrix) did not improve the deformability of the Nb. (3) This was only achieved

after an internal reduction during hot extrusion. An additional
component is added to the CuNb powder mixture, which has a larger
binding enthalpy for oxygen than Nb does (390 kJ/g-atom O: interstitial solution). A selection of various additives, together with
the binding enthalpies, is given in Table 1. Generally they are
added in elemental form, except Ca, where Cu_5Ca was preferred.
0.5-1 wt% of the additive is sufficient to lower the oxygen content
in the Nb to below 0.1 at.% after the hot extrusion. The microhardness of the Nb particles drops to $\sim$1200 MN/cm^2, which is about
the value of the surrounding Cu matrix. Additives, not consumed
in the internal reduction treatment, either dissolve in the matrix
(Al,Mg) or form compounds (e.g., Zr,Hf).

The oxide particles, whose distribution was studied,[13] and the
dissolved or precipitated additives do not impair the good deformability of the composites. The extruded compacts can be rolled or
wire drawn without intermediate anneals. The deformation of the
Nb particles to long filaments is ideal (Fig. 1a) and can be calculated from macroscopic degrees of deformation q (q=reduction in
cross-sectional area) of the wire. Depending on the initial powder
particle size and q, filament diameters between 0.1 and 2 μm can be
achieved easily. The filament cross-section does not stay circular
(Fig. 1b) due to the development of a fiber texture.

The tensile strength of the as-deformed Cu-20 wt% Nb composite[14]
increases with q and reaches values of 2000 MN/m^2 ($q=10^5$) at strains
between 2 and 3%. An annealing treatment at 550°C for 5 min reduces the tensile strength to $\sim$1400 MN/m^2, whereas the strains are
increased to $\sim$16%. Taking into account the texture of the composite
one can estimate a theoretical tensile strength, which is larger
than the experimental value by a factor of $\sim$2.4.

Thermal Stability of the Filamentary Structure

Prolonged annealing at relatively high temperatures, e.g.,
during the reaction treatment, results in a degradation of the
critical current densities of the wires. This can be ascribed to
a spheroidization[15-17] of the filaments, which disintegrate into

Table 1. Oxygen Binding Enthalpy in Oxides of
Different Additives

Additive	Al	Zr	Hf	Mg	Ca	La
Oxide	Al_2O_3	ZrO_2	HfO_2	MgO	CaO	La_2O_3
Binding Enthalpy kJ/g-atom O	536	540	570	603	636	640

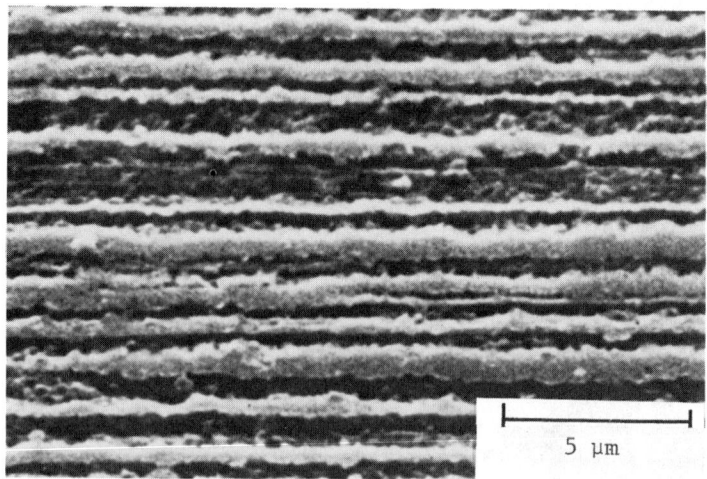

Fig. 1a. Longitudinal cross section of a Cu-30 wt% Nb+1 wt% Al
         composite (q=700). SEM.

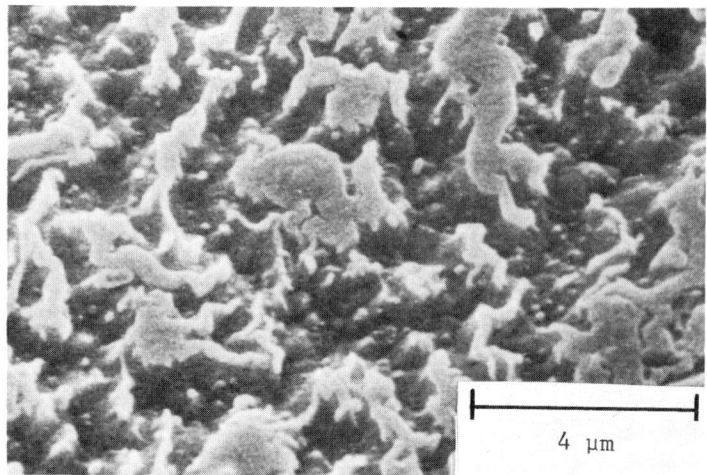

Fig. 1b. Cross section of a Cu-30 wt% Nb composite + additive
         (q=580).

rows of individual (finally spherical) particles. The time, t_{inc},
required to spheroidize a filament depends on the aspect ratio of
its cross section and - for cylindrical filaments of radius R - is
proportional to R^3 (Bormann[10]). It is to be noted that for the
composites investigated, the time t^*, required to completely convert Nb into Nb_3Sn, is found[10] to be shorter than t_{inc} at a given
reaction temperature.

Formation of the A15 Phase

Subsequent to the deformation, the wires are electrolytically
plated with Sn (or Ga) and diffusion and reaction treated between
550 and 650°C, generally long enough to convert all Nb or V into
Nb_3Sn or V_3Ga, respectively. The accompanying Sn (or Ga) diffusion
is rapid and seems to be assisted by grain boundary diffusion. The
ability to form, e.g., Nb_3Sn by the bronze process is a direct consequence[3] of the ternary Cu-Nb-Sn phase diagram, where equilibrium
tie lines (schematically shown in Fig. 2) exist between the α-CuSn
solid solution and the A15 phase and where the A15 field does not
extend far into the ternary phase diagram. Because of a low ter-
minal solubility of Sn in Nb, all Sn added to a CuNb composite is
initially consumed (c.f. broken line in Fig. 2) to produce Nb_3Sn,
which could be confirmed by measurements of the residual resistivity of CuNb+Sn composites.[9,10] Tie lines from the α-CuSn of different Sn concentrations are coupled with A15 phases of different
stoichiometries. Consequently the transition temperature T_c (and
also the upper critical field B_{c2}) of Nb_3Sn, which depend on its
stoichiometry, are determined by the Sn content of the bronze.
This will generally lead to a spread in the T_c values of the A15
layers if gradients in the stoichiometry are established during
the progress of the diffusion reaction. The microstructure on the
other hand, in particular the grain size distribution of the A15
phase, is determined by the reaction kinetics and can be influenced
by alloying additional elements to the composite (c.f. K. Tachikawa,
this Proceedings).

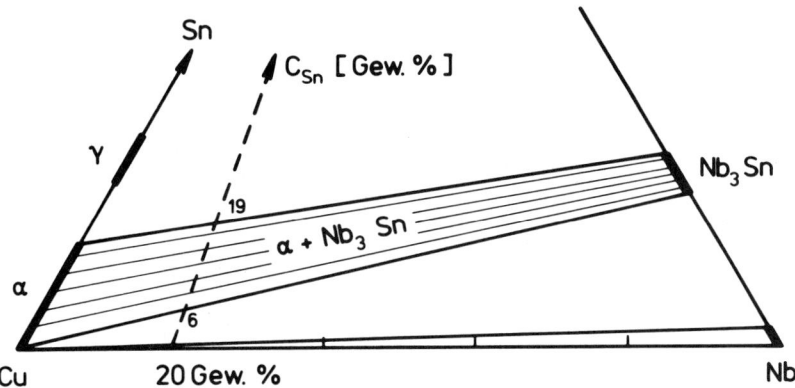

Fig. 2.   Schematic representation of the ternary CuNbSn phase
          diagram at 600°C.

SUPERCONDUCTING PROPERTIES

The transition temperature, T_c, of a Cu–Nb$_3$Sn composite reaches maximum values between 16.5 and 17 K. It is mainly determined by the stoichiometry and the dimensions of the A15 layer/ filament and by the stresses due to a difference in thermal expansion between Nb$_3$Sn and the bronze. For a Cu–30 wt% Nb composite T_c decreases for Sn concentration below 8 wt%.

Upper critical fields B_{c2} can be determined from a plot of $J_c^{1/2} B^{1/4}$ versus the magnetic field B (Fig. 3). Extrapolations of the straight lines in the field region between 13 and 15 T yield values of B_{c2} between 17 and 18.5 T, which (slightly) depend on tin concentration (Fig. 4). The high-field tail in Fig. 3, however, extrapolates to values above 20 T (see also ref. 18). Because of the differences in the ternary phase diagrams[2,3] Cu–V$_3$Ga composites exhibit a stronger variation of B_{c2} with Ga concentration.[10] B_{c2} changes caused by different treatments are brought about by variations of the homogeneity and stoichiometry of the A15 phase and by altering the thermal stresses. The physical or structural reason of the B_{c2} changes due to these effects is not yet understood.

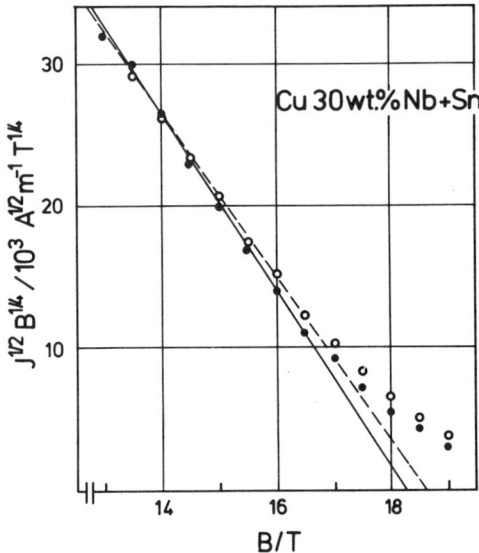

Fig. 3. $J_c^{1/2} B^{1/4}$ vs the magnetic field B for a Cu–30 wt% Nb+Sn composite reacted at 600°C without (●) and with (o) an additional heat treatment.

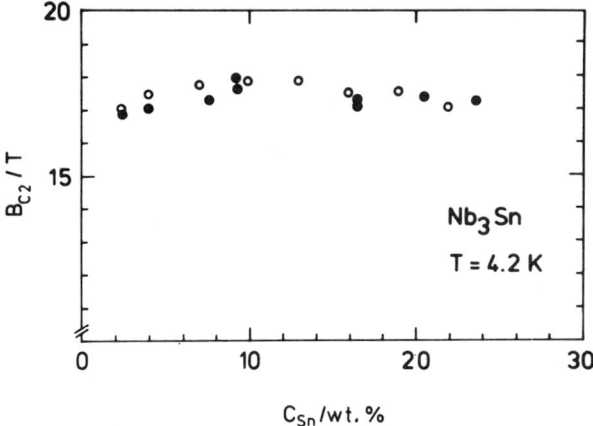

Fig. 4. Upper critical field B_{c2} vs tin content for Cu-20 wt% Nb
(o) and Cu-30 wt% Nb (●) composites (obtained from extrapolations from the field region 13-15 T; c.f. Fig. 4).

Overall critical current densities J_c of 2×10^8, 6×10^7, and
1.2×10^7 A/m^2 can be reached for unstressed Cu-30 wt% Nb+Sn wires
at 14, 16, and 18 T, respectively (Fig. 5). For Cu-30 wt% V+Ga
wire J_c at 16 T can exceed 4×10^8 A/m^2.

The critical current density J_c does not change appreciably
with the area reduction q beyond q ≈ 750. However, J_c (B=const.)
strongly depends on the volume percentage of the A15 phase formed
for Sn (or Ga) contents which are smaller than the amount needed
to convert all Nb (or V) into the A15 phase. J_c stays constant
above this threshold and decreases for very tin-rich alloys, as
expected from the phase diagrams. Whereas the critical current
at a given field remains approximately unchanged for reaction
times t > t* it deteriorates if spheroidization occurs (over-
aging t > t_{inc}), the earlier the higher the reaction temperature.

All investigations performed so far indicate that (for
strongly deformed composites with a well developed fiber structure and which are not spheroidized) the critical transport currents are determined by a bulk pinning of the flux lines due to
grain boundaries of the A15 phase, whereby the strength of the
pinning centers also depend on the reversible properties of the
material, in particular on the upper critical field (scaling

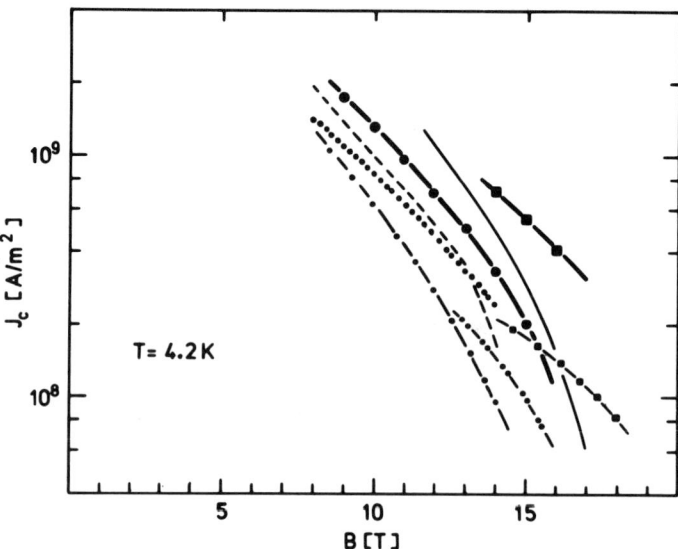

Fig. 5.   Overall critical current density vs external field in
          various composite materials
          Nb$_3$Sn: Cu 30 wt% Nb+Sn (this work),
                                  Cu 40 wt% Nb+Sn (Flükiger et al.),
                                  Cu 36 wt% Nb+Sn (Fihey et al.),
                                  Cu 17 wt% Nb+Sn (Bevk et al.),
                                  Cu 30 wt% Nb+Sn (Verhoeven et al.),
                                  commercial Nb$_3$Sn multicore conductor,
                                  25 wt% Nb (Vacuumschmelze),
          V$_3$Ga: Cu 30 wt% V+Ga (this work),
                                  Cu 14.5 wt% V+Ga (Bevk et al.).

laws) which, in turn, is determined by T_c and the residual resistivity of the A15 phase. Consequently, treatments which increase
B_{c2} will favorably influence J_c in the high field region. This
can be demonstrated if a tensile stress is applied to the wires
which counteracts the thermally induced straining of the A15 phase
(Fig. 6). J_c at 16 T reaches a maximum at

$$\varepsilon_{max} \overset{\sim}{\approx} 0.8\% \ (J_c(\varepsilon_{max})/J_c(\varepsilon = 0) \overset{\sim}{\approx} 2.4),$$

accompanied by a B_{c2} enhancement. This behavior is analogous to
what is observed for conventional multicore conductors, however,
the ε_{max} values and enhancement factors are different.

    If compared to commercial, in situ, and infiltration processes
used for preparing A15 composite conductors, the hot powder

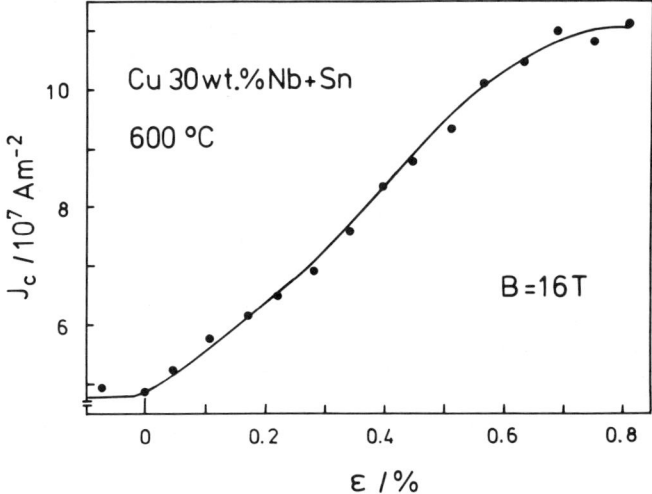

Fig. 6. Critical current density J_c vs strain ε at 16 T for a Cu-30 wt% Nb+Sn composite $(T_{ann}=600°C)$.

metallurgical approach offers certain advantages. The fractions of various constituents can be changed easily and further elements, which allow to optimize the conductor (Ta, Hf, Zr; In, Ga to Nb_3Sn. Si; In, Sn to V_3Ga), can be added readily. Because of the ideal compactation, sintering and purification during hot extrusion, the compacts can be drawn into wire down to area reduction ratios q of $\sim 10^5$ without intermediate anneals. At filament densities of the order of 10^5 (in the total cross section), final filament diameters ranging from 200-500 nm can be achieved (if desired). These small diameters of the filaments, in turn, allow the reaction treatment to be performed at relatively low temperatures (around 600°C), which generally favors small grain sizes in the A15 phase and, thus, large critical transport currents, in particular for V_3Ga conductors. The critical current densities are largely insensitive to the major preparation parameters. Consequently, additional optimization, e.g., with regard to ac losses, is rendered possible. The fact that the filaments are not continuous obviously does not influence the critical currents for highly deformed and well behaved composites. No principle difficulties are anticipated to upgrade the process to a technologically attractive scale.

REFERENCES

1. Metallurgy of Superconducting Materials in "Treatise on Mate-
    rials Science and Technology", T. Luhman and D. Dew-Hughes,
    eds., Academic Press, New York (1979).
2. R. H. Hopkins, G. W. Roland, and M. R. Daniel, Met. Trans.
    8A:19 (1977).
3. J. D. Livingston, General Electric Report 78 CRD 140 (1978).
4. C. C. Tsuei, Science 180:57 (1973).
5. J. Bevk and J. P. Harbison, J. Mater. Sci. 16:1457 (1979);
    J. Appl. Phys. 48:5180 (1977).
6. S. Foner, E. J. McNiff, Jr., B. B. Schwartz, R. Roberge, and
    L. J. Fihey, Appl. Phys. Letters 31:853 (1977).
7. E. P. Romanov, H. C. Freyhardt, and L. Schultz, Scripta Met.
    12:151 (1978).
8. J. D. Verhoeven, D. K. Finnemore et al., Appl. Phys. Letters
    33:101 (1978).
9. R. Bormann, H. C. Freyhardt, and H. Bergmann, Appl. Phys.
    Letters 35:944 (1979).
10. R. Bormann, Ph.D. Thesis, Univ. of Göttingen (1979).
11. R. Flükiger et al., IEEE Trans. on Magnetics MAG-15:689
    (1979); Appl. Phys. Letters 35:810 (1979).
12. M. R. Pickus, LBL 8580 Report, 1978, p. 88.
13. K. Mroviec, Diploma Thesis, University of Göttingen, to be
    published.
14. E. Goudah, Diploma Thesis (1979); F. E. Keunecke (1978), un-
    published, Univ. of Clausthal; H. Bergmann, B. L. Mordike,
    and R. Bormann, Proc. DGM Tagung Verbundwerkstoffe,
    Konstanz (1980).
15. H. E. Cline, Acta Met. 19:481 (1971).
16. J. van Suchtelen, J. Cryst. Growth 43:28 (1978).
17. S. Hansknecht (1980), unpublished.
18. J. Bevk, J. P. Harbison, and F. Habbal, Appl. Phys. Letters
    36:35 (1980).

CRITICAL CURRENTS OF Cu-$(Nb_{1-x}Ta_x)_3Sn$ IN SITU MULTIFILAMENTARY WIRES

R. Flükiger

Kernforschungszentrum Karlsruhe
Institut für Technische Physik
7500 Karlsruhe
Federal Republic of Germany

INTRODUCTION

Tantalum additions to the Nb core are known to considerably increase the critical current of bronze processed Cu-Nb_3Sn multifilamentary wires, particularly at magnetic fields above 12 T.[1,2] The present work is an attempt to reproduce this result on wires prepared by the in situ technique.[3,4] The addition of Ta as a third component, which is easily performed on bronze processed wires, introduces, however, a fundamentally new problem for the in situ process: the step from a two-component to a three-component system makes it impossible to predict the nature of the precipitates. The formation of additional phases can practically be excluded since Ta, like Nb, is nearly immiscible with Cu in the solid state,[5,6] while Ta and Nb are completely miscible. The main subsisting question is whether Ta would precipitate out of the melt in its elementary state, rather than dissolve in the Nb dendrites. The measurement of J_c at high fields allows to give a qualitative answer to this question.

SAMPLE PREPARATION

The starting materials were Grade I Nb and Ta powders and 99.999% Cu pieces from Johnson Matthey. The in situ samples were all melted at the University of Geneva. In order to study both the influence of the crucible material on the morphology of the dendrites as well as its possible influence on the quantity of Ta dissolved in the precipitated Nb, two different melting techniques were used, i.e., levitation melting (crucible-free)

and chill casting with graphite crucibles. The appropriate melting temperature for Cu-Nb samples with Ta additions varies between 1850 and 1900°C and depends on the Ta content. Unfortunately, two levitation melted samples containing Ta (x=0.05 and 0.10)
were not homogeneous, the rf power of the used generator being
not sufficient to reach the required high melting temperature.
The comparison between the dendrite morphologies will thus be
limited to pure Cu-Nb samples (x=0). The effects of Ta additions
will be discussed on chill cast samples only.

Levitation Melting

    Bevk et al.[7] first prepared small Cu-Nb samples by rf lev-
itation. In the present work, the samples were drawn immediately
after the levitation melting, which was carried out in a cold
Cambridge crucible, shown in Fig. 1. After melting approximately
20 g of material under an argon pressure of 4 atmospheres, the rf
power was rapidly turned off. The sample was quenched at the
contact with the water-cooled bottom of the crucible, situated
around 12 mm below the melting position of the sample. The Nb
dendrites in these samples are mainly oriented in the direction
of the temperature gradient, as shown in Fig. 2. The size of the
primary dendrites is always well below 1 μm, their aspect ratio
being very high. This dendrite morphology represents the ideal
case without any contaminations arising from the crucible material. In order to avoid too small final filament sizes, the
ingot was cut into several pieces, which were drawn to a final
diameter of 0.25 mm, with fiber diameters of the order of 0.1 μm.

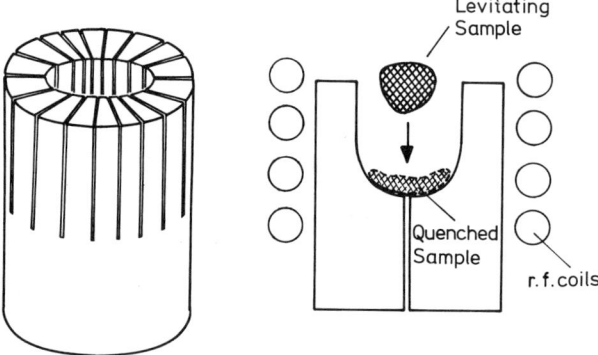

Fig. 1.   The cold Cambridge rf levitation crucible.

Fig. 2.   Cu-30 wt% Nb alloy, melted by levitation.   Deep etched.
          Primary dendrite size 0.6 to 1 μm.

Chill Casting

        The chill cast device used in the present work is very sim-
ilar to that reported by Roberge et al.[4] and permitted the formation of 20 to 30 g ingots having 10 mm diameter.   The temperature
was controlled by a quotient pyrometer and was chosen to 1800°C
for the samples without Ta and 1900°C for the samples with x=0.05
and 0.10 and x=0.20.   The pressure in the furnace was again 4
atmospheres of argon.   The dendrites produced by this method are
represented in Fig. 3 for x=0 and in Fig. 4 for x=0.10.   The main
problem of melting alloys containing Ta is the sample homogeneity: after a first melting, each sample was thus cut into two
halves along the cylinder axis, then remelted after turning up-
side down one of them.   The observed scattering in the measured
values of J$_c$ shows that a certain inhomogeneity in distribution
is still present.   The quenching rate is not identical to that
of levitated samples: the dendrites form at random orientations
(Figs. 3 and 4).   The aspect ratio of the dendrites is now much
smaller, their size being larger than in the levitation case,
i.e., between 2 and 3 μm.   These data confirm earlier experiments
of Verhoeven et al.[8] who studied the difference between Cu-Nb in
situ samples melted either in graphite or in thoria crucibles:
carbon impurities lead to coarser dendrites, with a smaller
aspect ratio than in the ideal case.   The dendrites obtained
after Ta additions are shown in Fig. 4 and show additional

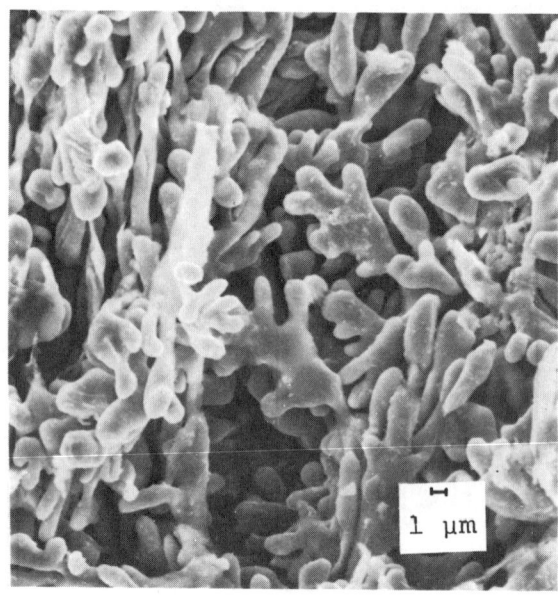

Fig. 3.   Cu-30 wt% Nb alloy, chill cast from graphite crucible.
          Deep etched.   Primary dendrite size 2 to 3 μm.

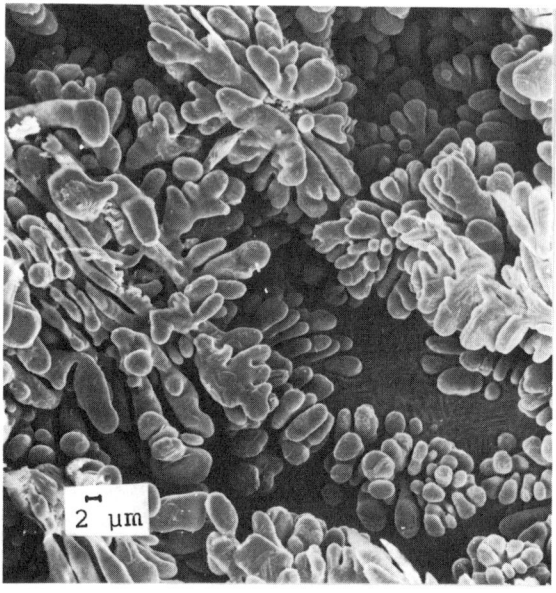

Fig. 4.   Cu-28.5 wt% Nb-1.5 wt% Ta alloy, chill cast from graphite
          crucible.   Deep etched.   Primary dendrite size 3 to 5 μm.

coarsening: their size is of the order of 5 μm. It is, however, difficult to determine whether this additional coarsening is really due to the presence of Ta or to an enhanced effect of carbon impurities due to the higher melting temperature.

After solidification, the chill cast Cu-Nb and Cu-Nb-Ta samples were treated 30 minutes at 800°C and subsequently drawn to wires of 0.25 mm diameter, with filaments varying between 0.1 and 0.3 μm size. The wires were then Sn coated and reacted to Cu-Nb$_3$Sn by the two-step process characterizing the external diffusion technique, the first step consisting in forming the Cu-Sn bronze at temperatures between 200 and 500°C, the second one in forming the Nb$_3$Sn filaments at temperatures between 650 and 700°C, for periods between 1 and 3 days. The Sn content in our samples varied between 10 and 14 wt% with respect to the Cu, which is somewhat lower than the maximum solubility limit, 15.4 wt%.[6] The J_c values reported in this paper are compared to the highest reported values of chill cast Cu-Nb$_3$Sn wires prepared by Roberge et al.,[9] with the composition Cu-36 wt% Nb-20 wt% Sn.

RESULTS

The critical current densities and the behavior of J_c under stress up to magnetic fields of 19 tesla were measured at the Francis Bitter National Magnet Laboratory. J_c was measured by a standard four-probe technique in transverse fields B_0. The critical current was defined at the 1 μV/cm level of detection. The area used to calculate J_c was the total cross-sectional area of the wire. Typical values of J_c versus the applied field B_0 for chill cast samples with compositions x=0, 0.05, 0.10 and 0.20 after two days at 650°C are shown in Fig. 5. These four wires are described by the formulas Cu-30%Nb-14%Sn, Cu-28.5%Nb-1.5%Ta-11%Sn, Cu-27%Nb-3%Ta-12%Sn and Cu-24%Nb-15%Sn (all values are in wt% relative to the Cu content). The dashed line in Fig. 5 represents the values for a fully optimized in situ wire with the composition Cu-36%Nb-20%Sn reported by Roberge et al.[9] The curves in Fig. 5 show a marked increase of J_c above 12 T as a consequence of Ta additions: in spite of the lower Sn and (Nb+Ta) content, the J_c values of the wire with x=0.05 exceed those of the wire of Roberge et al.[9] for fields above 15 T. The slope of this wire at high fields is comparable to that reported by Suenaga et al.[2] on a wire with x=0.03. This is an additional strong indication for the fact that Ta has effectively entered into the A15 structure. The wires with x=0.10 and x=0.20 in Fig. 5 have lower J_c values than that with x=0.05, but still show a slower decrease of J_c at high fields than that of the reference wire without Ta.[9] The value of J_c was found to increase slightly with the Sn content. The effect was most marked for the wire without Ta, where an increase from 11 to 14 wt% Sn enhanced J_c by 25% at 14 T. In the present work, the

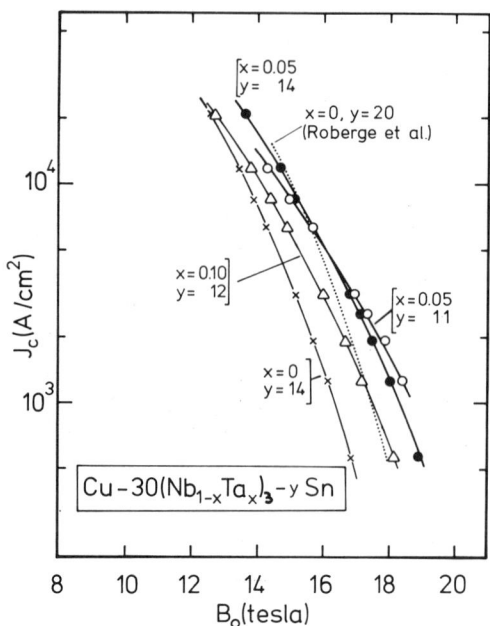

Fig. 5.   Overall critical current density J_c versus applied field
B_0 at 4.2 K for in situ wires Cu-30 wt% $(Nb_{1-x}Ta_x)_{3-y}$
wt% Sn, with x=0, 0.05, 0.10 and 0.20 and y=14,14,12 and
15, after 2 days at 650°C.   The dashed curve is a fully
optimized Cu-36 wt% Nb-20 wt% Sn wire by Roberge et al.[9]

reaction heat treatment was not optimized, 2 days at 650°C being
used for all samples.   In two cases (x=0.05 and 0.10), 1 day at
700°C was also tried, but resulted in slightly lower J_c values.
Like for wires without Ta, the small filament size (0.15 to 0.3 μm)
of in situ wires permits lower reaction temperatures than for the
bronze route: Suenaga et al.[2] found a maximum of J_c after 120 h at
725°C.

The size of the primary dendrites being very small (between
3 and 5 μm) microprobe analysis cannot be used for detecting the
amount of Ta dissolved in the Nb.   The present J_c measurements
allow to give a rough estimation for the solubility limit.   In
Fig. 6, the ratio $J_c(x)/J_c(x=0)$ is plotted versus the Ta content

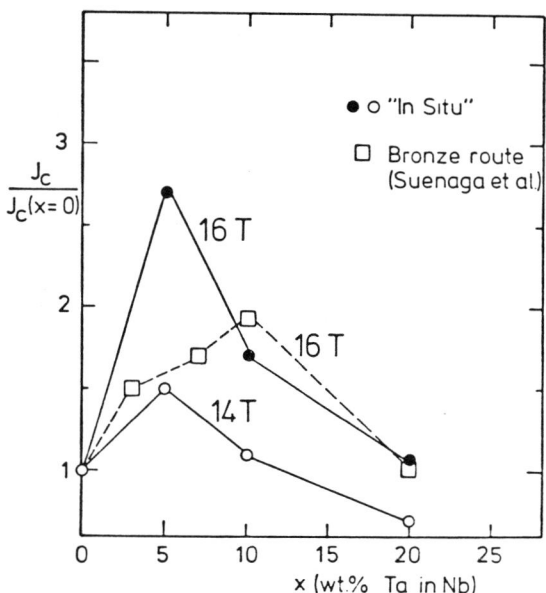

Fig. 6.   The ratio $J_c(x)/J_c(x=0)$ at the magnetic fields $B_0=14$ and
          16 T for in situ wires. For comparison, the data of
          Suenaga et al.[2] on wires produced by the bronze route are
          also illustrated. The maximum of J_c at lower Ta contents
          for in situ wires suggests that Ta is only partly dis-
          solved in Nb.

in Nb, x, for fixed magnetic fields, $B_0=14$ and 16 T. The corresponding data of Suenaga et al.[2] for intermediate annealing
times, 64 h at 725°C, have also been plotted in Fig. 6. The in
situ wires show a sharper peak at lower Ta contents. This figure
may be interpreted as follows. The amount of Ta in Nb precipitates
produced by the in situ process is determined by the initial Nb:Ta
ratio. If this ratio decreases, the amount of free Ta increases,
leading not only to a smaller volume fraction of A15 phase, but
also to interruptions of the Nb filaments by Ta filaments. Since
both effects reduce I_c it follows from Figs. 5 and 6 that
Cu-(Nb$_{1-x}$Ta$_x$)$_3$Sn wires should not contain more than 5% Ta:
$x \leq 0.05$.

     The critical current density, J_c, versus the applied uniaxial
stress for the wire with x=0.10 and y=10 is illustrated by Fig. 7.
At zero stress, J_c is somewhat smaller than for the measurement
reported in Fig. 5, which may be attributed to inhomogeneities in
the Ta distribution. The variation of J_c with stress shows a flat
maximum of J_c at 450 MPa, which is comparable to other in situ
wires[10] and to powder-metallurgically prepared wires.[11] Assuming

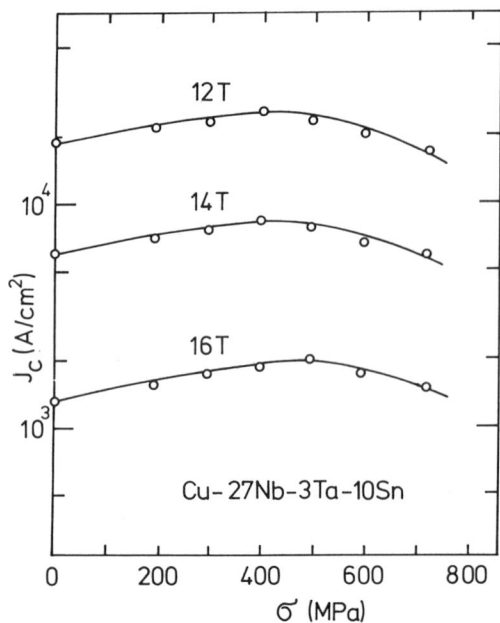

Fig. 7. Overall critical current density as a function of applied
         tensile stress at 4.2 K and fixed fields for the in situ
         wire Cu-27 wt% Nb-3 wt% Ta-10 wt% Sn after 3 days at 650°C.
         At all applied magnetic fields, the maximum of J_c, J_{cm},
         occurs at $\sim$ 450 MPa.

a similar stress-strain behavior than for other in situ materials,
the value of the strain at J_{cm} for this wire would also be of the
same order of magnitude, i.e., $\varepsilon_m \sim 0.6$ to 0.7.[10] This could be
expected from an investigation of Luhman et al.[12] on bronze processed wires, showing little effect on ε_m as a consequence of increasing Ta contents. The ratio J_{cm}/J_{co} in Fig. 7 is almost field
independent and is 1.5 at 16 T, which is much lower than for other
in situ wires where values between 2.5 and 4 have been reported.[10,11]
A probable reason for this behavior is the uncomplete reaction of
the filaments, the Sn content of this particular wire being too low.
Indeed, Foner et al.[10] have shown on a Cu-36% Nb wire that the ratio
J_{cm}/J_{co} decreases for lower Sn contents. On the other hand, the
reduced prestress could also arise from the presence of unreacted
Ta, randomly distributed in the bronze matrix.

CONCLUSION

We have shown that Cu−$(Nb_{1-x}Ta_x)_3$Sn composite wires can be prepared by the in situ technique. At least one part of the Ta precipitating out of a Cu−Nb−Ta melt is dissolved into the Nb dendrites, thus contributing to an enhanced current density, particularly at fields above 12 T. However, the random character of the Ta precipitation limits the applicability of the in situ process to Ta contents $x \leq 0.05$. Further increasing x leads to an accelerated decrease of J_c. Therefore, the enhancement of J_c relative to wires containing no Ta is markedly smaller than for the corresponding bronze route, which seems to be more suitable for controlled ternary or quaternary additions to the core. This is, of course, not valid for additions to the Cu matrix, e.g. Ga, Al, In, ... which can be easily dissolved in Cu by adequate heat treatments prior to the Sn plating.

For a possible large scale production of Cu−Nb_3Sn in situ composite wires containing Ta additions, it follows from the present data and from economical aspects that a Ta content below $x = = 0.05$ should be used. More research in this range should be undertaken.

The preparation of Cu−Nb ingots using either graphite crucibles or levitation melting (crucible-free) confirms that the presence of carbon favorizes the coarsening of the Nb (and also of the Nb−Ta) dendrites. This point could be of importance for an eventual application of the in situ process, where controlled additions to the graphite crucible could control the dendrite size.

ACKNOWLEDGMENTS

The author is grateful to A. Naula of the University of Geneva for his assistance with the preparation of in situ composites by both levitation melting and by chill casting. The use of high field magnets at the Francis Bitter National Magnet Laboratory and very helpful assistances by its staff are greatly appreciated. The author also wishes to thank Dr. S. Foner and E. J. McNiff, Jr. for the use of their stress apparatus.

REFERENCES

1. J. D. Livingston, IEEE Trans. Magn. MAG−14:611 (1978).
2. M. Suenaga, K. Aihara, K. Kaiho, and T. S. Luhman, ICMC Conf., Madison, WI (1979).
3. C. C. Tsuei, Science 180:57 (1973).

4. R. Roberge and J. L. Fihey, J. Appl. Phys. 48:1327 (1977).
5. C. S. Smith, Trans. AIME 215:905 (1959).
6. "The Constitution of Binary Alloys", M. Hansen, ed., McGraw-Hill, New York (1958).
7. J. Bevk and J. P. Harbison, J. Mat. Sci. 14:1459 (1979).
8. J. D. Verhoeven, E. D. Gibson, C. V. Owen, J. E. Ostenson, and D. K. Finnemore, Appl. Phys. Letters 35:270 (1979).
9. R. Roberge, S. Foner, E. J. McNiff, Jr., B. B. Schwartz, and J. L. Fihey, Appl. Phys. Letters 34:111 (1979).
10. S. Foner, R. Roberge, E. J. McNiff, Jr., B. B. Schwartz, and J. L. Fihey, Appl. Phys. Letters 34:241 (1979).
11. R. Flükiger, R. Akihama, S. Foner, E. J. McNiff, Jr., and B. B. Schwartz, Appl. Phys. Letters 35:810 (1979).
12. T. S. Luhman, K. Kaiho, and M. Suenaga, ICMC Conf., Madison, WI (1979).

MAGNETIC PROPERTIES OF MULTIFILAMENTARY Nb_3Sn COMPOSITES[*]

S. S. Shen

Oak Ridge National Laboratory
Oak Ridge, Tennessee 37830

INTRODUCTION

Magnetic behavior of composite superconductors depends strongly on the distribution and the properties of superconducting filaments and the matrix materials. This paper presents experimental results of magnetization measurements on both conventional internal-bronze wires and in-situ composites. The results are presented in M-H curves and analyzed in terms of filamentary magnetization and coupling magnetization. For comparison, results of a typical NbTi-Cu matrix wire are also presented. Techniques to maintain the individuality of each filament under external field change are reviewed and discussed in detail. ac loss analysis of each type of composite for certain possible technical applications is emphasized.

Two major problems often encountered in the development of new superconductors, namely, losses and stability, can be tested by a magnetic measurement. Moreover, "magnetization" - a directly measured quantity that describes the field distribution inside the conductor - can be fully correlated to the detailed structure of a multifilamentary conductor[1] in terms of superconductor distribution

[*]Research sponsored by the Office of Fusion Energy, U.S. Department of Energy, under contract W-7405-ENG-26 with the Union Carbide Corporation.

and matrix properties. The principal aim of this study is to
present such a technical approach to the testing of practial conductor such as multifilamentary Nb_3Sn composites.

It is well known that there are two distinguishable types of
magnetization generally observed in a multifilamentary conductor:
hysteresis and time-dependent shielding. The former results from
the intrinsic diamagnetic properties of individual superconducting
filaments while the latter comes about from the shielding effects
of matrix (eddy-current) and the collective reaction of filament
groups (coupling). To keep the hysteretic magnetization low,
filaments are generally very small (<50 μm). However, if care is
not taken, the coupling effect can be large enough to couple all
filaments and hence make a conductor behave like a bulk superconductor. In this situation the subdivision of the superconductor
into filaments is no longer effective or advantageous. The
consequences may be magnetic instability or high losses, depending
on the amplitude of the external field change. The two techniques
that have often been employed to avoid this are filament twisting
and the use of mixed matrix. Experimentally, if the mode of the
external field change is properly controlled, one can not only
determine the value of each component individually but also correlate them to the effects of these techniques. The information
thus obtained can be used for a conductor application like magnet
design or fed back to a conductor fabrication process. This paper
will show how such tasks are performed for three different samples:
(i) a typical NbTi conductor as reference, (ii) a conventional
Nb_3Sn multifilamentary wire for practical applications, and (iii)
an in-situ Nb_3Sn composite currently under development.

EXPERIMENTAL SAMPLES

Table 1 summarizes the major parameters of three types of
samples. Each sample over 1 m in length was wound into a single-
layer open coil with an inside diameter of 3.81 cm and placed in
the bore of a pulse magnet. All experiments were conducted in
liquid helium at 4.2 K.

Sample 1, a conductor with a low Cu/SC ratio (∿1.85) and a
large number of fine filaments (∿2000), is at present commonly
used for cable conductor elements. Sample 2 is the strand element
developed in an Intermagnetics General Corporation (IGC) development program sponsored by the Wright Patterson Air Force Materials
Laboratory.[2] The conductor has 20,538 fine filaments with 44%
copper provided as a stabilizer for operation at 8 K. The conductors are fabricated by a standard bronze technique in which a
tantalum shell is used as a reaction barrier to protect the copper
from contamination. The results to be presented on this sample
are a series of magnetization measurements on a progressively
heat-treated sample. Sample 3 is an in-situ Nb_3Sn composite

Table 1. Sample major parameters

	1	2	3
Description	Fermi Energy-Saver wire	Air Force phase-II wire	Ames in-situ composite
Type of super-conductor	NbTi	Nb₃Sn	Nb₃Sn^a
Wire size (mm)	0.4	0.75	0.25; 0.5
Sample winding density	0.66	0.45	0.40
Filament size (m)	5×10^{-6}	1.8×10^{-6b}	$<10^{-7}$
Matrix materials	Copper	Bronze plus copper	Bronze
Twist pitch (mm)	5.1	7.6	$2.5 \rightarrow \infty$

^{a}Nb₃Sn fibers are distributed in a shell of o.d. ∿ 250 μm and i.d. ∿ 130 μm.

^{b}Niobium core before reaction.

fabricated at Ames Laboratory by a "long-dendrite"[3] process. The special feature of this wire is its long Nb₃Sn fiber structure. The composites have exhibited excellent superconducting and mechanical properties.[3] To be presented are a number of M-H curves and ramp field losses for samples with different twist pitches and sizes.

EXPERIMENTAL TECHNIQUES

The electrical magnetic measurement system has been described in detail previously.[4] However, it is worth repeating that although results are generally given on a per-conductor-volume basis, it is the specific sample winding that is measured. The arrangement of the conductors in the winding, which may vary in density with conductor geometry or surface condition, can greatly influence the results of the measurement. This is especially true in the low-field excitations, where the demagnetization effect is important. Therefore, one should be very cautious in comparing results from different experiments. Also to be noted in regard to the measurement is the calibration of the voltage output of the pickup coils. This normally requires an accurate calculation of the mutual inductance of the pickup coils and the "sample winding." This paper presents only the results of an open-circuit sample winding subjected to a monopolar ramped transverse field.

DATA ANALYSIS

It is believed that measured magnetization well describes the flux distribution inside a sample winding and that it should therefore also provide information about the structure of the composite sample. This certainly would rely on correct interpretation of the results of the measurement and on appropriate application of the theory.[1] Generally speaking, the complete M-H curves for all field levels are hard to calculate, because the calculation requires a good knowledge of current distribution as a function of field and geometry. However, there are a few conditions under which the magnetization is readily defined. Three of these are: (for applied external field, H_e)

(i) For very small field ($\mu_o H_e \ll \mu_o H_p$, where H_p is penetration field), superconductor filaments should exhibit a Meissner effect, i.e., "complete shielding." The effective permeability for the sample winding can thus be expressed as[1]

$$\mu_\perp = \frac{1 - \lambda}{1 + \lambda} \text{ ,}$$

where λ is the fraction of superconductor. Therefore, the initial slope, i.e.,

$$\frac{dM}{dH_e} = - \frac{2\lambda}{1 + \lambda} \text{ ,} \tag{1}$$

should be a direct measure of the superconductor fraction in the winding and in the composites. In certain cases, λ also represents the maximum area enclosed by multiply connected or electromagnetically coupled superconductors.

(ii) For a field $\mu_o H_e \gg \mu_o H_p$, i.e., a "fully penetrated region" where a simple critical-state model applies, the magnetization of the composites can be expressed as

$$M = - \frac{2}{3\pi} \lambda J_c d_{eff} \text{ ,} \tag{2}$$

where d_{eff} represents the dimension of the "independent" superconductor region; this can vary from the filament size d_f to the composite size d_o depending on coupling conditions.

(iii) For a condition of changing magnetic field, the detailed field calculation would be even more difficult. Nevertheless, the circuit model has worked out quite well in understanding the phenomenon.[4] The technique used here is to compare a hysteretic and a transient M-H trace under discharge field. The results

will be discussed in terms of time constant and effective resistivity.

RESULTS AND DISCUSSION

NbTi-Cu Composite

For reference, the results obtained for a conventional NbTi
conductor are presented first. Figure 1 shows the hysteresis
losses per unit volume of the conductor for different external
field amplitudes. The results were obtained under the condition
that $\tau \gg \tau_o$ where τ is the time constant of the external field
change and τ_o is the time constant of the eddy current or coupling
current. It generally displays B_m^3 and B_m dependence representing
partial penetration and full penetration, respectively. The
absolute value of measured losses per superconductor volume in the
full penetration region has been shown to be in good agreement
with critical-state theory. However, in the partial penetration
region a wide discrepancy exists not only with theory but also
among various experiments. This is probably due to the sample-
winding effect as suggested earlier.

The flux penetration pattern can better be illustrated in an
original M-H curve such as that shown in loop 1 of Fig. 2. (In
the figure, it is assumed that the sample has already been cycled
to high field at least once.) Also shown in the figure are two
dynamic discharge traces recorded as the external field decreases
exponentially from 2 T. The difference between the static and
dynamic trace can thus be accounted for by the coupling-current
flows. Figure 3 plots the coupling losses (the additional area to
the hysteresis loop) versus the time constant of the discharged
field, which exhibits an inverse linear dependence as expected by
theory.[4] The time constant is found to be 7 ms, which in turn
yields an effective resistivity of 1.2×10^{-9} Ω-m for the composite.
In a practical large-current cable, the losses may be kept as low
as the sum of these elements if the interelement resistivity is
made larger in comparison with that of the strands.

Nb$_3$Sn Composite (Bronze Method)

Hysteresis loss. The sample wire, as shown in the insert of
Fig. 4, was reacted successively for increasing times at 650°C.
Magnetization measurements were conducted for samples with reaction
times of 0, 1, 16, 32, 64, 128, and 256 h, respectively. Figure 4
presents their hysteresis losses per conductor volume as a function
of external field level, and Fig. 5 illustrates M-H curves for
samples with different reaction times. The fact that both losses
and magnetization increase with the reaction time clearly indicates

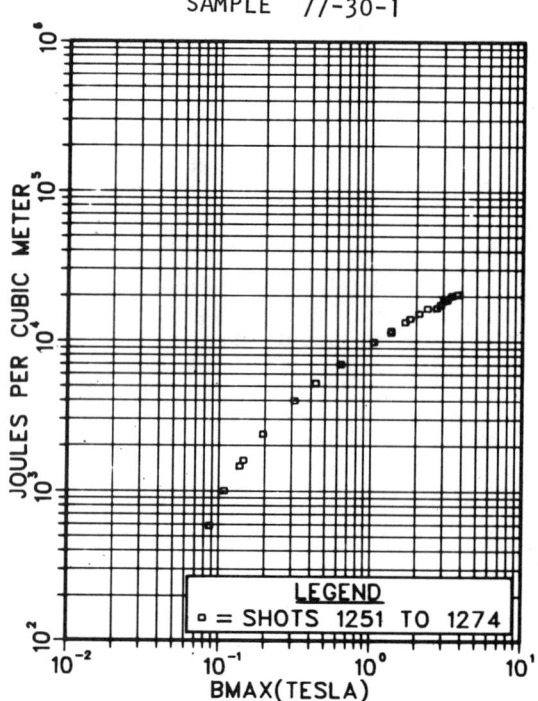

Fig. 1.   Hysteresis loss vs $\mu_o H_e$ for a NbTi-Cu composite.

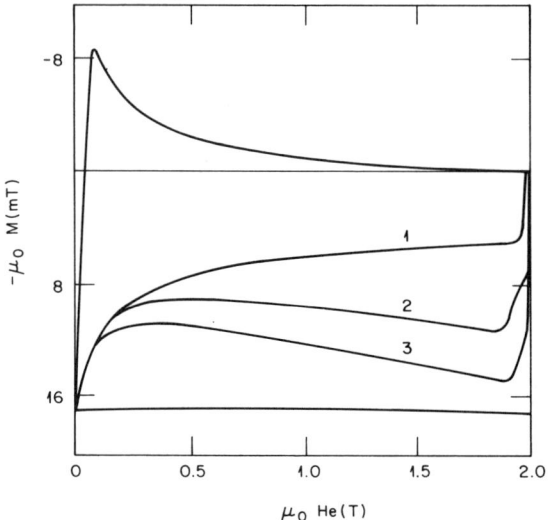

Fig. 2.   M-H curves for a NbTi-Cu composite: (1) hysteresis;
          (2) dynamic test $\tau = 0.18$ s; (3) dynamic test
          $\tau = 0.10$ s.

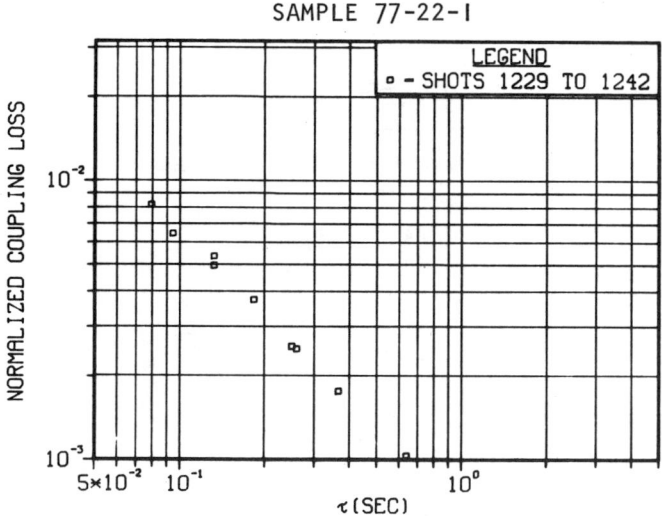

Fig. 3.   Normalized coupling losses vs τ for a NbTi–Cu composite.

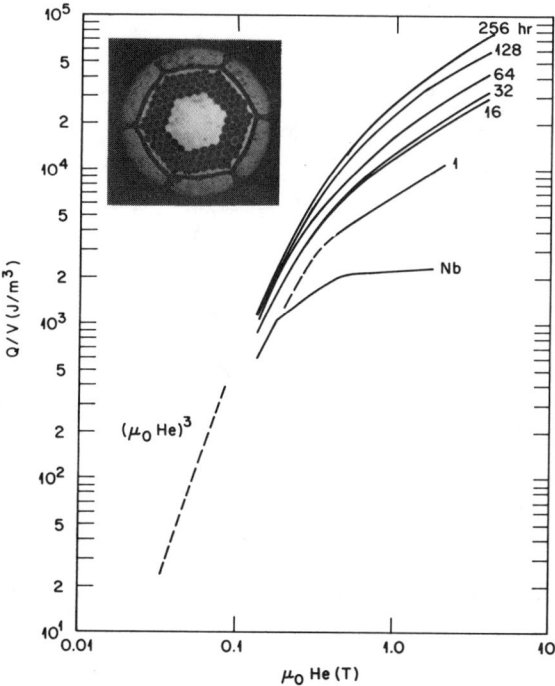

Fig. 4.   Hysteresis loss vs $\mu_0 H_e$ for a Nb$_3$Sn composite and its
cross section.

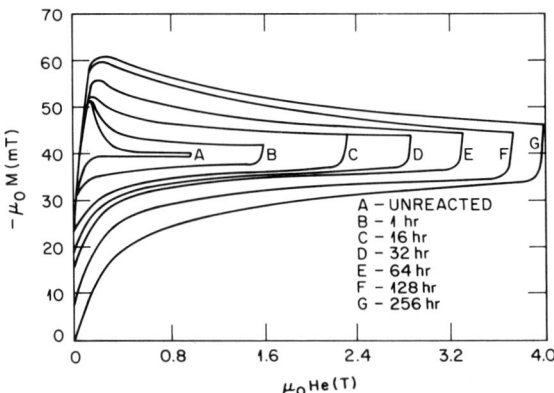

Fig. 5.   M-H curves for various reaction times.

that a Nb Sn layer gradually formed.   For such a Nb Sn shell
structure, Eq. (2) can be modified as

$$-M = \frac{4}{3\pi} \lambda J_c \, r_o \, (3x - 3x^2 + x^3) \, , \tag{3}$$

where $x = \delta/r_o$ and δ and r_o represent the Nb_3 Sn thickness and the
outside radius of the shell, respectively.

It is very important to note that because of the volume
expansion incurred in the reaction process, r_o can be appreciably
larger than the original radius of the niobium filaments (8.9 ×
10^{-7} m).   Calculation[5] has revealed that an 18% increase in the
radius is expected if the niobium filaments are completely reacted.

Since, in the practical conductor, the filament shape and
distribution are not uniform, the possibility that the reaction
might cause random interfilament contacts becomes a matter of
concern.

With the help of an independent critical-current measurement,
Eq. (3) is carefully examined at the full penetration region
($\mu_0 H_e$ = 4 T).   It is concluded that after a 64-h reaction, the
magnetization increased drastically, out of proportion with critical
current; this indicates that some contacts have occurred.   The
effective diameter for a 256-h sample was determined to be ∿3.2 μm
as compared with the 1.8-μm diameter of the original niobium
filament.

Another, more effective way to investigate this phenomenon is
to examine the initial slope (dM/dH) of M-H curves as functions of
reaction time (Fig. 6).   Using Eq. (1) for nonreacted niobium
wire, the fraction of niobium (presumably not in contact with

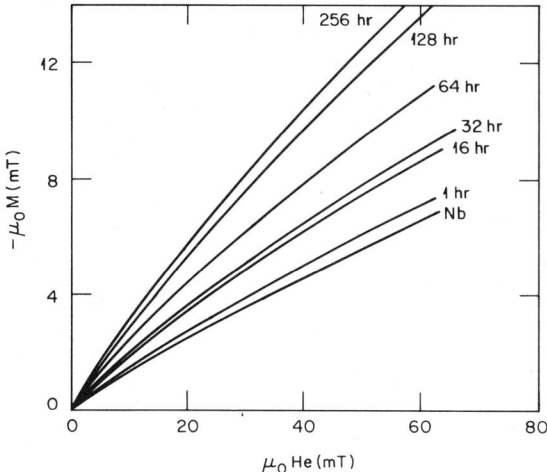

Fig. 6. Initial M-H curves for various reaction times.

other niobium filaments) in the composite is found to be

$$\lambda_{Nb} = 12.9\% \ ,$$

while data obtained from the billet assembly indicate a value of 11.5%. For a reaction time of 256 h, the fraction increases to

$$\lambda \sim 37\% \ ,$$

which corresponds to the area enclosed by multiply connected filaments; this is equivalent to the area covered by filament bundles.

Transient losses. In such conventional composites, filaments are embedded in bronze matrix; therefore its effective resistivity is commonly quite high ($\sim 10^{-8}$ Ω-m). (For conductors fabricated by other techniques, e.g., the "Nb-tube process," much lower effective resistivity may result.[6]) For operations with $\tau > 1$ s, the coupling effect can be neglected. Indeed, this was observed in the experiment conducted on a 16-h sample when the field was exponentially discharged from 2 T with $\tau = 0.14$ s. Only an additional 10% of hysteresis M-H area was measured. The time constant τ_0 of the wire was found to be 1.4×10^{-4} s. The composite would be suitable for any fast, large-amplitude, and high-field application.

Sample 3 is an in-situ prepared multifilamentary Nb_3Sn-Cu composite wire fabricated at Ames Laboratory.[3] The intent here was to determine if the formed superconducting fibers would act individually or collectively. Samples with different twist pitches and sizes were measured up to $\mu_0 H_e = 4$ T. Figure 7 shows typical M–H curves in which large hysteretic magnetization is observed. The effect of twisting is not enough to decouple the fibers but only to relieve some of the shielding.

Figure 8 summarizes ramp field losses per wire volume for all samples. Measured magnetization and losses are purely hysteretic. In other words, no time-dependent magnetization was detected when τ was varied from 0.1 s to 100 s. This supports the belief that the coupling is of a persistent nature.

Based on the analysis of the initial slope (dM/dH), the shielding strength ΔM at $\mu_0 H_e = 4$ T, the time-dependent test as mentioned above, and the magnitude of the hysteresis losses, one may infer that Nb_3Sn fibers are electromagnetically coupled. Therefore, the composites behave like a solid wire with equivalent

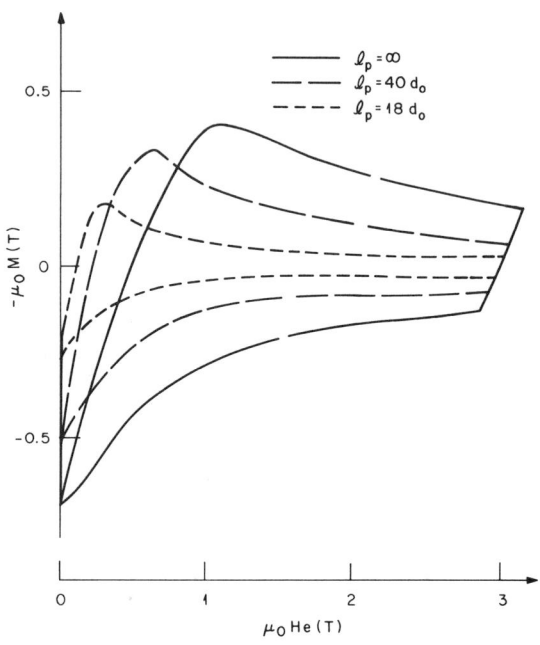

Fig. 7. M–H curves for in-situ processed composites.

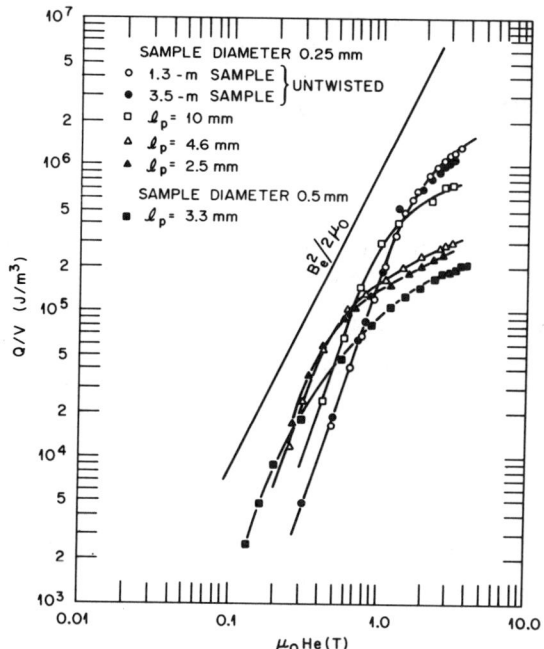

Fig. 8. Hysteresis losses for in-situ processed composites.

current density λJ_c and with an effective diameter of $d_{eff} \lesssim d_o$. For the sample of 0.25 mm with ℓ_p = 2.5 mm, it is found by Eq. (3) that d_{eff} = 40 μm.

Although this preliminary finding may seem to be a disappointment in direct comparison with the continuous filament bronze process wire, the potential value of in-situ material is too attractive to ignore, not to mention possible applications like those already demonstrated, such as dc or less-than-penetration-field operations.

It is strongly believed that much work still needs to be done to understand the microscopic superconducting mechanism in such a structure. Some of the areas needing special attention are the following:

. detailed characterization of the electromagnetically coupled mechanism;

. study of the effect on the magnetic properties of such metallurgical parameters as Cu/Nb ratio, tin content, reaction temperature, reaction time, and fiber size;

. measurements of magnetization at the high-field region ($\mu_o H_e \geq 8T$);

• study of the surface effects of such wire for low-field
operation and comparison with existing tape conductors; and

• for development purposes, exploration of stabilizing techniques[7]
for large-current applications.

CONCLUSION

A technical approach has been presented for gaining insight
into the superconducting mechanism in a multifilamentary structure.
The systematic probing technique has demonstrated the capability
to determine the superconductor distribution and matrix properties.
Hysteresis losses and the magnetization curve and its transient
effects have been presented for conventional bronze and in-situ
Nb_3Sn conductors. For the latter, research methods for possible
technological applications are outlined.

ACKNOWLEDGMENTS

The author wishes to thank the following persons and institu-
tions for their contributions during the course of the work:
D. K. Finnemore of Ames Laboratory; M. S. Walker and R. E. Schwall
of IGC; Wright Patterson Air Force Materials Laboratory; W. J.
Carr, Jr., of Westinghouse R&D Center; M. S. Lubell and W. A. Fietz
(for encouragement and helpful discussions); and J. P. Rudd (for
technical assistance).

REFERENCES

1. W. J. Carr, Jr., J. Appl. Phys. 46:4043 (1975).
2. G. R. Wagner, S. S. Shen, R. E. Schwall, A. Petrovich, and
     M. S. Walker, IEEE Trans. Magn. MAG-15:228 (1979).
3. D. K. Finnemore, J. D. Verhoeven, E. D. Gibson, and J. E.
     Ostenson, IEEE Trans. Magn. MAG-15:693 (1979).
4. S. S. Shen and R. E. Schwall, Proc. 7th Int. Cryogenic Eng.
     Conf., IPC Science and Technology Press, Surrey (1979).
5. W. A. Fietz, private communication.
6. S. S. Shen, "Magnetic Properties of Nb_3Sn Composites Made
     Without Intermediate Annealing", to be published.
7. J. D. Verhoeven, E. D. Gibson, C. V. Owen, J. E. Ostenson,
     and D. K. Finnemore, Appl. Phys. Letters 35:270 (1979).

ALTERNATING CURRENT LOSSES IN TWISTED IN-SITU

COMPOSITE WIRES

A. I. Braginski[*]   J. Bevk[+]

Westinghouse R&D Center   Harvard University
Pittsburgh, PA 15235   Cambridge, MA 02138

INTRODUCTION

The purpose of this study was to experimentally compare ac
losses and mechanical properties of twisted and untwisted in-situ
formed composite wires. Our earlier work concentrated on untwisted
in situ samples having remarkable mechanical properties, but
rather high losses. It is known, however, that conventional composites must be twisted to achieve acceptably low losses. Twisting
of in-situ wires could also be advantageous if mechanical properties were not degraded excessively. We will show that this degradation is, indeed, negligible but the loss reduction due to twisting is not large if the volume fraction of superconductor in the
composite is well above the percolation threshold.

AC LOSSES IN SUPERCONDUCTING COMPOSITES

In conventional filamentary composite wires, the continuous
superconducting filaments are separated from each other by a normal
matrix. In the case of A15 conductors fabricated by bronze diffusion processes, bronze is the typical matrix material.

Composite conductors operating in ac magnetic fields should be
twisted to reduce the eddy current, coupling losses in the normal
matrix to acceptably low levels predicted by theory.[1,2] Shielding
of the wire interior by interfilamentary ac coupling currents is
thus practically eliminated and full penetration of nearly all filaments occurs at a transverse magnetic field intensity, $B_{p\perp}$, defined

---

[*] Supported in part by the Air Force Office of Scientific Research
Contract No. F49620-78-C-0031.
[+] Supported by National Science Foundation, Grant No. DMR-76-01111.

by the filament diameter, $d \ll D$, where D is the wire diameter.
The hysteretic losses increase linearly with the ac field amplitude,
B_m, when the filaments are fully penetrated and as B_m^n with $n \geqslant 3$
when the penetration is only partial. The full penetration field is
$Bp_\perp = 4(10^{-7})$ J_c d/λ for each filament of diameter d in a properly
twisted composite having an overall critical current density, J_c,
and a superconductor volume fraction, λ. At a given B_m the hysteretic losses in the fully penetrated filamentary composite can be
thus orders of magnitude lower than in a penetrated bulk wire where
$Bp_\perp$ is D/d times higher. In contrast, longitudinal magnetic fields,
i.e., those parallel to the wire axis, will be partially shielded by
eddy currents as a result of twisting. Thus the penetration field
of the composite, $Bp_\parallel$ will increase over the single filament value.
Both, eddy current and hysteretic losses above $Bp_\parallel$ will be higher
than in an untwisted wire.

As discussed elsewhere in this volume the in situ formed composites have discontinuous filaments of $d = 10^2$ to 10^3 Å, two orders
of magnitude smaller than in conventional composites. The probability of lateral, random contacts between adjacent filaments increases
with the volume fraction of the superconductor in the matrix, λ, and
becomes very high above the percolation threshold situated at approximately 15-18 vol. %. Upon the application of dc magnetic field
supercurrents can thus set in loops formed by contacting filaments.
Even at lower concentrations, some supercurrents can conceivably
flow between the filaments due to the proximity effect. This occurs
when average spacings between filaments, $\sim d/\lambda$, are on the scale of
the proximity decay length k_n^{-1} of the matrix.[3] Since k_n^{-1} decreases
with increasing magnetic field intensity, B, the proximity links can,
however, be destroyed in higher fields.

Generally the presence of direct coupling supercurrents between
filaments can result in some dc shielding of the in-situ composite
interior in contrast to the conventional wire. In ac fields, additional shielding can be caused by ac coupling currents induced in
the matrix. We have found earlier that the dc behavior of untwisted
in situ composites with high $\lambda > 0.3$ and very high dc J_c's is essentially that of bulk superconductors.[4] This is the natural consequence of their sponge-like interconnected filament network
structure.

At low $\lambda \leqslant 0.1$ only partial dc shielding is observed.[4] It decreases significantly upon application of strong dc magnetic fields
$B_o > 1$ tesla as a consequence of the proximity effect suppression.[5]
The critical current density also decreases rapidly with increasing
B_o. In the range $0.2 < \lambda < 0.3$, where most of current work was
done because of acceptably high dc critical current densities, the
wire behavior is not quite that of a bulk superconductor but dc
shielding is considerable and dc fields up to 8.5 tesla do not

reduce it. The interfilamentary direct coupling is thus rather independent of the proximity effect.

From the above observations, one can infer that the loss reduction by twisting should become less effective with increasing λ, and
that sufficiently above the percolation threshold the direct interfilamentary coupling will dominate over the matrix-mediated ac coupling thus making the twisting useless.

SAMPLES AND EXPERIMENTAL PROCEDURES

The in-situ wire fabrication process was described elsewhere.[4,6] Wire samples with three different twist lengths, L, and a
diameter of $\sim$ 0.14 mm (5.5 mil) have been characterized. Their description is summarized in Table 1. The superconductor volume
fraction was in all cases $\lambda \simeq 0.23$. The section area reduction
ratio was $\sim$ 4500 thus producing an average filament length of 10 to
15 mm and an average ribbon-like filament cross section of 500 x
5000 Å. No filamentary texture was revealed by scanning electron
micrography of an etched wire section (with bronze removed), as
shown in Figure 1(a). The filamentary twist was clearly observed in
longitudinal photographs [Figure 1(b)].

Table 1. Measured Samples

Sample Number	Twist Length L, cm	$\frac{L}{D}$	Composition at %			Matrix Conductivity at $\sim$ 20 K σ_m, mho/m	J_c @ 0.2 T 4.2 K 10^{10} A m^{-2}
			Cu	Nb	Sn		
9	∞	∞	81.6	13.1	5.3	1.0×10^7	1.0
8	$\sim$ 0.280	20	81.5	13.1	5.4	0.8×10^7	no data
7	$\sim$ 0.125	9	80.2	12.9	6.9	1.0×10^7	0.6

Both, twisted and untwisted samples were tested in tension to
failure at room temperature ($\sim$ 293 K) and at 77 K. An Instron Tensile Testing Instrument was used. The crosshead speed was
1.27(10^{-4}) m/min, and the corresponding strain rate was $\dot{\epsilon}$ = 1%/min.
The ultimate tensile strength, σ_{UTS}, and the strain to failure, ϵ_T,
were then obtained from stress-strain curves.

The length of wire samples prepared for ac loss measurements
was in the range of 0.25 to 0.5 m. This limited the measuring
techniques to dc and ac magnetization methods. The dc magnetization hysteresis loops were determined using the P.A.R. Model 156
vibrating sample magnetometer with extremely low sweep rates, of
the order of 1 x 10^{-4} tesla sec^{-1}. At these sweep rates the eddy

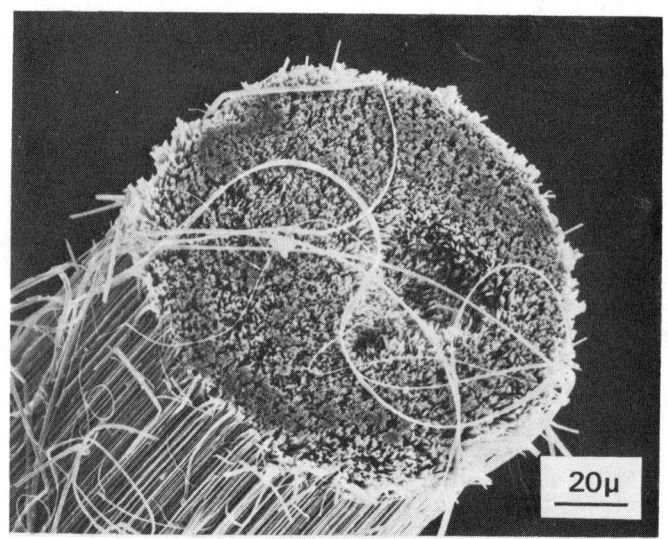

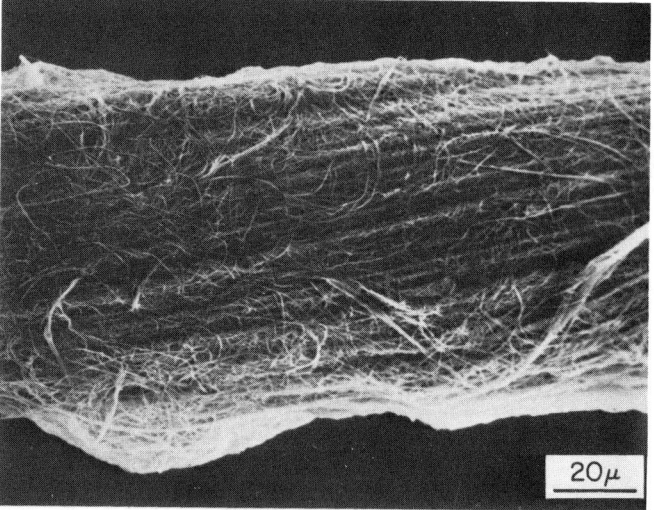

Figure 1.   Scanning electron microscope photograph of twisted
            sample No. 7 with bronze etched out.   (a) Cross section,
            (b) Longitudinal view.

current (coupling) losses could not be detected. The ac magnetization technique, aimed at determining coupling losses, was based on the use of electronic low power factor wattmeters: the Brookhaven National Laboratory Model IH-744 and a Westinghouse wattmeter operating at higher frequencies. Results reported were obtained at sine wave frequencies between f = 20 and 2000 Hz. Small sample lengths made the fabrication of longer sample solenoids impossible. Consequently, demagnetizing effects could affect the absolute values of measured loss. Comparison with calorimetric measurements of one untwisted sample at low frequencies, not exceeding 20 Hz, showed, however, a very good agreement.[7] The agreement was less satisfactory at high frequencies, $f \geqslant 1000$ Hz, and low field intensities, in the range of 10^{-4} to 10^{-3} tesla, where measured losses were very low.[4] Higher sinusoidal fields were not available at these frequencies. No dc bias fields were used. All measurements were performed at 4.2 K.

MECHANICAL PROPERTIES

The ultimate tensile strength and strain to failure data for the untwisted sample 9 and twisted sample 7 (L = 0.125 cm) are summarized and compared in Table 2.

Table 2. Ultimate Tensile Strength and Strain to Failure

Sample Number	L, cm	σ_{UTS}, MPa		ε_T, %	
		293 K	77 K	293 K	77 K
9	∞	780	920	1.9	1.8
7	0.125	730	920	2.3	1.8

At room temperature the twisted sample exhibits a slightly more plastic behavior (lower σ_{UTS}, higher ε_T). At 77 K the measured properties are identical within the accuracy of the method.

This somewhat unexpected lack of degradation upon twisting suggests that the ratio of the twist length to the average filament diameter is still sufficiently high ($\sim 10^4$) for the shear stresses to be small. Clean filament-matrix interfaces and fine filament sizes thus result in comparably high strength of both samples.[8] We expect that untwisted and twisted in-situ composites will also exhibit a similar behavior at 4.2 K and under cyclic stress.

LOSS RESULTS AND DISCUSSION

Hysteretic losses determined by magnetometer in transverse magnetic fields are shown in Figure 2 for one twisted and one untwisted sample (L = 0.125 cm and L = ∞). In the partial penetration regime, the twisted sample shows losses a factor of four or more higher than the untwisted although the ratio of critical current densities is only ∿ 1.7. This indicates that at any given $B_m < B_p$ the penetrated volume fraction is higher in the twisted wire. The determination of the penetration field is somewhat arbitrary as the transition from a cubic to linear field dependence is gradual. Assuming B = 0.3 tesla for L = ∞ and 0.12 tesla for L = 0.125 cm, we obtain the calculated effective filament diameters of d_e = 17 and 11 μm, respectively. Hence, twisting may reduce somewhat the dc shielding but certainly does not decouple the filaments fully. Indeed, the experimental average filament diameter d = $(500 \times 5000)^{1/2} \simeq$ 1600 Å is by two orders of magnitude smaller than d . The shielding should be attributed to random connections between filaments since no ac coupling currents could circulate in the matrix during the dc measurement. One can speculate that the observed reduction of shielding is due to either a redistribution of interfilamentary supercurrents or their reduction caused by breakage of some connections between filaments.

Low frequency (f = 20 and 60 Hz) loss measurements in transverse fields up to B_m = 0.3 tesla confirmed the B_p value obtained by the dc method for the twisted sample, as shown in Figure 3. In the untwisted sample B_p could not be exceeded, again in agreement with dc results. Above 0.1 tesla the exponent n = 4.5 obtained by the least squares fit to the experimental points of the P(B) curve is, however, higher than in the dc data, where n = 3.4. The reason for this discrepancy is not clear. At much lower fields the exponent decreases to n ≃ 2.5 thus suggesting that coupling currents contribute to the losses. In the twisted sample, this contribution is too small to be seen in Figure 3. It shows up, however, in higher frequency data.

The comparison of low frequency longitudinal field losses shown in Figure 4 confirms that twisting affects shielding to some extent. In this case the penetration field cannot be determined with any accuracy. It is, however, much higher in the twisted wire. Consequently, one observes a crossover of loss curves at $B_m \simeq 10^{-2}$ tesla. This behavior is qualitatively similar to that of conventional composites with ac coupling currents in the matrix. In the in-situ wire, however, interfilamentary supercurrents can produce the same behavior. Assuming that these supercurrents dominate at 20 Hz over the matrix currents the increase of longitudinal field shielding upon twisting suggests that breakage of interfilamentary connections is not a likely cause of the shielding reduction observed in transverse fields. The redistribution of interfilamentary supercurrents in more probable.

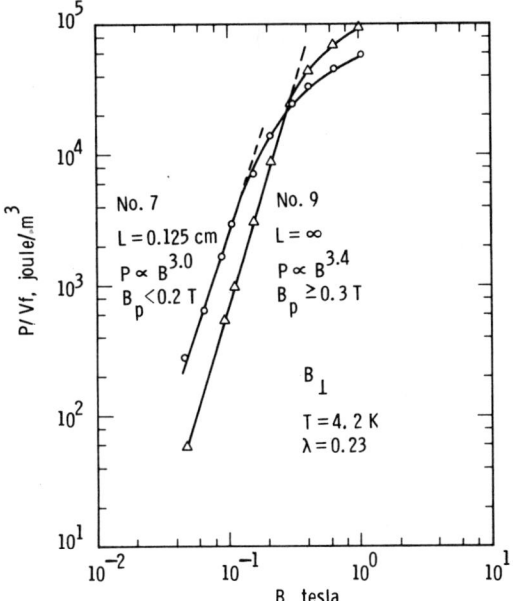

Fig. 2. Hysteretic losses
in transverse field for
twisted and untwisted sample;
measured by dc magnetization
technique.

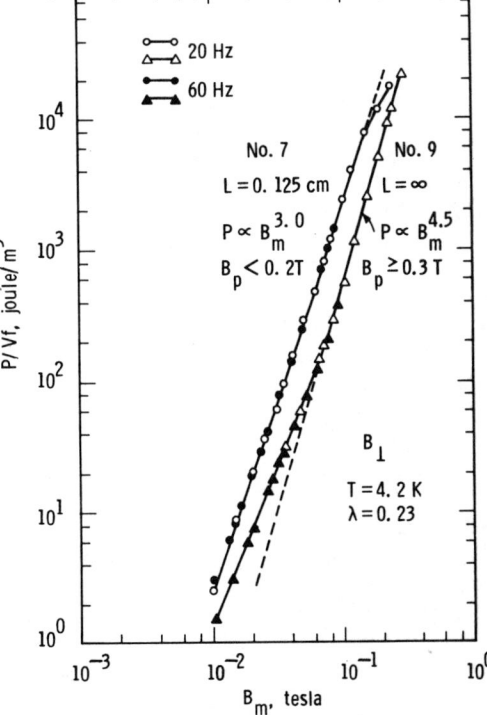

Fig.3. Low frequency losses
in transverse field for
twisted and untwisted sample.

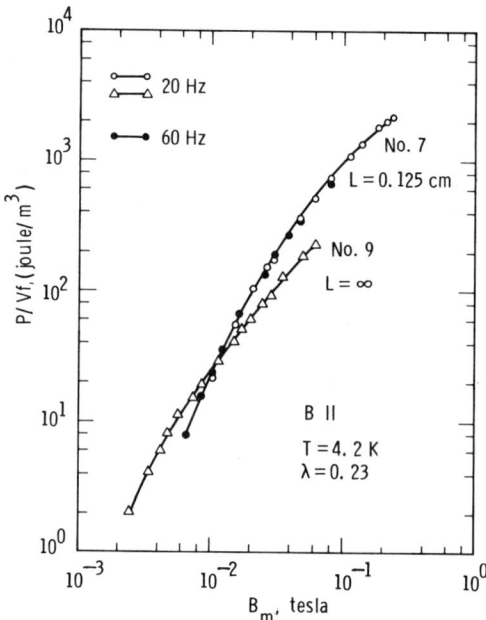

Figure 4. Low frequency losses in longitudinal field for twisted and untwisted sample.

The dc and low frequency results presented so far show, therefore, that in a twisted in-situ wire ($\lambda \simeq 0.23$) significant shielding persist and it cannot be attributed to ac coupling currents. Are the coupling (eddy) current losses of any importance? Earlier calorimetric and electronic ac loss measurements of an untwisted sample have shown clearly that at very low field amplitudes $B_m = 10^{-4}$ to 10^{-3} tesla, and frequencies between 200 and $\sim 10^4$ Hz, losses proportional to B^2 and f^2 dominate.[4] The calorimetrically determined reduced loss was $P/(V B_m^2 f^2) \simeq 500$ W/(m^3 T^2 Hz^2). The electronically determined value was by an order of magnitude lower but still non-negligible. We represented these losses by those of a conventional composite having a twist length comparable to the in-situ filament length.[4]

The reduced $B^2 f^2$ losses determined by wattmeter for the present samples at $B_m \leqslant 10^{-3}$ tesla and frequencies up to 2000 Hz are given in Table 3. The loss reduction with the twist length is less steep than the L^2 dependence predicted by the loss theory valid for conventional composites.[2] The experimental results are for twisted samples deceptively close to the theoretical low frequency loss calculated assuming L and σ_m from Table 1. One should stress, however, that the theory is not applicable to composites having their interior shielded also by direct coupling supercurrents. Neglecting all shielding that might be contributed by eddy currents, and using the dc results one can estimate the penetrated volume fraction of the conductor. For $B_m = 10^{-3}$ tesla it is of the order $2(10^{-3})/B_p \simeq 10^{-2}$. The measured eddy current losses may thus represent only a surface layer and not the volume of the wire.* All that can be deduced from Table 3 is that twisting reduces losses in that layer. The eddy current losses per unit surface are thus also given in Table 3. Attempts to arrive at an adequate characterization of volume losses by applying high frequency but low amplitude fields appear rather fruitless in the absence of applicable theory.

Table 3. Reduced Eddy Current Losses

Sample Number	L, cm	Experimental Losses		Calculated Loss[2]
		$P/(V\ B^2\ f^2)$, $W/(m^3\ T^2\ Hz^2)$	$P/(S\ B^2\ 2^2)$, $W/(m^2\ T^2\ Hz^2)$	Conventional Wire $P/V\ B^2\ f^2)$, $W/(m^3\ T^2\ Hz^2)$
9	∞	80 ± 20	$2.8(10^{-3})$	---
8	~ 0.280	40 ± 10	$1.4(10^{-3})$	32
7	~ 0.125	$25 \pm \frac{5}{10}$	$0.9(10^{-3})$	8

CONCLUSIONS

Mechanical properties of in-situ composite wires twisted to achieve a ratio of the twist length to the wire diameter of the order of 10 are similar to those of untwisted wire. At $\lambda = 0.23$ twisting somewhat reduces the shielding caused by random connections between filaments. Consequently, full penetration hysteretic losses are reduced. Surface eddy current losses are also reduced. The loss reduction is, however, much less significant than in conventional composites having a comparable twist. Twisting should be more effective in composites with lower superconductor volume fraction, λ, where the frequency of direct connections between filaments is reduced. Even at $\lambda = 0.23$, however, twisting should be helpful in low $\dot{B}$ applications.

ACKNOWLEDGEMENTS

We gratefully acknowledge the criticism and comments of W. J. Carr, Jr., M. Ashkin and G. R. Wagner. We thank H. C. Pohl for technical assistance and M. B. Warren for a careful typing of

---

*This was pointed out to us by S. S. Shen.

this text. We also thank the Francis Bitter National Magnet Laboratory for the use of high field magnets.

REFERENCES

1. M. N. Wilson, C. R. Walters, J. D. Lewin and P. F. Smith, J. Phys. D: Appl. Phys. 3:1518 (1970).
2. W. J. Carr, Jr., M. S. Walker, and J. H. Murphy, J. Appl. Phys. 46:4048 (1975).
3. G. Deutscher and P. G. de Gennes, in "Superconductivity", M. Dekker, Inc., New York, NY (1969).
4. A. I. Braginski, G. R. Wagner, and J. Bevk, Adv. in Cryo. Eng. 26 (in press).
5. A. I. Braginski and J. Bevk, Bull. Am. Phys. Soc. 25:385 (1980); (full text unpublished).
6. J. P. Harbison and J. Bevk, J. Appl. Phys. 48:5180 (1977).
7. J. Bevk, J. P. Harbison, F. Habbal, G. R. Wagner, and A. I. Braginski, Appl. Phys. Letters 36:85 (1980).
8. J. Bevk, J. P. Harbison, and J. L. Bell, J. Appl. Phys. 49: 6031 (1978).

A15 MULTIFILAMENTARY SUPERCONDUCTORS BY THE INFILTRATION PROCESS

M. R. Pickus, J. T. Holthuis, and M. Rosen

Materials and Molecular Research Division
Lawrence Berkeley Laboratory
University of California
Berkeley, CA 94720 (USA)

INTRODUCTION

The morphology of practical superconductors is an essential consideration that is directly related to the requirement of conductor stability. An extensive treatment of stabilization criteria has been provided by the Rutherford Laboratory Superconducting Applications Group.[1] The adiabatic criterion is perhaps the most basic. For intrinsic stability, it requires that a superconductor should not exceed a critical size that is determined by certain properties of the superconducting material. In view of the adiabatic and the other criteria, the optimal size is generally considered to be less than 10 μm. From this requirement for such small sizes stems the preferred morphology for an intrinsically stable conductor: a large number of superconducting filaments arrayed in a normal matrix. Multifilamentary conductors were readily achieved with the ductile niobium-titanium superconducting alloy, and these have been developed to a high degree of sophistication, involving twisting, braiding and cabling. However, as shown in Table 1, the superconducting properties of niobium-titanium do not compare favorably with those of the A15 compounds. Consequently a great deal of current research effort is devoted to the development of multifilamentary conductors based on the intrinsically brittle intermetallic compounds of the Nb_3Sn type.

The orthogonal chains of niobium atoms in close proximity are an important characteristic of the A15 structure. It is believed that the integrity of these chains strongly affect the superconducting properties of the A15 compounds. Disruption of

Table 1. Some High Field Superconducting Compounds (Niobium Based)

Compound	T_c (K)	H_{c2} at 4.2 K (Tesla)
$Nb_3(Al,Ge)$	21	410
$Nb_3(Al,Si)$	19.5	30
Nb_3Al	19	30
Nb_3Sn	18	23
NbTi	10	12

the chains by structural defects such as vacancies, or by tin atoms occupying niobium sites (in the case of Nb_3Sn), adversely affects the transition temperature (T_c) and the current carrying capacity. Thus in the design of processes for the fabrication of functional superconductors based on A15 compounds, two criteria have relevance in addition to the morphology constraints imposed by stability considerations. They are the stoichiometry and degree of order of the A15 phase. Most of the process development research presented in this report utilizes powder metallurgy techniques. There is one exception, involving the niobium-aluminum system, and it will be presented first.

THE NIOBIUM-ALUMINUM SYSTEM - A NON-POWDER APPROACH

This process[2,3,4] takes advantage of the increasing solubility of aluminum in niobium with increasing temperature. At room temperature niobium can dissolve only 9 at.% of aluminum, whereas at 1950°C the solubility of aluminum increases to 23 at.%. It was reasoned that a sample heated into the high temperature solid solution region could be quenched rapidly enough to retain a metastable solid solution and prevent any transformation to the stable A15 phase. After quenching, the solid solution might be sufficiently ductile to permit plastic deformation. In that event, traditional processing techniques could be employed. Moreover, having the material in a metastable state would provide another important advantage: at an appropriate stage in the processing sequence, it could be transformed to the superconducting phase by a low temperature aging treatment. It is desirable that the A15 phase be formed at temperatures not exceeding 1000°C. The superconducting properties of Nb_3Al are strongly dependent upon the degree of order of the niobium and aluminum atoms in the A15

lattice. For this reason, material formed by high temperature
diffusion processes is often subjected to low temperature anneals
of long duration. This restores much of the order which had been
disrupted by the high temperature treatment. Furthermore, the
critical current carrying capacity is related to the number and
efficiency of the flux pinning centers present in the superconducting phase. Flux pinning centers can be oxides, vacancies
or any other type of structural imperfection. Grain boundaries
are very effective pinning centers; consequently, the smaller the
grain size the better is the pinning effect. The grain size of
the phase is directly related to its formation temperature. The
lower the reaction temperature, the smaller is the grain size.
A low reaction temperature is especially important for Nb_3Al,
since no successful procedure has been developed to introduce a
dispersed oxide phase, such as ZrO_2, as pinning sites. For these
reasons the quench-age approach appeared particular attractive.

The process developed, which incorporates these ideas, is
shown schematically in Fig. 1. Niobium tubing was used to sheathe
a rod of grade 2011-T3 aluminum. This grade, which contains a
small percentage of copper, provided a good match with the mechanical properties of niobium. The composite (o.d. = 5.8 mm) was
first swaged and then wire drawn to a diameter of 0.5 mm. It was
then heated for 30 seconds at a temperature of 1950°C to form a
core of a saturated solid solution of aluminum in niobium. Direct
resistive heating was used for two reasons: the time of the reaction could be regulated easily from seconds to minutes as desired, with the sample reaching the high temperature very rapidly;
and the wire could be gripped in such a way that the ends were
cooled. This eliminated any possibility of aluminum escaping from
the ends. Quenching was accomplished by directing a jet of high
pressure helium gas on the sample. The cooling rate achieved was
approximately 550°C per second. Several observations confirmed
the fact that at this stage the core was a metastable solid solution. Its microhardness value of 385 VH is similar to that obtained in a Nb(Al) solid solution prepared by conventional techniques. No traces of the presence of the A15 phase were detectable
by either metallographic or x-ray examination. However, following
an aging treatment of one hour at temperatures from 850 to 1000°C,
the microhardness of the core increased to 890 VH, a value characteristic of the A15 compound. The presence of the A15 phase
after aging was further confirmed by x-ray analysis.

After quenching there are two options for further processing.
The composite can be aged to transform the metastable solid solution core completely to the A15 phase. The second option, which
is illustrated in Fig. 1, seems more interesting. Contrary to
what was expected, the saturated solid solution has only limited
ductility at room temperature, on the order of 7% reduction in
thickness. Although it was found that the ductility can be

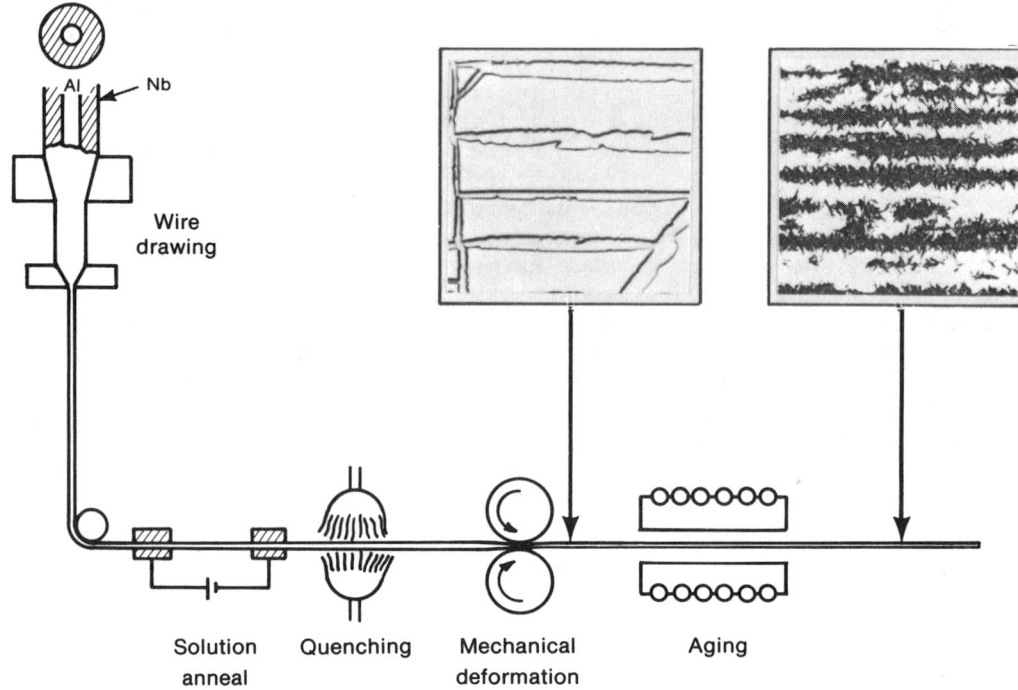

Fig. 1. A quench—age process for the fabrication of Nb(Al)
          superconductors.

doubled by working the material at 400°C, all indications suggest
that extensive mechanical deformation of the solid solution cannot
be readily performed. However, moderate deformation of the solid
solution results in the formation of a network of twins as shown
in the left photomicrograph of Fig. 1. These twins act as very
effective sites for nucleating the A15 phase during aging, as
shown in the photomicrograph to the right. Thus, an in situ multifilamentary morphology may be achieved.

The critical temperature of 3 samples which had been solution
quenched and aged for 1 h at 1000°C, 18 h at 950°C and 24 h at
900°C were measured inductively through the cooperation of Dr. R.
Hammond and the Stanford University Department of Applied Physics.
The onset of superconductivity for the samples was 17.6 ± 0.5 K
with a transition width of about 1 K. Although there was a 100°C
difference in aging temperature, no detectable difference in T_c
was observed. This value of T_c is a full 2 K higher than has been
previously reported for Nb_3Al formed at temperatures below 1000°C.

POWDER METALLURGY TECHNIQUES

Prealloyed Powders; [Nb_3Sn and $Nb_3(Al,Ge)$]

In any processing sequence that involves the formation of the
A15 phase by a diffusion controlled reaction there is always the
question of achieving optimal stoichiometry. It was thought that
this important question could be circumvented by preparing, as a
first step, a prealloyed powder having precisely the desired stoichiometry. A process of this type, if successful, would have tre-
mendous versatility. Not only would it be applicable to any binary
system, regardless of the individual properties of the component
elements, but also to ternary systems where the diffusion reactions
are far more complex. The niobium-tin system was selected for the
first feasibility assessments.

Niobium and tin powders of -325 mesh size were thoroughly
blended in a proportion corresponding to Nb_3Sn. The powder mix
was compacted in a steel die and then sintered at 1800°C for 1/2 h.
Metallographic and x-ray analysis showed the compact to be more
than 95% Nb_3Sn. The compact was then pulverized and ground to a
-400 mesh prealloyed powder. Niobium, copper and monel tubes were
used to jacket the powder. As one example, a powder filled niobium
tube was encased in a copper tube. The assembly was wire drawn to
a convenient diameter and then cut into a number of lengths which
were bundled in another copper tube. This procedure was repeated
to a point where a wire was obtained that had an outside diameter
of 0.030" and which contained 70 consolidated powder cores. A
metallographic examination of a cross section showed the cores to
be apparently well bonded. However, measurements of the current
carrying capacity were disappointing. It was surmised that there
must be a lack of true metallurgical bonding between the powder
particles. Another set of samples was prepared that had been heat
treated for 1 h at 900°C. No significant improvement was observed.
It was therefore decided to perform a systematic study of particle
bonding in an A15 compound.[5]

The niobium-aluminum-germanium system was selected for this
investigation. Niobium, aluminum and germanium powders were
weighed out to a nominal composition of 75 at.% niobium, 18.75 at.%
aluminum and 6.25 at.% germanium. After thorough blending, the
powder was pressed at 40 ksi in the form of small rectangular
compacts. These compacts were arc melted in water cooled copper
crucibles under an argon atmosphere, using a tungsten electrode.
Allowances were made when formulating the alloys to compensate
for a small weight loss, attributable mainly to evaporation of
aluminum during melting. The melted compacts were inverted and
remelted four times to ensure complete fusion and improved homogeneity. The arc melted buttons were further homogenized by

heating at 1840°C for 2 h in an argon atmosphere.

Phase identification by metallography was facilitated by
anodizing for five minutes in a 10 wt% solution of citric acid
at 25 volts. The characteristic colors of the four possible
phases that could occur are: niobium solid solution (blue-green);
A15 phase (gray-purple); and two minor solute rich phases (pink
and orange).[6] The sample consisted of more than 95% homogeneous
A15 phase, and less than 5% of solute rich phases which were
present mainly at the grain boundaries. The homogenized arc
melted buttons were crushed and then milled in a planetary grinder.
X-ray diffraction analysis of the powder confirmed the metallographic findings. The −400 mesh fraction of this powder was used
for the particle bonding studies.

The first procedure investigated was cold isostatic pressing
followed by sintering. Rubber molds containing the powder were
immersed in an oil bath which was pressurized to 60 ksi for 2
minutes. Compacts were sintered in a helium atmosphere for 1 h
over the temperature range 950-1650°C. Density measurements were
made using the water displacement technique. Accurate volume determination was facilitated by vacuum impregnation of the compacts
with epoxy. The density of an homogenized arc melted button was
found to be 7.72 g/cm^3, and this value was taken to be the theoretical density. The relative densities reported are the ratio
of the measured density to the theoretical density. The microstructures were in good conformity with the density measurements.

It appears from both the density measurements and the photomicrographs that the driving forces for sintering are insufficient
to achieve observable densification of the A15 compacts at temperatures up to 1550°C. Only at about 1650°C does the first stage of
sintering begin, with a small increase in the connectivity of the
powder particles. Increasing the compacting pressure or lengthen-
ing the sintering period had little effect. A number of additives
were tried in an attempt to activate the sintering process. Even
with copper, the most effective of the additives tried, only 75-
80% of the theoretical density could be obtained.

Hence, it may be concluded that the conventional pressing
and sintering technique is ineffective for densification of A15
powder. This may well be the principal reason for the poor superconducting properties exhibited by wires containing cores of precompounded Nb$_3$Sn.

To further elucidate this approach, it was decided to investigate the simultaneous application of heat and pressure via hot
pressing. The hot press employed was designed and constructed at
LBL. Essentially, it consists of a water cooled stainless steel
chamber containing an externally controlled revolving platform

which accommodates 8 graphite die assemblies. Thus eight press-
ings can be made without breaking the vacuum or waiting for the
cool down of individual assemblies. The heating unit surrounding
the die being pressed can be raised and lowered by an externally
controlled mechanism, thus permitting rotation of the platform.
Hot pressing was done in a vacuum of 5×10^{-5} mm of Hg with an
applied pressure of 6000 psi. The same $Nb_3(Al,Ge)$ powder used in
the press and sinter experiments was employed for the hot pressing
study. Samples were held under pressure for 1/2 h at selected
temperatures: 1100, 1200, 1300, 1325, 1370, 1400, 1450, and
1550°C.

The densification data for both the hot pressed and the cold
pressed and sintered samples are shown in Fig. 2. The contrast
in the two modes of processing is quite remarkable. Porosity was
substantially eliminated by hot pressing at approximately 1400°C,
whereas very little reduction in porosity was observed for cold
pressed samples that had been sintered at temperatures as high as
1650°C.

The pressure of 6000 psi used for hot pressing is close to
the maximum that can be applied due to the mechanical properties
of graphite. With this limitation, the onset of plasticity occurs
at 0.76 of the absolute melting point. It is very possible that
this value can be significantly reduced by the use of higher pressures. It is intended to further explore the potential of using
precompounded powders in the following manner. Wires would be
prepared containing one or many cores of precompounded powder.
After completion of all the mechanical processing, the wire would
be subjected to hot isostatic pressing (HIP). No die would be
required, and much higher pressures could be employed. Such a
procedure, if successful, would be applicable to other brittle
superconducting compounds, provided that they are thermodynamically stable at the HIP temperature required for full densification.

The Infiltration Process for Tapes; Nb_3Sn

The infiltration process was applied first to the niobium-
tin system because of the attention it has received in the literature. It was considered that the availability of extensive data
on the superconducting properties of niobium-tin conductors made
by other processes would facilitate the assessment of the merits
of a new approach. Since this is a diffusion controlled process
it is helpful to consider the equilibrium phase relationships.
The desired superconducting phase is the intermetallic compound
Nb_3Sn which has the A15 crystal structure. There are two additional intermetallic compounds, Nb_6Sn_5 and $NbSn_2$. Neither of
them is of interest for superconductivity, since their critical

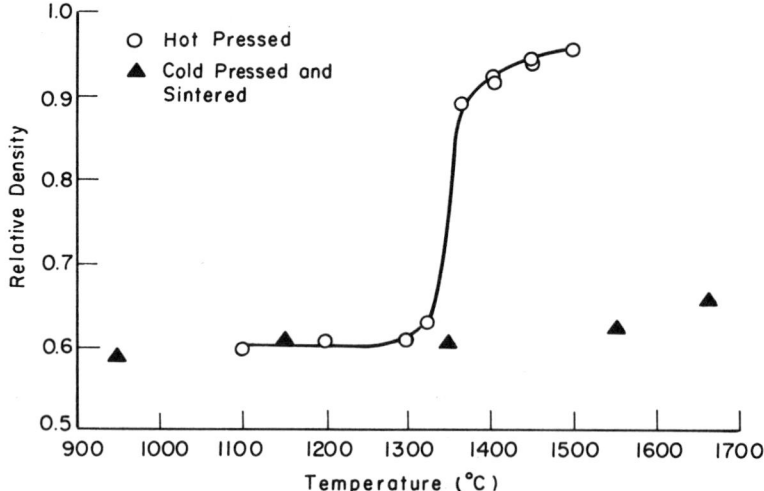

Fig. 2.   Temperature dependence of the relative density of
          Nb₃(Al,Ge) compacts for both hot pressed and cold
          pressed and sintered samples.

temperatures are in the order of 2 K compared with 18.1 K for
Nb₃Sn.   Like Nb₃Sn, both of these phases are brittle.   Of particular relevance is the fact that both are unstable at temperatures
above approximately 930°C.   To avoid them, the diffusion treatments
were therefore carried out in the temperature range 950–1000°C.

     The first application of the infiltration process was the
preparation of a flexible tape[7,8,9] that could be wound to form
solenoids after all processing operations had been completed.   The
morphology of the tape is that of a composite, with the brittle
Nb₃Sn phase in the form of thin, multiply connected filaments
arrayed in a ductile niobium matrix.   The filaments have a thickness of 5 microns or less.

     To obtain this composite, niobium powder was roll-compacted
and sintered to produce a tape that was permeated by a capillary
system of interconnected pores.   The porous tape was then infiltrated with molten tin in such a manner that all of the pores were
completely filled with tin.   Although a diffusion heat treatment
at this stage produced the desired Nb₃Sn phase, it was found preferable to subject the infiltrated tape to a substantial reduction
in thickness by cold rolling prior to the heat treatment.   With

this sequence a greater volume fraction and a more desirable morphology of the superconducting phase was obtained.

The niobium powder used throughout the development phase of
this program was supplied by the Teledyne Wah Chang Albany Corporation. It was produced by the hydride-dehydride process from slabs
of electron beam melted ingots. Its chemical composition, in parts
per million, was as follows: 1780 oxygen; 380 tantalum; 230 iron;
110 tungsten; and 70 carbon. The as-received powder contained a
wide range of particle sizes. A particle size analysis[10] gave the
following particle size distribution: 2 wt% -270 + 325 mesh (53-
44 μm); 26 wt% -325 + 400 mesh (44-37 μm); and 18 wt% for each of
the size fractions -400 + 500 mesh (37-30 μm); -500 + 750 mesh (30-
20 μm); -750 + 1000 mesh (20-15 μm) and -1000 mesh (15 μm). Work
was done using both the as-received powder and various sharply
graded size fractions. A scanning electron microscope study of
the powder showed that the particles were angular and highly irregular in shape. A particle shape of this type was an important
factor in achieving good green strength in the rolled tape.

The mill used for powder rolling has 2 inch diameter, hardened
steel rolls, the axes of which are in a horizontal plane. The mill
is powered by a stepless variable speed drive which provides a
range of roll speeds from 2 to 20 rpm. Grooves 1/8" wide and 1/8"
deep were ground into the rolls, leaving between the grooves the
desired width of rolling surface. Side retainers of 1/8" thick
plexiglass were machined to fit in the grooves and rest on the reduced roll diameters, thus confining the powder laterally throughout the roll bite. A plexiglass hopper, affixed to the side retainers, has an adjustable slide to regulate the powder flow and
maintain an adequately constant powder head. Uniformity of roll
compaction was significantly improved by roughening the roll surfaces.

The maximum tape thickness is determined mainly by the roll
diameter. The 2 inch diameter rolls employed imposed a maximum
tape thickness of approximately 0.025". Using the as-received
powder (-270 mesh), tape with a good surface appearance was produced at roll speeds from 2 to 20 rpm and with roll gaps from
0.010 to 0.018". With a given roll speed, a decrease in roll gap
led to a decrease in porosity and an increase in green strength.
When the roll gap was held constant, the effect of increasing the
roll speed was to diminish the green strength and increase the
porosity. A good balance between green strength and porosity was
obtained under the following conditions: roll speed, 5 rpm; roll
gap, 0.016". The green tape, with a thickness of 0.022", had a
porosity of about 30% and could support 830" of its own weight.

The sintering temperature and time period were constrained
to a rather narrow range by the simultaneous requirement of

ductility and porosity. A practical compromise was achieved by sintering for 3 minutes at 2250°C in a vacuum of 4×10^{-5} Torr. The tape thus processed had a porosity of about 25% and could sustain a bend of 0.1" radius.

Infiltration was accomplished by immersion of the sintered tape for one minute in a tin bath maintained at a temperature of 850°C. Although infiltration can be carried out successfully at temperatures as low as 350°C, problems were encountered in the cold rolling of the infiltrated tape. Because of the low flow stress of pure tin, some of it was squeezed out during the mechanical deformation, and this tin loss reduced the volume fraction of Nb_3Sn in the final product. Several methods were investigated for preventing the loss of tin. The simplest and most effective procedure was based on considerations relating to phase equilibria in the niobium-tin system. After infiltrating at temperatures in the range of 350-850°C, it is thermodynamically possible to have niobium, Nb_3Sn, Nb_6Sn_5, $NbSn_2$ and tin all present in the as-infiltrated tape. Nevertheless, the reaction kinetics at the lower temperatures for the short immersion times employed (one minute) are such that only niobium and tin are metallographically visible. Increasing the infiltration temperature to 850°C, however, produces a marked change in the reaction kinetics. A significant amount of Nb_6Sn_5, which is readily distinguished under the microscope by the characteristic reddish brown color imparted by anodic etching, is now present. It occurs as a dispersed phase and sufficiently hardens the tin to prevent any loss during the deformation. Thus, the maximum volume fraction of Nb_3Sn realizable from the amount of infiltrated tin is achieved.

As mentioned earlier, a diffusion reaction heat treatment, at temperatures above 930°C, carried out on the as-infiltrated tape does produce Nb_3Sn. However, there are marked disadvantages in doing so. Since the diffusion path is long (1/2 the original pore size), an extended reaction time is required to involve all of the tin. Moreover, an undesirable morphology is obtained. For example, an infiltrated sample was heat treated for 16 h at 1000°C. There was still some unreacted tin and the Nb_3Sn occurred in massive form with the result that the heat treated tape was extremely brittle and could be handled only with the utmost care.

Interposing a deformation by cold rolling after infiltration and before the heat treatment resulted in dramatic improvements. As the amount of deformation is increased the tin becomes elongated to a filamentary morphology, resulting in several benefits. The greatly increased interfacial area between the niobium and tin reduced the time for complete reaction of the tin from more than 20 h to a few minutes. As pointed out earlier, a filamentary structure is desirable for conductor stability. The filamentary structure also provides the tape flexibility required for coil winding. An appropriate morphology was obtained by cold rolling the infiltrated tape to a reduction in thickness of 75-85%.

The diffusion reaction can be carried out in the temperature range of 930-1350°C. However, the current carrying capacity degrades rapidly as the reaction temperature exceeds 1000°C. It is in fact desirable from the standpoint of grain size to carry out the reaction at as low a temperature as practicable. The tapes are normally heated for 3 minutes at 950-975°C.

The completed tapes exhibit high values for the important superconducting parameters: a critical temperature of 18.1 K, and an overall current carrying capacity, at 4.2 K, close to 10^5amp/cm^2 in an applied field of 10 tesla. The current carrying capacity was determined for a sample that had been formed into a coil of 0.5" radius. After testing, the sample was unsoldered from the contacts, uncoiled and then reverse coiled to the same diameter. A retest showed virtually no deterioration in superconducting properties.

The tapes have, however, a less than optimum filament morphology. Typically, the filaments have a cross section of 30 microns in width and 5 microns in thickness. According to the stability criteria discussed previously, no dimension in a filament section should exceed 10 microns. In the powder rolling process there is little control over the filament width, which is directly related to the original pore size. It was decided, therefore, to modify the process so that the conductor could be made in wire form. Wire would provide an additional advantage besides improved stability. Suitably clad, a single wire can be used as the basic element of a flexible multicored conductor or a multistrand cable to carry large currents.

The Infiltration Process for Wire; Nb3Sn

For producing wire, the concept of infiltrating a porous niobium form was retained, since the results were highly satisfactory. The powder rolling step, however, had to be replaced by a more appropriate mode of powder compaction. Two variations of pressureless sintering[11] were tried. Although successful, this approach was abandoned in favor of cold isostatic pressing which proved far more convenient.

Niobium powder with a sieve range of -250 + 400 mesh was siostatically compacted in rubber molds at a pressure of 207 N/mm^2 (30,000 psi) to produce rods. Vacuum sintering at 2250°C provided the desired combination of porosity and ductility. In order to control the level of interstitials, especially oxygen, the time at sintering temperature depends on the mass of the rod. For the

small diameter rods (3/16") intended for laboratory use, a sintering time of 30-60 minutes was satisfactory. A scanning electron
micrograph, Fig. 3, shows clearly the ductile niobium matrix and
the system of interconnected pores.

For infiltration, the sintered rods were immersed in a tin
bath maintained at a temperature in the range of 350-400°C. These
lower temperatures were preferred in order to prevent the formation
of brittle intermediate phases which could impair the ductility of
the composite. At this point a cladding of one or more metals is
applied to the infiltrated rod to accomplish several purposes. An
exterior cladding of monel, for example, greatly facilitates wire
drawing. An inner jacket of tantalum, in principle, serves as a
diffusion barrier to prevent undesired reactions between the infiltrated core and the exterior jacket. Fortunately, as a result
of the very short reaction times required in the infiltration
process, it was found that the diffusion barrier could be eliminated

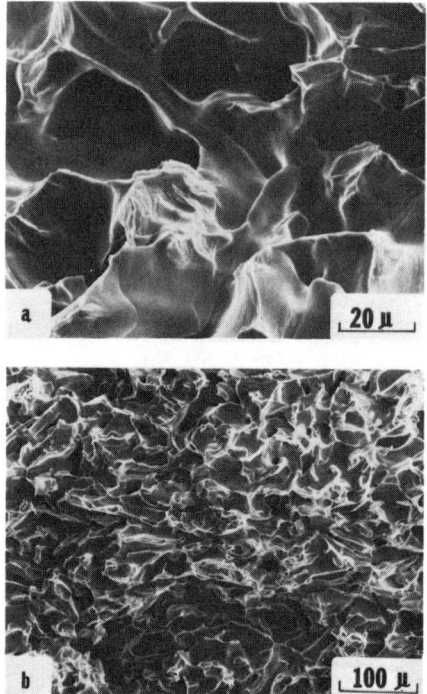

Fig. 3.   Scanning electron micrograph of porous niobium rod
          sintered for 30 minutes at 2250°C.

with no adverse effects, thus greatly simplifying the procedure.

The composite rods were mechanically reduced to fine wire by a combination of form rolling and wire drawing. Indicative of the ductility achieved, a reduction ratio in cross sectional area of 4000:1 was consistently obtained. Heat treating the wire for 2-3 minutes at 950°C results in all of the tin reacting with niobium to produce multiply connected filaments of the superconducting compound, Nb_3Sn. The filaments are well within the desired size range for adiabatic stability, having transverse dimensions of the order of 1 micron. A schematic representation of the entire process is shown in Fig. 4.

Properties of Wire Made by the Infiltration Process

The critical temperature, taken as the midpoint of the inductance change during the transition from the superconducting to the normal state, is 18 K. This value remains substantially constant for reaction temperatures from 900 to 1200°C. Data from both pulsed (rise time, 8 milliseconds) and steady field measurements were used to determine the dependence of the critical current

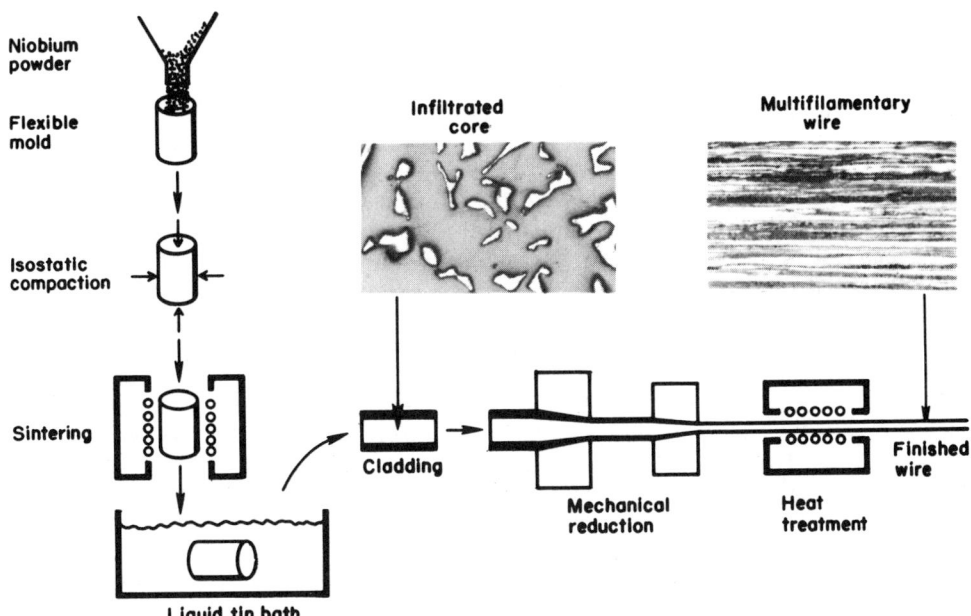

Fig. 4. The infiltration process for producing multifilamentary superconducting wire.

density on the applied magnetic field.[12] The area used in the
computations was that of the entire cross section of the Nb–Nb$_3$Sn
core. Steady field data were obtained through the courtesy of the
Accelerator and Fusion Research Division of the Lawrence Berkeley
Laboratory, the Brookhaven National Laboratory and the Francis
Bitter National Magnet Laboratory (M.I.T.). They were determined
at a resistivity of 10^{-12} Ω-cm. The results (Fig. 5) show good
agreement between the two sets of data at high fields, but lower
values for the pulsed field determinations at fields below 5 tesla.
This discrepancy is attributed to Joule heating at the current
contacts of the very short samples used in the pulsed field tests.
To further elucidate the current carrying capacity at very high
fields, steady field tests were made recently at the Francis
Bitter National Magnet Laboratory over the range from 17 to 22
tesla. The data points fell smoothly along the steady field curve
of Fig. 5. The critical density was 1×10^4 amps/cm^2 at 20 tesla
and 4×10^3 amps/cm^2 at 21 tesla. Wires made by this process appear
to be especially suitable for applications requiring fields of the
order of 15 tesla. In another series of tests, it was found that
the critical current density increases with decreasing filament
size down to 1 micron, below which there appears to be a diminish-
ing effect. The minimum diameter to which the reacted wires can
be bent with no degradation of the critical current density is 2
cm. Here too, there is an effect of filament size and it is anticipated that with finer filaments even smaller bend diameters
can be tolerated.

POROUS NIOBIUM RODS FROM POLYMER COATED POWDER

     The infiltrated rods used for most of the laboratory studies
just described had a diameter of 5 mm. In order to produce long
lengths of superconducting wire, the diameter of the isostatically
pressed compacts would have to be increased substantially. Con-
comitant with the studies of the size effect, it was decided to
investigate an alternative approach[13,14,15,16] that could produce
indeterminate lengths of a given diameter.

     The particles of niobium powder were coated with a layer of
the thermoplastic, polystyrene. The polymer content was varied
from 7.5 to 10% by weight. A significant result was the fact
that the coating imparted to the polymer–metal powder mixture the
characteristics of a thermoplastic, and conventional plastic forming methods such as extrusion could be performed readily. It was
found that plasticizing the polymer by the addition, in the amount
of 1/5 of its weight, of the eutectic mixture of diphenyl and
diphenyl ether not only reduced the required extrusion pressure,
but also assured consistent extrudability.

The polymer coated niobium was heated to 165°C and extruded through a 0.48 cm (3/16") diameter orifice. The extrusion pressure varied from 27,000 psi for a 7.5 wt% polymer content to 7000 psi for a 10 wt% polymer content. After cooling to ambient temperature, both the surface finish and strength of the extruded rods were quite satisfactory. Samples were then subjected to a programmed heating cycle. During a dwell period at 400°C the polymer was completely expelled by volatilization. Although the shape was undistorted, the strength at this stage was virtually nil. However, after increasing the temperature to 1350°C, the strength was more than adequate for any handling required prior to final sintering. After sintering, the infiltration, mechanical deformation and diffusion heat treatment were carried out as described in the previous section. The residual porosity after sintering varies nearly linearly with the wt% of polystyrene in the polymer-metal powder mixture. The porosity was approximately 25% for a 7.5 wt% polymer content and about 40% for a 10 wt% polymer content.

A15 TERNARY COMPOUNDS

$Nb_3(Al,Ge)$

The A15 ternary compounds offer high promise for substantial improvement of the superconducting properties over the presently available conductors, like Nb_3Sn and V_3Ga. Of the stable superconductors with high critical temperatures the ternary compound $Nb_3(Al,Ge)$ has the highest critical field. It was reported to possess a T_c of 20.7 K[17,18] and a H_{c2} of about 41 tesla.[19]

$Nb_3(Al,Ge)$ has received significant attention since its discovery by Matthias et al.[17] For a variety of reasons associated with the phase relationships and inherent mechanical properties difficulties were encountered in preparing $Nb_3(Al,Ge)$ in the form of a flexible multifilamentary conductor. Several methods of preparation have been reported in the literature.[20-24] The sputtering process yields promising qualities in terms of the current carrying capacity,[20] but this approach is limited by the relatively low rate of deposition. Feasibility of the powder metallurgy process for producing tape of $Nb_3Al_{0.75}Ge_{0.25}$ has been demonstrated.[21] The Kunzler technique was used whereby powders of Nb, Al, and Ge were mixed in the appropriate proportions, sealed in a cupronickel tube and cold worked by swaging and rolling. The tape was subsequently reacted to form the A15 compound. The critical current densities, in steady transverse magnetic fields of 15 tesla, were found to be about 10^4 amp/cm^{-2}.

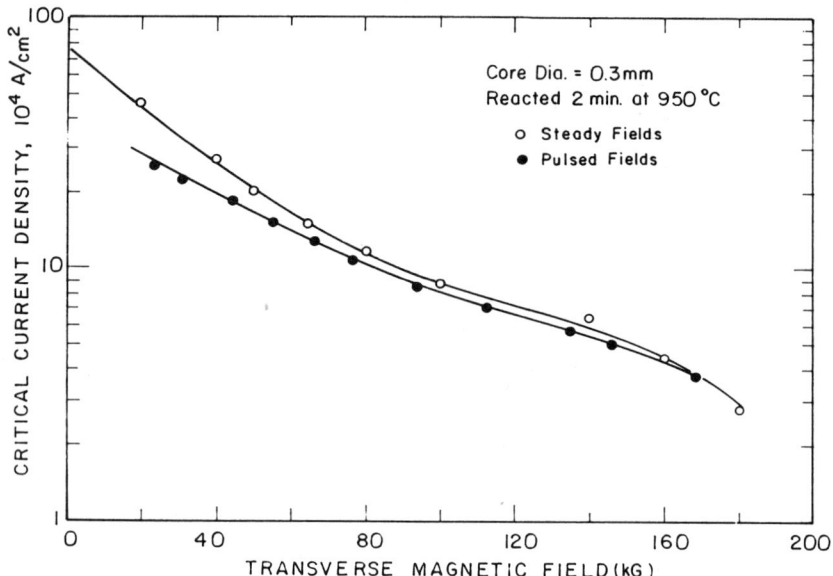

Fig. 5.   Critical current density of the Nb–Nb$_3$Sn core as a
         function of the magnetic field under pulsed and
         steady field conditions.

An investigation was undertaken at Lawrence Berkeley Laboratory[5] to evaluate the factors affecting densification of precompounded powder of Nb$_3$(Al,Ge). This study has been already
described in the present paper. The salient feature of the results was that a hot pressing process, of 6000 psi at 1400°C,
substantially eliminates the porosity. It can further be assumed
that by application of higher hydrostatic pressures the temperature
for plastic flow could be reduced below 1000°C. Considerable plastic flow, at ambient temperature, was observed in samples of V$_3$Ga
in response to high pressure in the 3000 ksi range.[25]

The infiltration route was adopted in order to obtain a relatively ductile material that can be mechanically deformed into a
fine wire, containing an interconnected network of filaments of
the order of a few microns in diameter. Subsequently a diffusion
heat treatment was applied in order to convert these filaments
into A15 Nb$_3$(Al,Ge).

Additional advantages of the infiltration process are due to
the fact that the composition of the infiltrant (Al-Ge or Al-Si
alloys and eutectics) can be readily controlled. Appropriately
programmed diffusional anneals may form the desired composition
and microstructure of the compound. Moreover, an optimized
thermomechanical process may yield the desired combination of

high critical current density and critical temperature. This
objective can be achieved by appropriate control of the grain
size and long range ordering, and by determining conditions for
the approach to optimal stoichiometry.

The Al-Ge system forms a low melting point eutectic (424°C)
at the composition of 70% Al-30% Ge. The inherent brittleness of
the Al-Ge eutectic requires particular attention as to the temperature of the infiltrating bath, the temperature difference between the porous sintered niobium rod and the bath, and the rate
of infiltration. The formability of the Al-Ge eutectic, entrapped
and solidified within the pore volume, during form rolling or wire
drawing can be significantly improved through the application of
superplasticity principles.[26] Superplastic behavior requires a
fine duplex microstructure which is stable at the deformation temperature. The ductility of the Al-Ge alloy confined in a network
of pores in the niobium matrix was investigated.[27] In the bulk
eutectic a lamellar structure was associated with extreme brittleness, whereas when confined in a pore undergoing the controlled
cooling regime of the infiltrated rod, e.g., directional solidification or quenching, the composite after infiltration yielded
relatively ductile microstructures. Two desirable morphologies
of the solidified eutectic were observed--spheroidized particles
and a fine lamellar structure. An infiltrated niobium rod containing Al-Ge eutectic of such morphology can be plastically deformed into fine multifilamentary wire, as shown in Fig. 6.

The Al-Ge eutectic was prepared by induction melting the
constituent mixture of the exact eutectic composition under an
argon atmosphere and casting in a copper-chilled mold. The
molten eutectic bath was maintained at about 600°C during immersion. A close temperature control is necessary in order to prevent
freezing of the eutectic in the outer pores of the sintered rod
and thus blocking further infiltration. The helium gas pressure
above the molten eutectic, the immersion time, and the rate of
cooling of the infiltrated rod to the ambient temperature are
important parameters. They depend on the pore size and shape,
viscosity of the eutectic and the desired microstructure of the
solidified eutectic in order to assure good formability. Furthermore, the infiltration temperature should be maintained at a practical minimum so as to avoid, at this stage, any diffusion reaction between the eutectic and the niobium.[28] At higher infiltration temperatures, and with excess Al in the eutectic bath, there
is a tendency towards formation of an Al-solid solution and an
Al-rich intermetallic phase $Nb(Al,Ge)_3$. It was found that this
intermetallic phase nucleates and grows at relatively low temperatures.[29] Holding an infiltrated rod for 12 h at 400°C resulted in complete conversion of the eutectic into $Nb(Al,Ge)_3$.
These phases are brittle and have, therefore, a detrimental effect
during the subsequent plastic deformation of the infiltrated
composite.

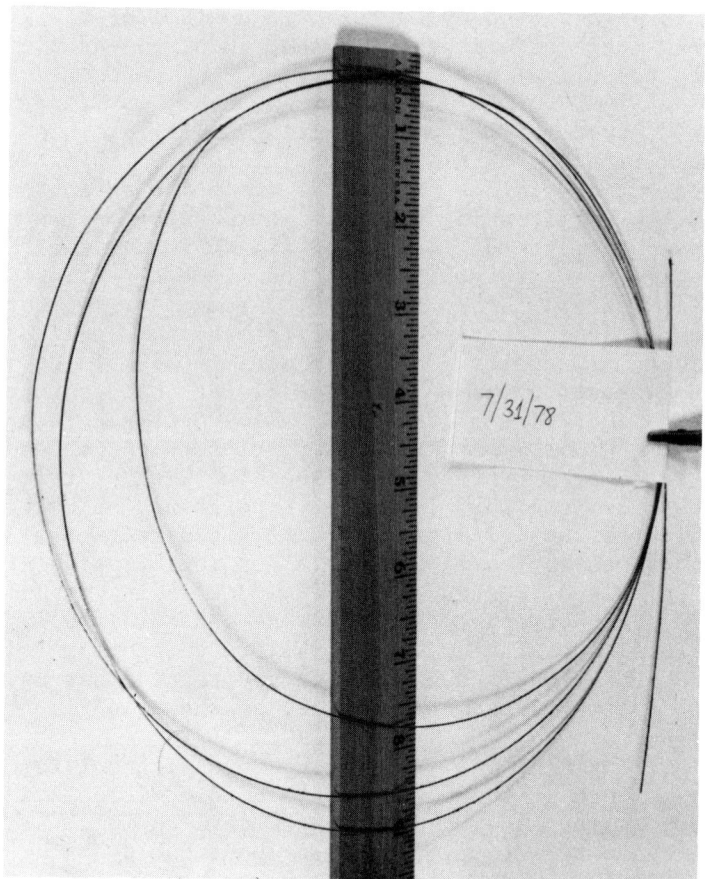

Fig. 6. Nb-Al-Ge multifilamentary wire of 0.18 mm diameter.

The kinetics of the A15 $Nb_3(Al,Ge)$ phase formation was investigated over a wide range of temperatures (1250-1700°C), reaction times (15-900 sec) and post-reaction anneals.[28-31] Data are
presented in Table 2, for the critical temperatures of $Nb_3(Al,Ge)$
after various heat treatments.

The multifilamentary superconducting A15 phase is formed as
a result of a diffusion reaction. Its microstructure and superconductive properties are determined by the reaction kinetics.
It was found[30] that the A15 phase grows rapidly during the early
stages of the reaction. The growth rate depends on the reaction
temperature. In the early stages of the reaction at 1700°C for
different phases coexist: the Ge-rich, the Al-rich, the A15

Table 2. Critical Temperature of $Nb-Nb_3(Al,Ge)$
Multifilamentary Superconductors

Reaction Anneal		Post-Reaction Anneal		
Temp. (°C)	Time (sec)	Temp. (°C)	Time (h)	T_c (K)
1250	60	750	96	17.2
1300	30	750	96	17.2
1300	30	750	96	17.5
1300	30	750	250	17.4
1300	60	...	...	15.6
1300	60	750	48	16.3
1300	900	...	...	15.4
1400	30	750	96	18.2
1500	30	750	96	18.5
1600	15	750	96	19.1
1700	15	750	96	18.7
1700	60	...	...	17.4
1700	60	750	48	18.7
1700	60	750	96	16.8

$Nb_3(Al,Ge)$ and the A2 phase which is a Nb solid solution. The Ge-rich phase forms the innermost layer next to the pore and the Al-rich sigma and the superconducting A15 phase form outside of it. These three phases are embedded in the Nb solid solution matrix. As the reaction proceeds, the Al-rich phase gradually transforms into A15. After 15 minutes holding time at 1700°C the Ge-rich phase transforms also, leaving the A15 in a matrix of Nb solid solution. However, all four phases were found to exist after one hour reaction at 1300°C.[30] The generally observed trend of higher T_c for higher reaction temperatures[29] (Table 2) may suggest that the presence of either one or both Ge-rich and Al-rich phases, in the specimens reacted at lower temperatures, may cause a reduction in the critical temperature. However, if this were the case, one would expect an increase in the critical temperature with an increase in the reaction time. Such a behavior was not observed.[29] The reason for this apparent inconsistency may be due to deviation from the optimal stoichiometric composition with increased diffusion times. The present data indicate that high reaction temperatures and short reaction times improve the T_c value. Figure 7 shows the results of the current carrying carrying capacity of $Nb_3(Al,Ge)$ samples, prepared by different procedures.

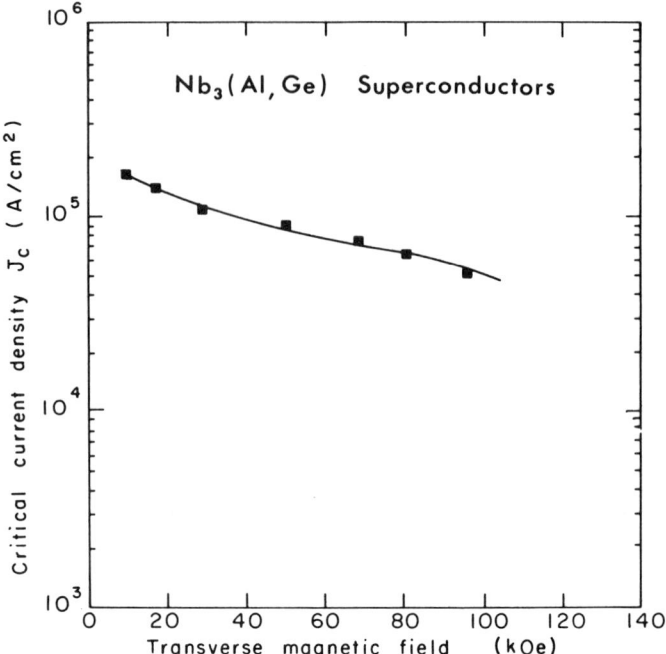

Fig. 7. Variation of the critical current density with magnetic
         field of $Nb_3(Al,Ge)$.

In contrast to the behavior of T_c with respect to reaction
times and temperatures, the critical current density J_c is significantly improved by decreasing the temperature of the reaction
anneal and its duration.[29] But this effect is diminished in absolute terms, by the reduced volume fraction of the A15 phase formed
under these conditions. An additional negative contribution to the
reduced critical current density in the high temperature reaction
anneals is due to the excessive grain growth at the elevated temperatures. Grain growth causes a significant reduction in the number of pinning centers, thus reducing J_c.

The full potential of the ternary A15 $Nb_3(Al,Ge)$ for the preparation of flexible multifilamentary stable superconductors can be
realized by controlling the kinetic conditions for the diffusion
reaction with the purpose of obtaining an optimal combination of T_c
and J_c. The problem of formability, for the fabrication of long
lengths of multifilamentary conductor, was shown to be soluble by
applying suitable thermomechanical procedures.

Nb₃(Al,Si)

The infiltration process was applied for the preparation of flexible, multifilamentary superconducting $Nb_3(Al,Si)$ wire.[32,33] The infiltrant employed in the preparation of $Nb + Nb_3(Al,Si)$ is the binary Al-Si eutectic with a melting point of 577°C. The structure and properties of this eutectic have been reported in the literature.[34] The Al-Si eutectic exhibits poor ductility, but the ductility can be substantially improved by unidirectional solidification of the eutectic. Care was therefore exercised to slowly cool, unidirectionally, the infiltrated sintered niobium rod. Similar to the procedure adopted for the Al-Ge eutectic described before, the immersion time and temperature for the Al-Si eutectic had to be optimized in order to prevent formation of brittle intermetallic phases that reduce the ductility of the infiltrated composite. The optimal immersion temperature and time were found to be 580°C and 30 seconds, respectively. This temperature is only 3°C above the melting point of the eutectic. Therefore, the temperature of the sintered niobium rod had to be closely controlled in order to assure complete infiltration.

One major advantage of the Nb-Al-Si system is the ease with which it reduces to wire. Infiltrated Nb rods of 5 mm in diameter were easily reduced to 0.150 mm diameter wire at room temperature. The average filament size was found to be about 1 micron. The extremely fine filamentary microstructure achieved with the Nb-Al-Si system allows the formation of A15 at reaction temperatures as low as 850°C.

The most favorable diffusional anneal for the formation of A15 $Nb_3(Al,Si)$ was carried out in two stages.[33] The Al-Si eutectic in the form of elongated interconnected fine filaments will readily melt and react at A15 reaction temperatures above 1000°C. Therefore, a two stage procedure was adopted by which formation of an intermediate compound prevents migration of the Al-Si eutectic.

a) Primary heat treatment: After infiltration, the niobium rod was held for 1 h at 650°C in order to form, by a diffusional process, an intermetallic barrier around the pores, or elongated filaments. It was found that the composition of the barrier corresponds with the region of $NbAl_3$ in the Nb-Al phase diagram. This phase is very stable at 650°C.

b) Secondary heat treatment: Following the primary heat treatment, the specimen was heated to temperatures in the range of 1000-1700°C for the A15 formation. The appropriate holding time at 1700°C was found to be 30 seconds. During the secondary heat treatment Al diffuses and reacts with Nb and Si to form $Nb_3(Al,Si)$. This compound is very stable at 750°C. Its composition remains unchanged after 96 h of annealing at this temperature. Longer holding times at low reaction temperatures, below 1000°C, resulted in extremely thin layers of the A15 compound.

The highest T_c value for $Nb_3(Al,Si)$, thus far obtained in flexible wire samples, was 18.8 K. The reaction occurred at a temperature of 1700°C for 30 seconds, followed by an annealing heat treatment for 96 h at 750°C. This post heat treatment proved to be beneficial for improving the T_c value.

In conclusion, the present work demonstrates the applicability of the infiltration process to prepare multifilamentary superconductors based on both binary and ternary A15 compounds, and indicates that for the ternaries in-depth studies of the reaction kinetics are required in order to realize their full potential.

ACKNOWLEDGMENTS

The authors are deeply indebted to numerous individuals and laboratories for substantial contributions.

Thanks are due to Drs. M. P. Dariel, J. L. F. Wang, W. Gilbert, K. Hemachalam and Mr. J. Jacobsen of LBL, Dr. R. Scanlan of LLL, Dr. M. Suenaga of BNL, Mr. Craig Wojcek of Teledyne-Wah Chang, and to the members of the National Magnet Laboratory (MIT), especially Drs. L. Rubin and S. Foner.

This work was supported by the Division of Materials Science, Office of Basic Energy Sciences, U.S. Department of Energy under Contract No. W-7405-ENG-48.

REFERENCES

1. M. N. Wilson, C. R. Walters, J. D. Lewin, and P. F. Smith, J. Phys. D 3:151 (1970).
2. R. L. Ciardella, Report LBL-4174, 1975 (unpublished).
3. R. L. Ciardella, M. P. Dariel, J. L. F. Wang, and M. R. Pickus, IEEE Trans. on Magnetics MAG-13:832 (1977).
4. M. R. Pickus and R. L. Ciardella, U.S. Patent 4,088,512 (May 1978).
5. D. P. Modi, Report LBL-4173, 1976 (unpublished).
6. J. A. Cave and T. J. Davies, Metal Sci. 8:28 (1974).
7. M. R. Pickus, V. F. Zackay, E. R. Parker, and J. T. Holthius, Int. J. Powder Met. 9:3 (1973).
8. M. R. Pickus, K. Hemachalam and B. N. P. Babu, Mater. Sci. Eng. 14:265 (1974).

9. M. R. Pickus, E. R. Parker, and V. F. Zackay, U.S. Patent 3,815,224 (June 1974).
10. B. N. P. Babu, Report LBL-437, 1971 (unpublished).
11. K. Hemachalam and M. R. Pickus, J. Less Common Met. 46:297 (1976).
12. K. Hemachalam and M. R. Pickus, IEEE Trans. on Magnetics MAG-13: 466 (1977).
13. M. R. Pickus and M. Wells, Powder Metallurgy 8:16 (1965).
14. M. R. Pickus, Int. J. Powder Met. 5:3 (1969).
15. D. R. Nordin, Report LBL-7346, 1978 (unpublished).
16. A. Noman, Report LBL-8501, 1978 (unpublished).
17. B. T. Matthias, T. H. Geballe, L. D. Longinotti, E. Corenzwit, G. W. Hull, J. P. Maitu, and R. N. Willens, Science 156:645 (1967).
18. J. Ruzicka, Z. Physik 237:432 (1970).
19. S. Foner, E. J. McNiff, B. T. Matthias, T. H. Geballe, R. N. Willens, and E. Corenzwit, Phys. Letters A 31:349 (1970).
20. S. D. Dahlgren and D. M. Kroeger, J. Appl. Phys. 44:5595 (1973).
21. R. Lohberg, T. W. Eager, J. M. Puffer, and R. M. Rose, Appl. Phys. Letters 22:69 (1973).
22. J. Ruzicka, Cryogenics 14:434 (1974).
23. U. Zwicker, H. J. Miericke, and H. J. Renner, Z. Metallkunde 66:669 (1975).
24. A. Muller, J. Less Common Met. 42:29 (1975).
25. Y. D. Martynov, B. J. Beresnev, I. A. Baranov, V. Y. Meais, A. E. Fokin, S. P. Chizhik, and Yu N. Ryabinin, Fiz. Met. Metalloved. 24:522 (1967).
26. G. J. Pech, Report LBL-7300, 1977 (unpublished).
27. C. Rutan, Report LBL-8502, 1978 (unpublished).
28. J. J. Granda, Report LBL-5772, 1976 (unpublished).
29. M. R. Pickus, M. P. Dariel, J. T. Holthuis, J. L. F. Wang, and J. Granda, Appl. Phys. Letters 29:810 (1976).
30. E. Kannatey-Asibu, Report LBL-6266, 1977 (unpublished).
31. K. Douglas, Report LBL-7629, 1978 (unpublished).
32. G. C. Quinn, Report LBL-6999, 1977 (unpublished).
33. B. Phung, Report LBL-8500, 1978 (unpublished).
34. H. Steen, A. Hellawell, Acta Met. 20:363 (1972).

REPORT OF MAGNETIC FUSION ENERGY GROUP

D. N. Cornish

Lawrence Livermore National Lab.
Livermore, CA

M. N. Wilson

Rutherford Lab.
Chilton, Didcot, UK

Group Members: A. I. Braginski (Westinghouse R&D Center), D. N.
Cornish - Cochairman - (Lawrence Livermore National Laboratory),
J. File (Princeton Plasma Physics Laboratory), W. A. Fietz (Oak
Ridge National Laboratory), R. Flükiger (Kernforschungszentrum
Karlsruhe), J. E. C. Williams (Francis Bitter National Magnet Lab-
oratory), M. N. Wilson - Cochairman - (Rutherford Laboratory).

The group's discussion was principally concerned with the
relatively near-term problems of the next generation of large
fusion devices, which are expected to be constructed within 10-15
years. In this regard, it was generally felt that the major problems will be of a mechanical nature and that these will limit the
maximum fields to $\sim$10-12 T. For such magnets, the critical current
densities, fields and temperatures now being achieved in the best
composites presently available, were felt to be quite adequate.
Mechanical properties were thought to be much less satisfactory,
however, and the group would welcome further improvements in strain
tolerance. As a typical example, requirements for the conductor in
the proposed ETF Tokamak might be a manufacturing strain of 0.8%
and a cyclic operating strain of 0.2%. It may thus be seen that
improved handling capability during coil manufacture is the major
requirement. This could be brought about either by further improvements in materials properties or by better technologies for conductor fabrication or coil winding.

In the fusion devices presently envisaged it seems likely that
most of the operating stress will be taken by an external structure
rather than by the coil itself. For this reason, the most important
variable in assessing conductor performance will be strain rather
than stress and the group wished to encourage the materials community to present any future results on critical current degradation

in terms of strain. The effects of cyclic strain were also felt
to be important and more work is clearly needed in this area.
Structural materials were recognized as a very important area. Although this area is not directly connected with A15 conductor development, it is already clear that the future utilization of A15
materials in fusion will be more dependent on structural strength
than on any other single factor.

In developing novel techniques for the production of A15 conductors, it will be important to bear in mind the ever-present need
for stabilization and protection. Copper has generally been used
for this purpose but its high magnetoresistance at fields of 12 T
may present difficulties in attaining the necessary current density.
If a lower resistivity alternative, such as dispersion-hardened
aluminum, could be developed to a satisfactory level of performance,
it would be very useful.

Economic studies of conceptual reactor designs have made it
clear that substantial reductions in conductor cost will be needed
if fusion power is to be competitive with alternative sources.
Conductor designs should, therefore, be as simple as possible. In
this regard, it is worth noting that fine filaments are not generally necessary for stability in large conductors when they are
cryogenically stabilized. Tape conductors could be just as effective and may be cheaper. Even the presence of ac fields need
not necessarily preclude the use of tape conductors, provided the
amplitude of the pulsed field is sufficiently low for the conductor
to work in a self-screening mode.

Finally, the group was anxious to point out that, although it
had concentrated on the most pressing short-term problems, it did
not wish to discourage in any way the longer term work on new materials. There can be little doubt that higher temperature operation
will eventually be required, particularly when helium supplies
start to be restricted. Higher fields may also become desirable
as a result of failure to attain high values of β in present design
concepts or the development of radically different concepts. Basic
work on materials development must therefore continue with the
objective of achieving substantial improvements in all the critical
parameters.

REPORT OF HIGH ENERGY PHYSICS GROUP

W. B. Sampson

Brookhaven National Laboratory
Upton, NY

Group Members: Y. Furuto (Furukawa Electric Company, Ltd.), P. Genevey (CEN-Saclay), W. S. Gilbert (Lawrence Berkeley Laboratory), D. Hagedorn (CERN), W. Hassenzahl (Lawrence Berkeley Laboratory), M. Kuchnir (Fermi Laboratory), C. Scott (Rutherford and Appleton Laboratories), W. B. Sampson (Brookhaven National Laboratory), C. Taylor (Lawrence Berkeley Laboratory).

The discussion by this group centered on the potential of A15 materials such as Nb_3Sn for the next generation of high field accelerator magnets. Dipole coils capable of fields as high as 10 T are generally regarded as the goal despite the considerable difficulties being experienced with current magnets which are only required to produce fields between 4 and 5 tesla.

Considerable concern was voiced about the mechanical properties of the intermetallic compounds since it is virtually impossible to design a dipole magnet which does not require bending the conductor around a radius smaller than that permitted by the maximum strain limitation. There was no agreement reached on how practical a "wind and react" scheme might be for such long coils nor on the possibilities of developing a way to construct coils from prereacted conductor. It was the author's impression, however, that the general attitude towards making magnets from conductors of such limited ductility had improved considerably in recent years and there was unanimous agreement that methods could be worked out if conductor with impressive current carrying capacity was available.

Experience with multifilamentary A15 composites in magnets is minimal, only a few non-solenoidal coils having been constructed from this material. The results from these coils have been quite encouraging but for the most part they have been rather small and

357

wound from conductor of questionable quality. Magazine advertisements to the contrary A15 composites are not readily available and this effects the development of magnets.

It was pointed out by a number of participants that the construction of high field dipoles requires very high current density conductor. Geometrical consideration dictates that current density at 10 T be of the same order as that presently being achieved at 5 T in NbTi, in order for a practical design to be developed. Such high current windings lead to large stored energies and severe problems with quench protection.

The effect of superconductor magnetization on the field quality at low field was also discussed. In this respect the in-situ materials may prove especially troublesome due to coupling between the "filaments". The general ac characteristics of such composites were questioned and it seems that a test of this type of material in a reasonably sized coil is long overdue.

The economic situation was discussed briefly as it plays a large part in the usefulness of a conductor for large scale applications. Special devices for specific experiments can be justified even if the conductor is very expensive but a project such as a large accelerator or storage ring which may require of the order of one half million pounds of conductor would be priced out of existence at present small lot prices. The elimination of tantalum diffusion barriers would probably greatly improve the cost situation since the other new materials are comparable in price to those used in alloy composites.

The willingness to consider A15 materials for future applications has increased dramatically in recent years. This is no doubt due to the substantial progress being made in understanding the basic properties of these new composites. Magnet development will certainly progress more rapidly when conductor becomes readily available.

REPORT OF SOLID STATE PHYSICS GROUP

M. R. Beasley

Stanford University
Stanford, CA

Group Members: M. R. Beasley - Chairman - (Stanford University),
J. E. Evetts (Cambridge University), E. J. Kramer (Cornell University), D. O. Welch (Brookhaven National Laboratory).

The solid state physics group addressed the question what
should be done in the future to improve the properties or increase
the understanding of the problems in A15 filamentary superconductors. In their deliberations the group sought to avoid simply
advocating more basic research in a blanket way or holding out the
promise of a mythical new superconductor ($T_c > 30$ K, $H_{c2} > 10^2$ T,
$J_c > 10^8$ A/cm^2, and outstanding mechanical properties) as the solution to today's practical problems. There was the general percep-
tion, however, that recent advances in the basic understanding of
the science of A15 superconductors relevant to practical conductors needs to be transmitted to the practical composite community.
Moreover, it was likewise evident that, particularly for the newer
very fine multifilamentary in situ and powder fabricated composites, new regimes of behavior were being encountered, the physics
of which has not been completely elucidated heretofore. Some
specific points felt generally worthy of mention both for their
own sake and as illustrations of the type of work that would be
helpful in the future are outlined below. Logically they divide
conveniently into intrinsic properties (e.g., T_c, H_{c2}, strain dependence, etc.), extrinsic properties (e.g., flux pinning) and
areas to watch for developments of possible relevance.

Intrinsic Properties

Recent studies of the dependence of the superconducting properties of the A15 superconductors on order and composition are of
direct relevance to the composite maker. These are changing the

orientation from which these materials should be viewed, clarify-
ing the bounds of superconducting behavior possible, and providing
qualitative and quantitative guidance on how to optimize physical
properties. For example, well-made A15 materials are clean, not
dirty superconductors with $\xi_0/\ell \simeq 0.5$ ($\xi_0 \simeq 50$ Å; $\ell \simeq 100$ Å).
Thus, they differ from the more familiar alloys such as Nb-Ti.
Moreover, it is now clear that slight disordering of Nb_3Sn can
raise $H_{c2}(0)$ to as high as 30 T with only a slight reduction in
T_c. Hence, Nb_3Sn can go much further than it has been applied to
date, and in seeking optimization one should not necessarily be a
slave to pursuing the highest possible T_c. The possibilities of
achieving improved performance by means of tertiary substitutions
is also attractive and in need of systematic work on well controlled materials. V_3Ga, which is severely Pauli limited, raises
even more interesting questions and possibilities. Since $dH_{c2}/dT|_{T_c}$
is so large ($\simeq$ 30 kOe/K) in this material, introduction of spin-
orbit scattering substitutions to counteract the Pauli limiting
would appear to hold the promise of really dramatic increases in
H_{c2}. Moreover, theories of pinning designed for non-Pauli-limited
materials (such as the well-known Kramer theory) don't obviously
apply in V_3Ga, and a proper theory needs to be developed to provide guidance to the composite maker.

The importance of strain effects in Nb_3Sn is sobering and
proper phemonenological models of these effects, solidly based on
the microscopic properties and theories of the materials, need to
be developed. Present engineering models, while probably good at
calculating strain, make very crude assumptions about the dependence of the physical properties on that strain. A thorough
understanding of why V_3Ga is so much more strain tolerant might
be very enlightening.

Jim Livingston in his summary of the metallurgical issues of
importance in the A15 materials has stressed the need for really
good basic chemical free energy and kinetic information about this
class of materials. In a similar vein we would mention the poor
understanding of diffusion, defect properties, and internal friction, etc. from the solid state physics point of view.

Extrinsic Properties

It is clear that the understanding of flux pinning in the A15
superconductors is incomplete. For example the temperature dependence of J_c in multifilamentary Nb_3Sn is different from that of
Nb_3Sn tape, multifilamentary V_3Ga, and multifilamentary Nb_3Sn with
Ga additions. Moreover, in the more advanced in situ and powder
fabricated composites the filaments are getting small enough that
finite size effects (filament size less than the penetration depth
λ) and surface pinning are becoming relevant. Also since the

filaments are only a few grains wide, non-statistical theories
of flux pinning may be required. Finally, the nature and role of
coupling in the discontinuous composites has not been fully elucidated. Indeed, the full implications (coupling, effects on boundary pinning, etc.) of the proximity effect on very fine filamentary
conductors remains vague.

Some Areas to Watch

There are also some areas of current superconducting research
which do not presently impact on composite fabrication and properties but which might in the long run. They are areas to watch
for possible breakthroughs. Amorphous superconductors are currently of great fundamental interest. They can exhibit excellent
mechanical properties and critical fields. Critical temperatures
and currents are still low but research aimed at improving them
goes on. The Chevrel phase superconductors (e.g., $PbMoS_8$) have
very high critical fields and fair critical currents but share
poor mechanical properties with the A15 superconductors. Interest
remains high in the materials, however.

The general area of inhomogeneous superconductors - that is
coupled superconducting elements (spheres, rods or planes) im-
bedded in a nonsuperconducting matrix) - is one of growing interest
because of the novel superconducting properties (including very
large critical fields and currents) they exhibit. The present in
situ and powder produced composites are a not-very-idealized example of an inhomogeneous superconductor. The general understanding that is sure to develop about such systems may be of use to
the composite community.

Finally, there are always new materials. This field of research has traditionally played a crucial role in the development
of superconductivity and will certainly continue to do so. Dramatically different and better superconductors will not be there when
needed if they are not sought now in earnest - it is as simple as
that.

REPORT OF METALLURGY GROUP

J. D. Livingston

General Electric Research Center
Schenectady, NY

Group Members: S. Foner (Francis Bitter National Magnet Laboratory), D. C. Larbalestier (University of Wisconsin), J. D.
Livingston – Chairman – (General Electric Research Center), R. M.
Rose (Massachusetts Institute of Technology).

The discussion by this group centered on multifilamentary
Nb_3Sn and the further metallurgical work necessary to improve our
understanding of these materials.

Multifilamentary Nb_3Sn is usually formed by the diffusion reaction between Nb and Cu(Sn) solid solution, but the basic thermodynamics and kinetics pertaining to this reaction are inadequately
characterized. The Cu–Nb–Sn equilibrium phase diagram – specifically, the detailed shape of the A15 phase field and its tie lines
to the Cu(Sn) field at various temperatures – remains to be established quantitatively. Another need is for measurements of diffusion coefficients in Nb_3Sn, as a function of temperature and
stoichiometry, and establishment of the relative importance of
grain boundary and volume diffusion. Such thermodynamic and
kinetic data are necessary to understand the composition (and
hence property) gradients to be found in Nb_3Sn reaction layers.
More characterization of such gradients (p. 143) is needed. Although diffusion in Cu(Sn) is better characterized in the literature than that in Nb_3Sn, secondary effects such as Kirkendall
voids have been inadequately understood and controlled (p. 91).
Interesting ternary diffusion effects, such as the influence of Ga
gradients on Sn diffusion (p. 1), are worthy of more study and
possible exploitation.

Much published data on superconducting properties of multifilamentary Nb_3Sn composites are of questionable scientific value because of the probable presence of such composition and property gradients across the layers. Impurity effects (p. 143) may also be important. Residual stresses also have influenced much of the data in the literature, but understanding of stress effects is now fairly advanced, as evidenced by the papers in session II. However, with regard to residual stresses in Nb_3Sn composites, x-ray measurement of such stresses would be of interest. Quantitative interpretation also requires more data on the temperature dependence of the flow properties of Cu(Sn). More work is also needed to establish whether stress effects can be interpreted entirely in terms of changes in T_c and H_{c2} or whether changes in pinning centers need also be considered.

Although critical currents are generally believed to be controlled primarily by flux pinning by grain boundaries, the relative importance of surface pinning in fine-filament composites needs to be established. The possible importance of impurity segregation in grain-boundary pinning could perhaps be examined by Auger analysis of samples exhibiting intergranular fracture. The presence or absence of the cubic-tetragonal martensitic transformation in multifilamentary A15 composites has not been established. If present, it could play a role in influencing the dependence of superconducting properties on stress, composition, etc.

Finally, other metallurgical areas worthy of more study include thermal coarsening of superfine filaments, micromechanics of polyphase materials, high-speed casting, and high-purity powder processing.

CONTRIBUTOR INDEX

SUBJECT INDEX